Shipping in Arctic Waters

A Comparison of the Northeast, Northwest and Trans Polar Passages

Willy Østreng, Karl Magnus Eger, Brit Fløistad,
Arnfinn Jørgensen-Dahl, Lars Lothe, Morten Mejlænder-Larsen
and Tor Wergeland

Shipping in Arctic Waters

A Comparison of the Northeast, Northwest and Trans Polar Passages

Springer

Published in association with
Praxis Publishing
Chichester, UK

PRAXIS

Willy Østreng
University of Alaska Fairbanks
and Ocean Futures
Oslo
Norway

Karl Magnus Eger
Ocean Futures
Oslo
Norway

Brit Fløistad
Ocean Futures
Oslo
Norway

Arnfinn Jørgensen-Dahl
Ocean Futures
Oslo
Norway

Lars Lothe
Ocean Futures
Oslo
Norway

Morten Mejlænder-Larsen
Det Norske Veritas
Høvik
Norway

Tor Wergeland
Copenhagen
Denmark

http://www.arctis-search.com/

SPRINGER-PRAXIS BOOKS IN GEOPHYSICAL SCIENCES
SUBJECT *ADVISORY EDITOR*: Philippe Blondel, C.Geol., FGS., Ph.D., M.Sc.,
Senior Scientist, Department of Physics, University of Bath, UK

ISBN 978-3-642-16789-8 ISBN 978-3-642-16790-4 (eBook)
DOI 10.1007/978-3-642-16790-4
Springer Heidelberg New York Dordrecht London

Library of Congress Control Number: 2012947034

Cover design: Jim Wilkie
Project copy editor: Rachael Wilkie
Typesetting: David Peduzzi

Printed on acid-free paper

Springer is a part of Springer Science+Business Media (www.springer.com)

Contents

List of figures

List of tables

About the authors

Arnfinn Jorgensen-Dahl is a senior Research Fellow at Ocean Futures, Oslo. He has occupied teaching, research and visiting positions at universities and research institutes in Australia (University of Tasmania, University of Queensland, Monash University), National University of Singapore (NUS), and at the Fridtjof Nansen Institute in Norway. He received the degree of Magister Artium (Mag. Art) in Political Science from the University of Oslo, a PhD in International Relations from the Australian National University (ANU), and a Diploma in International Relations (with Distinction) from the Institute of Social Studies, The Hague. His research includes northern and Arctic natural resources, political and jurisdictional issues and affairs, Southeast and East Asian regional political and economic cooperation integration and development, and global population and environment issues. Prior to joining Ocean Futures, he spent ten years on the staff of the United Nations Population Fund (UNFPA) dealing with the Fund's environment and population policies and activities.

Karl Magnus Eger is Adviser at Analys & Strategi (part of Muticonsult) and former Research Fellow at the Research Institute Ocean Futures. He has a Masters degree in Political Science from the University of Oslo, and also studied Entrepreneurship and Business Development at Boston University, School of Management. Eger has experience from several interdisciplinary projects at the intersection of technology, policy, governance and strategy. Recently, Eger has been working with an impact assessment on behalf of the Norwegian Ministry of Petroleum and Energy, focusing on infrastructure and logistics relevant for petroleum activities in the Barents Sea South-East, the Norwegian Sea North-East and the sea areas around Jan Mayen.

Brit Fløistad has a Masters degree in Political Science and in Law from the University of Oslo. She has been a researcher at the Fridtjof Nansen Institute, dealing primarily with matters related to coastal states' jurisdiction and international fisheries cooperation and management. She is an affiliate of Ocean Futures in Oslo. She has also been head of the Parliamentary Library and of the Parliamentary Research Service of the Norwegian Storting.

Lars Lothe is a Norwegian legal counsel in private practice in Oslo, and a research fellow of Ocean Futures, where he is also chairing the Board of Directors. He was a scientific assistant at the Fridtjof Nansen Institute until 1990, and was then affiliated with the Scandinavian Institute for Maritime Law in Oslo until 1991. From 1992 until 1996, he worked with Vogt & Co in Oslo, a law firm specializing in maritime law. He then worked as legal counsel for Norsk Hydro ASA until 2004, including one year a the company's office in the United States, and was then legal counsel for Yara International ASA until entering his present legal practice in 1999.

Morten Mejlaender-Larsen graduated from the Technical University in Trondheim in 1983, and started to work for Det Norske Veritas, DNV, in 1985. At DNV, he has worked in different positions related to ships and structures. Since 2004, he has been responsible for Arctic Operations and Technology in DNV and is today Discipline Leader for this group. The activity includes management of and participation in projects involving R&D, rule development and JIP, as well as advisory services related to shipping in polar waters. He is also a member of several steering committees for different R&D projects related to Arctic shipping.

Willy Østreng is President of the Norwegian Scientific Academy for Polar Research and affiliated faculty at the University of Alaska Fairbanks. He was Scientific Director at the Centre for Advanced Study at the Norwegian Academy of Sciences and Letters, 2003–2009, Director General of the Fridtjof Nansen Institute, 1978–2003, Adjunct Professor of Political Sciences at the Norwegian University of Science and Technology, 1994–2003 and Senior Research at Ocean Futures, 2005–2012. He has been Visiting Senior Fellow to the Harvard University and the University of California Berkeley, Visiting Professor to the University of Alaska-Fairbanks, and Special Advisor to the Royal Norwegian Ministry of Foreign Affairs. He is an elected member of the Royal Norwegian Society of Sciences and Letters, and of the International Academy of Regional Development and Cooperation in Russia. From 1993–1999, he was head of the International Northern Sea Route Programme (INSROP) between Russian, Japan and Norway, and from 2009 to 2012 he was project leader of Shipping in Arctic Waters of Ocean Futures.

Tor Wergeland is an independent consultant living and working out of Copenhagen and he is also associated senior advisor to Marintek, Trondheim. For more than 20 years he was Processor in Shipping Economics at the Norwegian School of Economics and Business Administration in Bergen, Norway. He has been Professor at what is now the University of Agder, Kristiansand, Norway, as well as at the Copenhagen Business School where he started the MBA in Shipping and Logistics in 2001. He has also been involved in a Maritime MBA at Euromed Management, Marseille. He is co-author of two MBA-targeted textbooks – *Shipping* (1997) and *Shipping Innovation* (2008) – written jointly with Niko Wijnolst. He is associated with Ocean Futures in Oslo relating to shipping in Arctic waters, and with the Copenhagen Institute for Futures Studies.

Foreword

The overall objective of this study is to compare the shipping conditions along the Northeast, Northwest and Transpolar passages on the basis of a state-of-the-art analysis. To fulfill this objective three core premises were applied. First, the concept of sailing conditions was defined at the intersection of a variety of interacting factors such as geopolitics, military affairs, global warming, sea ice melting, international economic trends, resources, competing modes of transportation, environmental challenges, logistics, ocean law and regulations, corporate governance, jurisdictional matters and rights of indigenous peoples. On this premise, the study is multi- and interdisciplinary in approach and substance. Second, the report is to serve applied purposes – to provide a fact basis both for real-life decisions and for developing a comprehensive multiyear research program to fill in, expand, deepen, broaden and supplement existing knowledge of relevance to shipping in Arctic waters. For this reason, the study is empirical and highly fact-oriented, not least revealing the shortcomings of existing knowledge. Third, the study is to a large extent based on secondary scientific material stemming from research programs like INSROP, ARCOP, JANSROP, AMSA, etc., all of which have a singular focus on the Northern Sea Route. To compensate for this bias and provide a satisfactory database for all three passages, the research team had to draw on a much broader range of literature – official as well as research based – and to produce its own data through interviews, not least in Canada. In this way, the study became much more of a research program in its own right than originally planned for.

To secure user relevance and the multidisciplinary character of the report the research team conducted 32 interactive seminars over a period of a year and a half. To overcome the "Tower of Babel" the team members had to learn each other's way of thinking, methodologies, terminologies and approaches. To help out in this respect, multiple experts from research, industry and professional organizations took part in the seminars and contributed valuable insights to the benefit of the outcome. The research team would like to acknowledge the value and usefulness of this participation, which involved the following individuals: Felix H. Tschudi, Ulf Hagen, John Edvard Sundnes and Henrik Falck (Tschudi Shipping Company), Kristin Omholt Jensen (Center for High North Logistics), Egil Rensvik (MARINTEK), Ottar Hermansen, Eivind Ingdahl and Truls Herland (Innovation Norway), Yuki Yoshida (Embassy of Japan), Bengt Ramberg and Atle Nordli (BI Norwegian School of Management), Kristine Offerdal (Norwegian Institute for Defence Studies), Daniel Fjærtoft (Econ Pöyry), Irina Fotcsjenko (Scandinavian Institute of Maritime Law), Tor Husjord (Maritimt Forum Nord), Robert Hermansen and Svein Helge Orheim (The Barents Institute), Oddgeir Danielsen (The Norwegian Barents Secretariat), Tom Tjomsland and Hennig Smidt (Norwegian Shipowners' Association), Johan Petter Barlindhaug and Andrei Mineev (Bodø Graduate School of Business), Knut Espen Solberg, Steven Sawhill and Lars Ingolf Eide (Det Norske Veritas), Øystein Kristiansen (University of Oslo) Gunnar M. Gamlem (Wilh. Wilhelmsen Group).

Special thanks go to Ottar Hermansen (Innovation Norway) and Knut Espen Solberg (Det Norske Veritas) who attended more seminars than any other participants and made particularly valuable contributions. Equally important is the contribution of Jan Magne Markussen, Director of Ocean Futures, for his strenuous efforts in making Shipping in Arctic Waters a commissioned project under the auspices of Ocean Futures. Karl Magnus Eger and Øystein Kristiansen deserve recognition for compiling and organizing multiple pieces of scattered texts into a coherent readable script of structure and logic.

To secure a satisfactory database and to get a realistic backdrop on how to interpret Canada's position on the Northwest Passage, Karl Magnus Eger and the undersigned had several meetings in Ottawa with experts in government and private institutions and with knowledgeable individuals. These interviews and conversations gave much needed and highly valuable insights to the benefit of the study. In random order we would like to express our gratitude to the following persons and institutions for sharing with us their profound expertise in Arctic matters: Victor Santos-Pedro, Ross MacDonald, and Laura Hoy (Transport Canada), Harley Trudeau (Yukon Government), Kari Hoel, Else Berit Eikeland and Jo Sletbak (Royal Norwegian Embassy in Ottawa), Steven C. Bigras, John Bennett and Jean Marie Beaulieu (Canadian Polar Commission), Peter Johnson (University of Ottawa), Peter Jackson, Fiona Robertson and Barb O'Connell (Canadian Coast Guard) Udloriak Hanson and Laurie M. Pelly (Nunavik Tunngavik), John Merritt (Inuit Tapiriit Kanatami), Terry Fenge (Terry Fenge Consulting), Donat Pharand (University of Ottawa), Robert C. Kadas, Manfred Auster and Allison Stewart (Foreign Affairs and International Trade Canada) and Joseph Gough (Secretary to Senator William Rompkey, P.C.).

Our appreciation also goes to Professor Edgar Gold, Dalhousie University, Canada/ University of Brisbane, Australia for setting time aside and sharing with us his unique insights on insurance matters relating to Arctic shipping. The same appreciation goes to Terry Halliday who has contributed a most detailed and useful index covering the complexity of the book in a way making sense across disciplines and sectors.

Last but not least, I express my personal appreciation and admiration to all the members of the research team. Without their interdisciplinary and inter-sectorial curiosity and solid disciplinary in-depth knowledge, this report would have suffered in quality and comprehension. Combined, the team members have invested some of their leisure time to make the study more comprehensive than the funding allowed for. For that they deserve acknowledgement. I will miss our fruitful weekly discussions and, not least, our joint laughter – a most underrated factor of creativity.

It is the hope of the authors that the readers, funders and users of the result of this study will enjoy the report as much as we enjoyed writing it. Any shortcomings related to the content are, of course, the sole responsibility of the authors

Oslo, 25.03.12

Willy Østreng

Project Leader, Ocean Futures

Ocean Futures and the study: *Shipping in Arctic Waters*

About Ocean Futures

Ocean Futures is an independent and interdisciplinary research institute focusing on the High North, the oceans and the deep seabed. A holistic approach is an important feature and aim of all research. All projects are internationally oriented. Ocean Futures has among its staff some of Norway's most experienced researchers in the fields mentioned above.

Ocean Futures has written three reports on the High North for the Storting (Norwegian Parliament): (1) on the border issues in the Barents Sea; (2) the Svalbard Treaty; and (3) the natural resources of the High North. Ocean Futures has presented several reports on oil and gas in the high North in various forums, and safety at sea in the Arctic is a continuing subject of study. Ocean Futures' series of eight Barents Sea maps has won the BCS (British Cartographic Society) Award. Ocean Futures recently published a report on the Framework Conditions for the Norwegian Fishing Fleet. Ocean Futures has IPY (International Polar Year) approval.

About the study

The *Shipping in Arctic Water* study was initiated by Ocean Futures in 2008. There were three main reasons why Ocean Futures wanted to make a comprehensive report on Arctic shipping: (1) the increasing international focus on resources in the High North; (2) climate change; and (3) technical and logistical changes in the high north.

The study was developed with financial support from and cooperation with the Center for High North Logistics (CHNL) in Kirkenes, Norway. CHNL is a foundation, funded by the Tschudi Shipping Company, which acts as an international center for the collection of high-quality and up-to-date information that will contribute to economically viable, do-able, and environmentally- friendly transport and logistics solutions for the High North.

The information is broadly disseminated and will, for example, provide shipping and logistics companies with the best available information in order to make informed business decisions about their future business activities in the Arctic. CHNL has established a Logistics Information Office (LiO) in Kirkenes.

CHNL financed one third of the study. The remaining two-thirds came from Innovation Norway – Finnmark and Ocean Futures respectively.

The study *Shipping in Arctic Waters* has a holistic/interdisciplinary approach and focuses on the Northeast Passage, Trans-Arctic Ocean shipping routes and the Northwest Passage. In addition to transit sailings, the study also covers destinational shipping and intra-Arctic shipping. Six of Ocean Futures' researchers and a researcher from DnV have been involved in the study: Senior Research Fellow Professor Willy Østreng, Ocean Futures (project leader); Researcher Karl Magnus Eger, formerly Ocean Futures, presently advisor at Analyse & Strategy, Multiconsult; Senior Research Fellow Brit Fløistad, Ocean Futures; Dr Arnfinn Jørgensen-Dahl, Senior Research Fellow, Ocean Futures; Senior Research Fellow Lars Lothe, Ocean Futures; Segment Director Arctic Operations and Technology Morten Mejlaender-Larsen, DNV; and Professor Tor Wergeland, Ocean Futures (affiliated). Other researchers from Norway, Russia, Canada, the United States, and Japan also contributed to the study.

This book, *Shipping in Arctic Waters*, is an updated version of the study which was completed in May 2010. Ocean Futures wishes to thank the Port of Narvik, the Norwegian Shipowners Association, and Maritime Forum North for financial support in connection with updating of the report.

In a speech in Kirkenes on 3 February 2011, the former Norwegian Minister of Foreign Affairs, Mr Jonas Gahr Støre, said that "…the report *Shipping in Arctic Waters*, written by Ocean Futures, has become a reference work on Arctic shipping" (see www.regjeringen.no).

Oslo, 12 April 2013

Jan Magne Markussen

Managing Director, Ocean Futures

Summary: Shipping in Arctic Waters

A Comparison of the Northeast, Northwest and Transpolar Passages*

This study compares the shipping conditions along the Northeast Passage (NEP), Northwest Passage (NWP) and Transpolar Passage (TPP) on the basis of a state-of-the-art analysis. The concept of sailing conditions is defined at the intersection of a variety of interacting and interdependent factors such as geopolitics, military affairs, global warming, sea ice melting, international economic trends, resources, competing modes of transportation, environmental challenges, logistics, ocean law and regulations, corporate governance, jurisdictional matters and rights of indigenous peoples. Thus, the study is multi- and interdisciplinary in orientation and approach, opting to define the complexity of operational parameters at work in these unique waters. The comparisons are conducted to provide an imagery of what the state-of-the-art of regional shipping is for the Arctic Ocean as such, for the individual passages and for each of the interacting navigational factors. Within the Passages, three types of sailing routes are addressed: Intra-Arctic routes, i.e., sailing lanes between locations within the Arctic; destination-Arctic routes, i.e., sailing lanes between harbors inside and outside of the region; and transit routes, i.e., sailing lanes between harbors in the Pacific and the Atlantic via the Arctic Ocean. The study also comprises fourtransportation corridors in southern waters connecting the Arctic Passages to world markets: the Northern Maritime Corridor (NMC), connecting the Northeast Passage to the European continent and to the east coast of the USA; the Northern Pacific Corridor (NPC), connecting the NEP, the TPP and the NWP to Asian markets and the western coast of North America; the Fram Corridor (FC) between Greenland and Svalbard, connecting the TPP to the North Atlantic and the western branch of the NMC; and the Davis Corridor (DC), connecting the NWP to the western branch of the NMC.

These are some of the main findings:

- The study defines *Arctic waters* to comply with the IMO Guidelines of 2002, including the waters of the Barents and White Seas. Broadly-based definitions like the ones used

* This chapter has been compiled and written by Willy Østreng, Ocean Futures.

in the AMSA study serve political rather than operational purposes for shipping in the Arctic Ocean.

- Most Arctic resources, in particular oil and gas, are to be found in the Russian Arctic. About 91% of her natural gas production and about 80% of her natural gas reserves are in the Arctic. It is estimated that 90% of Russia's offshore reserves of hydrocarbons are located in the High North. According to the United States Geological Survey 22% of the world's undiscovered oil and gas are to be found in the Arctic and most of it is thought to be in offshore fields in Russian-controlled Arctic waters.
- The ice-edge will steadily migrate northward along with a further thinning and weakening of sea ice. Over the last 30 years, sea ice thickness in the Central Arctic Ocean has decreased by 42%, a decrease of 1.3 meters – from 3.1 to 1.8 meters, with an accompanying reduction of some 73% in the frequency of deep pressure ridges. The influx of multiyear ice from the Central Arctic Ocean to the coastal areas has decreased by 14 percent from 1978 to 1998. This decrease greatly benefits economic activities in coastal waters. Model experiments suggest a further decrease in sea ice thickness of some 30%, and an ice volume decrease between 15 and 40% by 2050. One postulate is that summertime disappearance of the ice cap is possible in the course of this century and that significant areas of the Arctic Ocean may become permanently free of sea ice in summer.
- Improvements in one operational condition for ships in Arctic waters – retreat of sea ice – is countered by the introduction of a new natural challenge – *polar lows,* which are small-scale cyclones moving northward as the sea ice melts. The NEP and NWP may be affected by polar lows in the future.
- There is an obvious and at time considerable distance advantage involved in using the Arctic Ocean between ports in the Pacific and those in the Atlantic, as compared to the Suez and Panama Canals. Most of the North American West coast, the Russian east coast, Japan, Northern Korea and China are all closer to the European Union/European Economic Area through the Arctic than by way of the Mediterranean. Thus viewed, the Arctic Ocean is indeed an "industrial Mediterranean" – a shortcut between the most industrialized regions of the world.
- In pure geographical terms, the TPP provides the shortest transportation distance of the three Passages between the Atlantic and Pacific Oceans. As long as sea ice conditions prevail in the Arctic Ocean and ships have to tack between ice floes to find a suitable navigation channel, none of the passages has a definite distance advantage over the others. In the relation between the passages, the TPP can only provide a distance advantage if the ocean is free of ice. Geopolitically speaking, the shipping industry is bound to focus on a complexity of interacting factors – sea ice conditions, depths, icebreaking capabilities, etc. – to calculate the length of any Arctic transportation voyage.
- Natural parameters provide for cohabitation between military and civil actors in the Arctic Ocean, and for both sectors to fulfill their obligations on their own terms without necessarily interfering too heavily in the activities of the other. The shallow shelf areas seem to be "reserved" by nature for the civil sector – shipping and resource mining – whereas the Central Arctic Basin, providing thousands of meters of depths, is the prime

operational space of strategic submarines. This natural separation scheme is in no way absolute, but it seems to suffice the needs of each sector to conduct most oftheir respective activities without continuous and undue interference from the other. If, in the future, the TPP is used for regular freighting purposes, the shipping industry will share ocean space with strategic submarines. This, however, does not provide an insurmountable risk for clashes of interests in that freighter traffic across the North Pole area will only create additional noise to the noise created by moving sea ice, providing added protection for submarines against detection from enemy sonar.

- The NEP is by far the most attractive of the three passages when it comes to offer spaces of maneuverable ice. This implies that Russia, controlling the waters of the NSR, has a key role in regional shipping. If the sea ice disappears – seasonally or permanently – the power position of the Russian government will be weakened when it comes to transit sailings. In such a scenario the TPP will assume a more prominent position of utility.
- The NMC is by far the safest and most developed of the three connecting corridors. It was established by the EU in collaboration with Russia and Norway. Regional tensions and lack of pressing needs explain why the countries of the North Pacific have not yet formally established what we for the purpose of this study label the "Northern Pacific Corridor". Indications are that this corridor may be established in due time, possibly with China as a prime mover.
- What mode of transport will become the most important in the Euro-Arctic area is an open question. As far as oil and gas are concerned, the choice is between shipping and pipelines. Gas is the dominant resource in the Russian Arctic, especially offshore. How it will be transported also depends on the location of the principal markets. If these markets are mainly in Europe and Asia, pipelines are likely to be the favored mode. Oil and gas for markets outside Europe and Asia will be sent by ship.
- Three factors points to the NEP as the sea route that will be favored in the foreseeable future. Great stretches of the route go along or through parts of the Arctic where great amounts of resources are found or likely to be found both onshore and offshore. The second factor that favors the NEP is the already established and continuously developing infrastructure along the route. And the third factor is the less challenging ice conditions and the longer sailing season during summer.
- With the prospect of less ice-covered waters in the Arctic in the future, it is expected that cruise operators will follow the edge of the ice. The ships are self-contained and most of them carry zodiacs and even hovercrafts, enabling the cruise passengers to explore shallow waters. The chances are high that cruise traffic will interfere with the traditional hunting life of the indigenous peoples, mainly along the NEP and the NWP, so the cruise industry must show social responsibility in order to avoid direct conflicts.
- Several vulnerable Large Marine Ecosystems are identified on the three Arctic transportation passages. No significant negative impacts have been documented for the ecosystems on the NEP, NWP and TPP caused by shipping activities.

- As of the present, there is not an adequate preparedness for oil spills, either on the NEP or the NWP. The lack of sufficient infrastructure, response and necessary combating equipment make the ecosystems vulnerable if accidents should occur.
- From a navigational point of view, the NWP will be the last area where the multiyear ice will disappear. Shipping through these waters will remain risky even in summer in the foreseeable future. It is not likely that the TPP will be used as a regular transport corridor, even in a long time perspective.
- At present, there are no internationally binding requirements when it comes to standards for ice classing of vessels, equipment and crew training to Arctic operational conditions. National regulations apply to Arctic coastal waters, but they are not harmonized, compatible nor adequate. Several nations support an initiative to establish a set of mandatory requirements to ensure safe shipping in the Arctic in the future.
- There are limited aids to navigation and traffic service systems in support of navigation along all passages. Available Search and Rescue facilities diverge between the NEP, NWP and the TPP. Global Positioning Systems (GPS) are adequate for the coastal and marine routes of the NSR, but of restricted value to the NWP. The GPS coverage is not optimal in high latitudes which means that position accuracy for the TPP is insufficient.
- There are no deepwater ports along the NWP. The most adequate ports are found along the NEP. The ports of the NSR are in need of significant improvements in order to meet the standard required by the international shipping industry.
- The NEP is the only one of the passages where sailing fees have already been introduced. The main one is an icebreaker fee for the passage through the NSR. The system of fees has changed several times, where the principle has been that the total traffic should cover the total costs. In a period of declining traffic this led to an increase in fees per ton, and figures from 2003 indicate fee levels that will not attract transit traffic. Currently the situation is non-transparent, and one may conclude that fees are subject to negotiations. If the Russian government wants to attract transit traffic through the Northern Sea Route, the fee levels must be commercially reasonable. The NWP has currently no fee system and the Canadian government will most likely not introduce any fees as they will discourage the economic viability for any transportation on the passage.
- The international insurance market is willing and able to underwrite risks in terms of shipping on the three Arctic transport corridors, but to a very different price. On the basis of the risk factors identified in this study, it is likely that the insurance premium will be higher for the NWP and TPP than for the NEP.
- The entire Arctic Ocean is subject to applicable international ocean law, most prominently UNCLOS, but also other instruments of international law. In addition, norms for good corporate governance and IMO-Guidelines for Arctic shipping will have an influence for how shipping is carried out. The NEP and NWP are in addition governed by national legislation, most importantly passed by Russia and Canada. In the NWP there is a jurisdictional controversy between Canada and the US concerning the status of the

waters in the Canadian Arctic Archipelago and a resembling dispute between the US and Russia pertaining to Russian ice-covered straits along the coastal route of the NSR.

- With regard to the NSR there are some uncertainties when it comes to the definition of Russia Arctic waters, indirectly related to possible implication/interpretation of the Sector principle laid down in a Russian Decree of 1926. Those uncertainties may, under certain specific conditions, be claimed to have a bearing on the legal status of sea ice and the freedom of the High Seas of the Arctic Ocean.
- Indigenous peoples in the Arctic have worked steadfastly towards self-determination to preserve their own culture and way of life as well as having a share in the resource development in the territories on which they have lived for centuries. In such development schemes, ships may on the one hand encroach on indigenous peoples' traditional way of life when going through areas used for hunting and fishing or causing cultural disturbances. On the other hand, shipping is an integrated part of the economic development of the Arctic, welcomed also by indigenous peoples.
- When comparing the cost-saving aspects of using the three Arctic passages instead of the Suez Canal between Yokohama and Hamburg and between Shanghai and Hamburg, the Arctic alternatives turn out to be the most economic (under certain specific conditions). Of the three Arctic routes, the TPP comes out as the most economic, simply because crossing the pole is the shortest route if ice is not too much of an obstacle. This is also by far the least likely route to become accessible without icebreaker support in the near future. The NSR seems marginally better than the NWP, but then there is the question of icebreaker fees. The average Russian fee rose to $23 per ton in 2003, and this level will kill the economics for both cases. The fee must be way below $15 per ton for commercial activities to seek an Arctic solution, given current oil prices. Most experts seem to believe we are facing much higher oil prices in the future, which will make the use of Arctic passages more attractive. The analysis includes practical experiences learned from transits through these waters in the period 2009–2011.

Introduction: An operational view of Arctic waters*

At present, the Arctic attracts economic attention for three main reasons: its abundance of strategically important and exploitable resources, its rapidly dwindling sea ice, and its shortcut location between three of the most industrialized and developed continents of the world. Arctic waters are getting bluer, more accessible, exploitable and increasingly attractive for economic utilisation. This may in due time open up a new chapter in Arctic shipping, which is the topic of this book.

1 Purpose and conduct of study

The overall purpose of this study is to identify determining factors influencing shipping in Arctic waters. It comprises onshore and offshore activities that may shape and affect present and future sailing conditions as they manifest in a complex web of interacting factors – geopolitical, economic, logistic, social, cultural, military-strategic, environmental, political, legal, technological and natural. The main focus is on the challenges and opportunities of shipping along the Northeast Passage of which the Northern Sea Route (NSR) is the main part, the Northwest Passage (NWP) and the Transpolar Passage (TPP) running through the high seas of the Arctic Ocean. Three types of sailing routes are addressed: Intra-Arctic routes between ports within the Arctic, Destination Arctic-routes between ports inside and outside of the region, and Transit routes between ports in the Pacific and the Atlantic Oceans via the Arctic Ocean. The two latter connect with transportation corridors running southward from the Arctic to markets in the south. As these routes are functionally integrated with the three passages of the Arctic Ocean, their operational conditions are part of the study. The approach is empirical, applied and holistic, aiming at defining the state-of-the-art when it comes to shipping in these waters, which, in turn, may be useful

* This chapter has been written by Willy Østreng, Ocean Futures.

in forging an operational and applied research agenda for the future. The book comprises eight chapters dealing with different interrelated aspects of arctic navigation.

In this chapter the reader is introduced to the definitional challenges of the notion of Arctic waters with the aim of arriving at a definition that meets the needs and challenges of shipping in ice-infested waters.

Chapter 1 discusses two types of interconnected transportation routes: the *routes of the Arctic Ocean* – the NEP, NWP and TPP, and *routes in southern waters* connecting the Arctic to world markets, i.e., the Northern Maritime Corridor (NMC), the Northern Pacific Corridor (NPC), the Fram Corridor (FC) and the Davis Corridor (DC). The navigability of these routes is discussed on the basis of their geographical features, natural conditions, political significance, legal characteristics, usage pattern and historical background.

Chapter 2 presents an overall imagery of the geopolitical features of the Arctic as expressed in its location in between three continents, assumed abundance of strategically-important resources, continuous sea ice reductions, unique environmental fragility, regulatory affinity to global ocean conventions and international power politics.

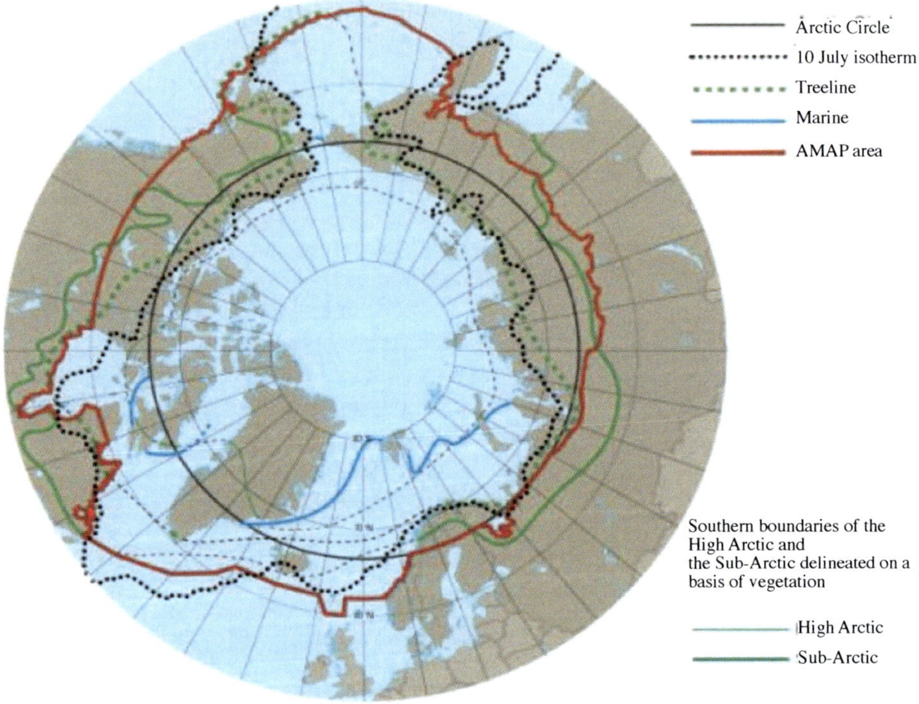

Source: AMAP (1998)

Figure 1 Different definitions of the Arctic.

Chapter 3 discusses the volume, harvesting and location of regional petroleum, mineral, forest and living resources in the context of world economic developments. The address is on offshore as well as onshore resource deposits. The discussion is related to modes of resource transportation, i.e., shipping, rivers, roads, pipelines and railways. The cruise and tourism industry is an object of special attention in the chapter.

Chapter 4 focuses on the environmental challenges and other operating factors affecting shipping in Arctic waters. The purpose is to identify the most important aspects of natural polar conditions to navigation, to assess shipping as an independent source of pollution and the role of sea ice melting in ship-based transport.

Chapter 5 discusses the operational aspects of Arctic shipping along the three transportation passages in terms of security, icebreaking technology, crew competence and education, safety, rescue, marine insurance, ports, cost structure, etc.

Chapter 6 addresses questions related to regional governance as expressed in existing international ocean law, soft laws, domestic laws, human rights (indigenous people) and jurisdictional issues.

In Chapter 7 the navigability of the individual passages and routes are compared to highlight the varying operational conditions of the coastal and central areas of the Arctic Ocean. The analysis ends with a commercial shipping assessment of the transportation passages of the Arctic Ocean.

The operational relevance of the book depends to a large extent on the preciseness of the concept of Arctic waters. As will be demonstrated, the term has throughout history been defined in multiple ways, serving political as well as different operational purposes. In this introduction, the term is defined to fit the operational needs and constraints of the shipping industry.

2 The elusive Arctic

Many criteria have been used to delineate the Arctic from the temperate zone. None of them has achieved common approval or acceptance.

The oldest and most common definition focuses on solar radiation, defining the Arctic as the area north of the Arctic Circle (66 degrees 32 minutes North), which encircles the area of the midnight sun. The area is about 21 million square kilometers in extension – making up about 8 percent of the Earth's surface.[1] This definition excludes sub-regions like the Bering Strait, the White Sea, the southern part of Greenland and the Hudson Bay (see Figure 1). In most expert opinions, these areas are as arctic in natural conditions as most of the areas situated north of the Circle. In 1888–1889, Fridtjof Nansen crossed the inland ice of Greenland using a route to the south of the Arctic Circle.[2] Polar historians are unanimous that this daring event is an undisputable part of the exploratory history of the Arctic. On grounds like these, the noted Canadian researcher, Moira Dunbar has characterized the Arctic Circle – a purely astronomical concept – to be meaningless from most points of view.[3] A more reasonable approach, in her mind, would be to define the Arctic by physical criteria that are domestic to the region.

In line with this thinking, multiple alternative criteria have been suggested and used: the *tree line*, the *10 degree C July isotherm*, the *continuous permafrost,* the *sea ice cover,* the *marine boundary between cool and warm waters*, etc. (see Figure 1). A fairly popular and long-lived assumption has been that the limits of the tree line, which in certain areas coincide with the 10 degree July isotherm, are more meaningful from the point of view of human activity than any of the climatic ones.[4] Although the terrestrial and marine environments vary considerably throughout this area, they nevertheless share certain Arctic characteristics that distinguish them from temperate regions, including cold temperatures, extensive snow and ice cover, large seasonal fluctuations in solar energy, continuous permafrost, polar darkness and lack of nutrients.[5] Thus, in terms of natural features, this definition makes the Arctic stands out as a relatively coherent and uniform region. But also these criteria suffer from weaknesses that act counter to common perceptions of what the Arctic is and is not. For instance, the blue oceans of the North Atlantic and North Pacific are here defined as constituent parts of the Arctic along with the ice-covered waters of the Central Arctic Ocean (see Chapter 1). This definition also excludes certain coastal areas of the Northeast and Northwest Passages from the Arctic.

In short: the application of most single definitional criteria does not suffice to *include* all geographical areas hosting polar characteristics in the concept of the Arctic or, for that matter, to *exclude* areas featuring the characteristics of the temperate zone. The differences in aerial extension between the various definitions amount to thousands of square kilometers. The competing delineations depend upon the perspective from which one approaches it. This is not problematic when working within the confines of single scientific disciplines; rather, the difficulty arises when one seeks a definition to suit all subject areas – the multi- and cross-sections of the sciences as well as of industries and politics.

In the multilateral process of defining the Arctic in a way that is relevant for all areas of science, the Arctic Monitoring Assessment Program (AMAP) of the Arctic Council decided not to define the region in too precise terms. Instead, the Program provides guidelines about a core geographical area to be used by all scientists irrespective of disciplinary orientation. Although the boundaries are not fixed once and for all they are to be located somewhere in the latitudinal area between 60 degrees N and the Arctic Circle. This approach provides flexibility in boundary drawing without repudiating borders as a useful measure for multi- and interdisciplinary science. The flexibility achieved also shows in the modifications of borders suggested by AMAP. One such modification is that the Hudson Bay and the White Sea – which were excluded as Arctic territory by the Arctic Circle – are defined as integral parts of the Arctic. A second modification concerns the terrestrial environment of the eight Arctic states. Here the southern boundary of the region shall not be determined by any concrete natural criteria but by the individual countries in locations of their own liking between the Arctic Circle and 60 degree N (see Figure 1).[6] In this definition, parts of the land territories of all the eight Arctic states[7] have been defined as belonging to the Arctic. Previously, only the five littoral states to the Arctic Ocean were thought of as having land territory in the region. What is more, by this definition the blue warm waters of the Norwegian Sea have become part of Arctic waters, which again deviates from the common perception of what is characteristic of polar natural conditions. By this extension, the Northern Maritime Corridor becomes an integral part of Arctic waters along with the three passages contained within the ice-covered part of the Arctic Ocean (see Chapter 1 and

Figure 1). In AMAP terms, the Arctic covers an area of approximately 33.4 million square kilometers, out of which 60 percent is defined as Arctic water.[9] Here, the AMAP applies multiple scientific, political and pragmatic criteria that have been blended together to reach consensus across sectors and between the states. However, the ambiguity and flexibility of this definition causes overlaps with the Sub-Arctic, which in principle should be the transition zone between the Arctic proper and the temperate zone (see Figure 1).

The Sub-Arctic is even more difficult to define than the Arctic proper. In fact, no completely satisfactory definition has ever been put forward, and it has become necessary to compromise on some arbitrary climatic boundaries to sketch out what the Sub-Arctic is. The complexity of the matter is shown in Figure 1 where the boundaries of the Sub-Arctic at certain stretches in North America and Siberia are located north of the southern boundary of the Arctic proper. In other places the borderline is situated south of the Arctic as it should be according to theory. The AMAP definition has become a "bastard" between two definitional traditions and the Sub-Arctic. How then does this broad scale definition relate to recent studies on shipping in these waters?

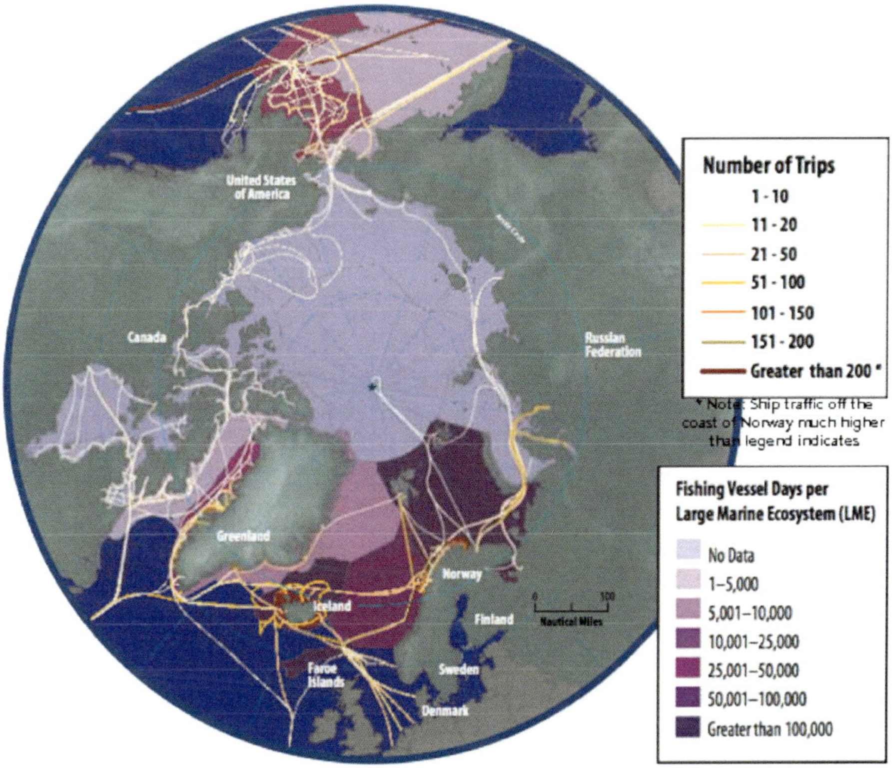

Source: AMSA (2009), p. 73

Figure 2 Overview of all vessel activity in Arctic waters for 2004, including fishing vessels.

3 The shipping industry and broad-based definitions

Traditionally, and based purely on scientific criteria, the boundaries of the exit areas – or the open sea borders – of the Arctic Ocean have been drawn along the Barents shelf edge from Norway to Svalbard, across the Fram Strait, down the western margin of the Canadian Archipelago and across the Bering Strait. Thus defined, and including the Canadian Archipelago, the total area of the Arctic Ocean is 11.5 million square kilometers.[10] In AMAP terms, the "marine Arctic" or "Arctic waters" have been extended far beyond the Arctic Ocean to denote an area that includes Baffin, Hudson and James Bays; the Labrador, Greenland, Iceland, Norwegian, and Bering Seas (see Chapter 1 and Figure 1).[10] This definition has been embraced by some of the most cited Arctic shipping assessments of the present.

The *Arctic Marine Shipping Assessment* (AMSA) of the Arctic Council has adopted the AMAP definition lock, stock and barrel, and allowed the five coastal states to define their own Arctic waters for the purpose of data collection efforts. In this way, the AMSA definition stands out as the "least common (political?) denominator" of the eight Arctic States, covering an extensive area of a variety of operational conditions. Shipping is here defined to cover all types of marine transport, including tankers, bulk carriers, offshore supply vessels, passenger ships, tug/barge combinations, fishing vessels, ferries, research vessels and government and commercial icebreakers.[11] The *Canadian Arctic Shipping Assessment* (CASA) follows suit when claiming that shipping "... is best defined in the broadest terms, whilst recognizing that there will be differences that depend on the issues under consideration".[12] Also in the CASA study, shipping is taken to include: cargo vessels, passenger vessels, military/defence vessels, navigational support vessels, oil/production vessels, support vessels and fishing vessels.[13] These lists suggest that Arctic waters comprise blue waters as well. Several of these vessels – among them fishing boats – have no sea ice capability and are functionally bound to operate in open waters only. As shown in Figure 2, the most busy "Arctic waters" of vessel activity in 2004 were, according to the AMSA, the Norwegian and Bering Seas. Fishing vessels were responsible for an important part of this total.

In this definition, Arctic waters span a variety of different sea areas – cold, warm, ice free, ice covered, salty, less salty, etc. – waters of no, few and multiple polar characteristics. By implication, AMSA defines these waters along three continuums. The first continuum is, as discussed above, *definitional* extending from temperate warm waters in the southern part of the Norwegian Sea via the cold but ice free waters of the Barents Sea to the permanently ice covered waters of the Central Arctic Ocean. Only the northern part of this long ocean stretch contains true Arctic characteristics. The second continuum is *seasonal* where operational conditions vary throughout the year according to the annual cyclical process of melting and freezing of sea ice. This continuum relates to the Arctic Ocean only (as defined by scientific criteria, see above). The third continuum is *generational* and concerns the man-made long-term melting and reduction of sea ice in the Arctic Ocean. This continuum spans from the continuous reduction of sea ice in the Central Arctic Ocean to the geographically expanding ice free stretches of the Northeast and Northwest Passages in summer, but not in winter (see Chapters 1 and 5). The latter continuum has moved the shipping industry to take a second look at the sailing conditions of the Arctic Ocean. If all three continuums are put together – as the AMSA study does – the challenges encountered by ships in these waters are gradual as one sails northward from the southern part of the Norwegian Sea and into and through the Arctic Ocean. In the south, the ships may encounter storms that are domestic to the temperate zone; when reaching the ice free parts of the Barents Sea some of the operational conditions of the Arctic such as icing from sea spray and polar darkness in the winter will pose a challenge to shipping. Further north and east the ship will enter into areas of light sea ice conditions in addition to icing, polar darkness, fog, etc., followed by more and more heavy ice conditions.

The geographical broadness of the AMSA definition has inspired politically motivated definitions also by states. In 2011, the Icelandic government claimed publically that Iceland is "the only country" among the Arctic eight that is located both "entirely within the Arctic region" and "in the centre of the North Atlantic Ocean".[14] This move sided firmly against the ministrial meetings of May 2008 and March 2010 between the five littoral states to the Arctic Ocean in which Iceland was not invited (see Chapter 2). Later, and in support of the government, the Icelandic Parliament passed a Resolution to secure "Iceland's position as a coastal State within the Arctic region", promoting the view that the Arctic "should not be limited to a narrow geographical definition but rather be viewed as an extensive area when it comes to ecological, economic, political and security matters".[15] What Icelandic authorities opt for is to redefine the southern borders of the Arctic Ocean for Iceland to become an Arctic coastal state and to extend the group of Arctic five to become the Arctic six (see Chapter 2). The Icelandic Parliament is of course right when pointing out that the North Atlantic connects and interacts with the Arctic Ocean both in terms of ecology, economy, politics and security. However, the same is true with the relationship between the North Atlantic and the South Atlantic, or for that matter between all oceans of the world which are physically connected through straits and broad ocean areas. Undoubtedly, this fact poses a political challenge and calls for international cooperation, but is not in its own right a definitional criteria to delimit one region from another. Planet Earth is the only geographical term comprising all features and interactions between regions and world oceans. Thus, the Icelandic move is serving political purposes inspired by the AMSA definition.

Due to global warming, the Arctic itself is moving north. The tree lines and isotherms move towards the North Pole on a continuous basis, albeit at varying rates depending on location. The sea ice edge follows suit. In the long term, the Arctic Ocean may even be ice free in summer. As has been observed, "As an icy milieu the Arctic is shrinking. The north is being denorthified".[16] Whereas nature makes the Arctic shrink in areal extent, politics makes it expand.

On these grounds, there is something basically wrong with broadly-based politically-influenced definitions. They all comprise temperate waters that have been sailed for centuries and where the operational parameters are well known to the shipping industry. What makes the Arctic a true shipping challenge is its domestic characteristics – sea ice, cold temperatures, spraying, whitening, pitch darkness in winter, etc. In this perspective, there is a need to get away from "politically motivated" definitions of Arctic waters like the ones applied by the AMSA and CASA studies and to focus on the issue specific needs of the shipping industry in the Arctic Ocean proper. The AMSA borders are much too liberal to make the concept of Arctic waters a uniform geographical and operational entity. As seen from the point of view of shipping, delineations between oceans have one function only: *to encircle areas of operations that present relatively uniform challenges*. As is well known, ships built for sea ice operations are performing inefficiently in blue waters.

Today, some define the Arctic as the physical and biological environment spanning across the northern reaches of Asia, Europe and North America, and the expanses of sea and ocean in between them.[17] Here the Arctic is thought of as a group of concepts and attributes unique to a region far from the industrial centers of the world.[18] Definitional attempts of this kind share a common feature: The region is defined as borderless and portrayed as an aerial entity fading away somewhere in the south without telling *where* and *why* it happens. In principle as well as in operational terms, lack of defined borders creates on the one hand uncertainties about the extension of the region, allowing for definitions like the ones suggested by AMSA and CASA. On the other hand, it opens up for an issue-specific border setting, leaving it to the mining and shipping industry to create borders that make sense from their respective operational points of view. One issue-specific definition that has achieved broad international acceptance in the shipping industry is launched and applied by the *International Maritime Organisation* (IMO).

4 Issue specific definitions relating to shipping in Arctic waters

In December 2002 the IMO issued *Guidelines for Ships Operating in Arctic Ice-covered Waters* (IMO Guidelines), in which it made sea ice concentrations the only definitional criteria to define Arctic waters. The IMO definition states that waters that have "... sea ice concentrations of 1/10 coverage or greater.. and which pose a structural risk to ships" are Arctic in character (see Chapter 5).[19] This definition excludes the southern part of the Barents and White Seas, and Hudson Bay, but includes the east coast of Greenland down to its southernmost point (see Figure 3, making Fridtjof Nansen's crossing of the Greenlandic ice sheet an Arctic endeavour, see above). Since the Denmark strait between Iceland and Greenland is part of this definition, Iceland may rightfully claim to border on Arctic waters, if not on the Arctic Ocean. In this definition, and counter to the definitions applied by the AMSA and CASA studies, a polar "ship is defined as any vessel required to complying with the 1974 SOLAS Convention" (see Chapter 5). This excludes from the area of application fishing vessels, pleasure yachts, wooden ships of primitive build, cargo ships of less than 500 gross tonnage and naval vessels, whereas passenger ships and cargo ships of 500 gross tonnages or more engaged on international voyages are subject to regulation.[20] In this definition the three passages, that can be delineated by fairly exact historical coordinates, and the three types of routes located in the ice covered waters of the High North are all safe within the confines of the IMO definition. At the same time, connecting blue water transportation corridors extending from the Arctic to southern markets are, with the exception of parts of the FC and DC, singled out on the basis of their own ice free operational characteristics (see Chapter 1).

The IMO definition applies just one polar criterion – sea ice concentration – to delineate its waters of operations. In so doing, it is Arctic specific and therefore useful in regional shipping operations. In this study we take the IMO definition as our point of departure but

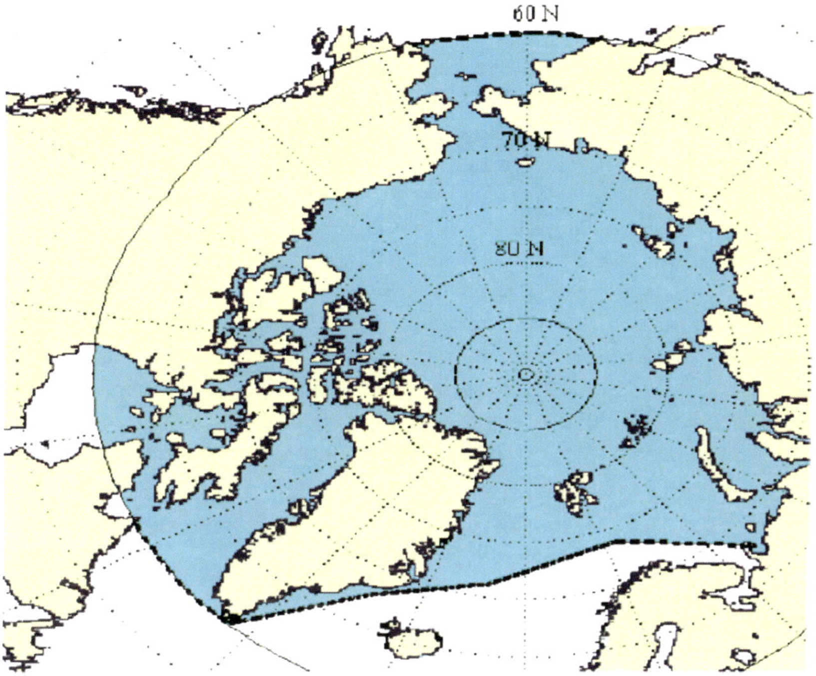

Source: IMO Guidelines (2002)

Figure 3 The IMO definition of Arctic waters.

add two more sea areas – the White Sea, which freezes over in winter and the southern part of the Barents Sea – which host some of the other natural characteristics of Arctic waters – sea ice spraying, cold temperatures, darkness in winter, etc., and even polar lows, which are discussed in Chapter 1. The extension of sea ice in the IMO definition basically complies with the sea ice coverage of the Arctic Ocean in winter, which in most climate models is projected to be the same in the long term (despite global warming).

IMO used *60 degree N* to define Arctic waters in North America (see Figure 3). The same parallel can be used to identify those parts of the Canadian and Alaskan land territories that in due time may prove to have a bearing on shipping in these waters. Thus, in functional and operational terms we regard these land territories, including Greenland, to be part of the Arctic region. On the Eurasian side, the land territories north of the *tree line* in Russia and Norway are part of the Arctic because these areas already have a function in Arctic shipping (see Figure 1).

5 Some concluding remarks

In short, all of the different criteria used to reach a general agreement on the delineation of the Arctic from southern zones are arbitrary and only partially satisfactory. Although the AMAP, AMSA and CASA definitions have reached a certain degree of international acceptance, the business of reaching an overall consensus on what the Arctic is or is not is still a matter of functionality and debate. As of the present, competing definitions live side by side and are used interchangeably as if they cover identical territories. Representatives of different industries are therefore well advised to be aware of the definitional confusions applying to the region, not least in multisector projects like shipping.

The only definitions to be used consistently for operational purposes seem to be functional – issue-specific in character – reflecting the uniqueness and pragmatic needs of the respective industries. In this study, the concept of Arctic waters comprises the areas of the IMO definition plus the White Sea and the southern part of the Barents Sea. These waters are in turn part of the Arctic region which also comprises the land areas north of the tree line in Eurasia and the 60th parallel in North America.

Notes

1 AMSA (2009), p.19
2 Vogt (2011), p. 65
3 Dunbar (1966), p. 4
4 Dunbar (1966), p. 4
5 AMAP (1998), p. 117
6 AMAP (1998), p. 7
7 Canada, Denmark (Greenland and the Faroe islands), Finland. Iceland, Norway, Russia, Sweden and the United States (Alaska)
8 AMAP (1998), pp. 7 and 10
9 ACIA (2005), p. 26
10 AMAP (2007), p. 26
11 AMSA (2009), p.11
12 CASA (2007), p. 3
13 CASA (2007), p. 3
14 Heininen (2011), pp.33-35
15 Althingi (2011) , p. 1
16 Griffiths (2012), p. 2
17 Polunin (1951) and Barry and Ives (1974)
18 Rogers and Rowley (1978)
19 IMO Guidelines (2002), G-3.2.2
20 IMO Guidelines (2002), G-2.3

1 The transportation passages of the Arctic Ocean and connecting corridors in southern waters*

The purpose of this chapter is twofold.

First, we define the transportation passages of the Arctic Ocean on the basis of their geographical features, natural conditions, political significance, legal characteristics and usage pattern, displaying their distinctions, interrelations and eventual overlaps. The main focus will be on the *Northeast Passage (*NEP), of which the *Northern Sea Route* (NSR) is the main part, the *Northwest Passage* (NWP) and on the *Transpolar Passage* (TPP) running through the Central Arctic Ocean outside of coastal state jurisdiction (see Figure 1.1). This focus will allow us to address three types of sailing routes within each passage: *intra-Arctic routes*, i.e., sailing lanes between locations within the Arctic; *destination Arctic routes*, i.e., sailing lanes between harbors inside and outside of the region; and *transit routes*, i.e., sailing lanes between harbors in the Pacific and the Atlantic via the Arctic Ocean.

Second, we connect Arctic passages to world markets by interconnect transportation routes through southern waters. Here the focus will be on four routes: the *Northern Maritime Corridor* (NMC), connecting the NSR/NEP to the European continent and the east coast of the USA; the *Northern Pacific Corridor,* connecting the NSR, TPP and NWP to Asian markets and the western coast of North America; the *Fram Corridor* (FC) between Greenland and Svalbard, connecting the TPP to the North Atlantic and ultimately to the NMC; and the *Davis Corridor* (DC) connecting the NWP to the western branch of the NMC and the east coast of North America. Thus, the function of this chapter is to serve as an overarching definitional and organizing framework for the ensuing chapters.

* This chapter has been written by Willy Østreng, Ocean Futures. Input on logistics has been provided by Karl Magnus Eger, Ocean Futures

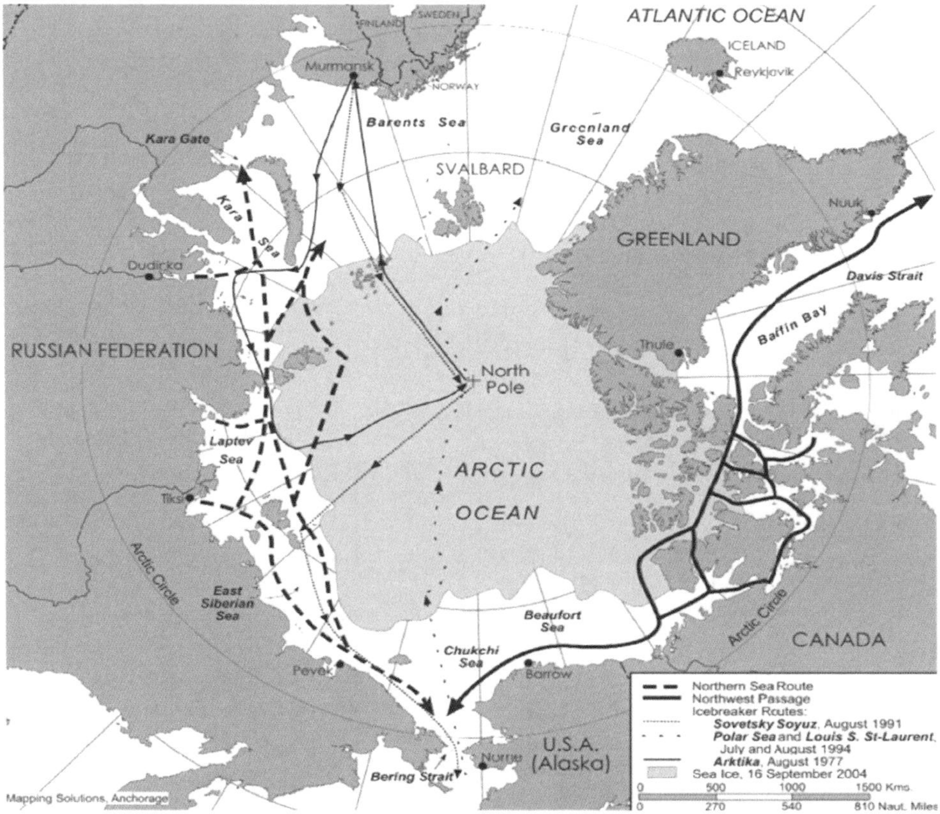

Source: Mapping Solutions, Lawson Brigham, USARC Anchorage 2006

Figure 1.1 The Arctic Ocean with transportation corridors.

1.1 The transportation passages of the Arctic Ocean

The assumption that the Arctic Ocean has three partly separated transportation passages dates back approximately 500 years. In 1527, the English merchant Robert Thorne suggested the existence of three alternative sea routes through these icy waters, "all of which were supposed to lead to the spice markets of the East Indies": the NWP, the NEP and the "open polar sea-route via the North Pole".[1] This idea stood the test of time and was gradually and imaginatively developed, first by the famous Russian scientist Mikhail Lomonosov who in the later part of the 17th century claimed that the Central Arctic Ocean was ice free and easily navigable between the Far East and Europe. A hundred years later, the German geographer August Peterman followed suit, explaining that the ice free part of the Arctic Ocean was due to the warming effect of the Gulf Stream. The assumptions of an ice free Central Arctic Ocean had no scientific justification or foundation, but have attracted renewed interest in light of global warming. Today, all of Thorne's three passages have been tested and navigated by surface ships.

The Northeast and Northwest passages are often perceived as coastal sea lanes, whereas the Transpolar Passage is assumed to be a mid-ocean route across the North Pole to and from ports in the Pacific and Atlantic. This perception is far from accurate. Due to the presence of sea ice, neither of these transportation passages can offer ships a single set channel to follow. In practice, ships are forced to follow the channel that offers the best ice and navigational conditions at any one time and place. Thus, each of them is more like a broad transportation corridor stretching out in the north–south direction, containing several alternative navigational channels and fairly huge expanses of ice-infested waters. The corridor feature of these passages implies that they occupy broad stretches of waters that under certain specific circumstances and on occasions make them overlap and interact. In sum, the three corridors occupy the whole of the Arctic Ocean, which covers an area of 14.75 million sq. km and carries a volume of 18 million cubic km. of water. The lack of accurate geographical coordinates in the north–south direction leaves it to future politics and international law to delimit them from each other.

1.1.1 The Northeast Passage and the Northern Sea Route

Politics were involved when the Russian government introduced the concept of the Northern Sea Route as constituting the main part of the NEP in the early 1930s. Therefore we will start our definitional exercise of the routes with the relationship between the NEP and the NSR.

Two approaches are often applied to determine the coordinates of the NSR: an *official definition* as found in Russian laws and regulations, and an *unofficial Russian functional definition* based on a mixture of organizational, operational and geopolitical criteria. The former restricts the route geographically to Arctic waters claimed to be national and under the exclusive national jurisdiction of the Russian Federation, whilst the latter extends it geographically to include additional expanses of international waters in the Atlantic and Pacific Oceans (see definitional difficulties in Chapter 1).

The official Russian definition of the NSR

According to political perception and legal regulations in Russia,[2] the NSR stretches from Novaya Zemlja in the west (meridian 168 degrees 58 minutes and 37 seconds west) to the Bering Strait in the east (parallel 66 degree north). The establishment of the NSR as a separate part of the NEP was decided by the Council of People's Commissars of the USSR on 17 December 1932, which marks the beginning of the NSR as an administered, legal entity under full Soviet jurisdiction and control.[3] It comprises the main part of the NEP which, with the addition of the waters of the Barents Sea, connects the Atlantic and Pacific Oceans along the entire length of the northern coast of Eurasia.

The NSR is a series of different sailing lanes, and ice conditions at any one time and place will decide the sailing course to be set. The route covers some 2,200 to 2,900 nautical miles of ice-infested waters (see Figure 1.2). It consists of a series of marginal seas – the Kara Sea, the Laptev Sea, the East Siberian Sea and the Chukchi Sea – which are linked by some 58 straits running through three archipelagos – the Novaya Zemlja, the Severnaya

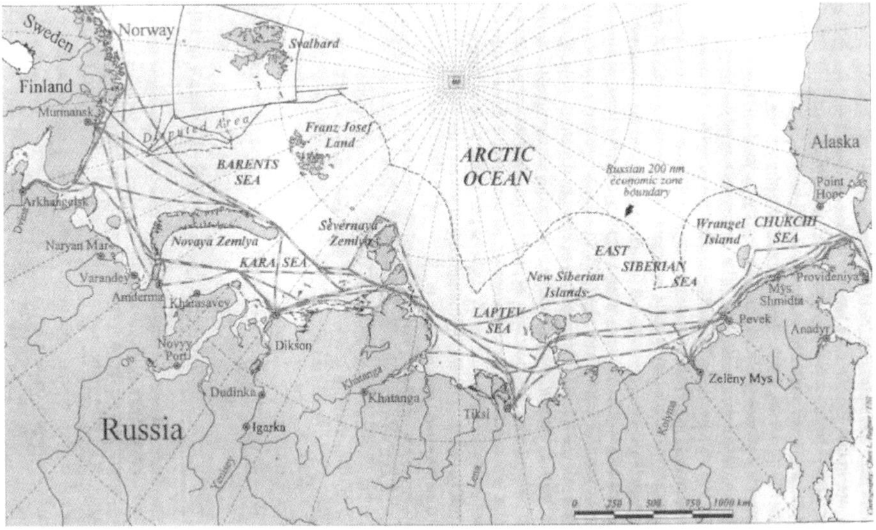

Source: Løvås & Brude, INSROP GIS, 1999

Figure 1.2 The northern sea route.

Zemlja and the New Siberian Islands. At times, surface vessels operating in convoys are forced to proceed due north of the large island masses as a result of the accumulation of pack ice in the straits,[4] which may be clogged with sea ice. Ice conditions are in general more difficult along the eastern extremity of the route than in the west. In the Laptev, East Siberian and south-western Chukchi seas five *ice massifs* – large areas of close and very close ice – are identified (see Figure 1.3 and Chapter 4). These massifs often block the entrances to important navigational straits along the route, among them the Long Strait and Vilkitskii Strait. Although, some of these ice massifs are relatively stable, on rare occasions they disappear at the end of the melt season, but reoccur again in winter.[5] The eastern sector is also the part of the route with the most shallow shelf areas. The East Siberian Sea has an average depth of 58 meters and the Chukchi Sea of 88 meters. The shallowness of the shelf is the most pronounced in the straits, with minimum depths of 8 meters (see Figure 1.6 and Chapter 5).[6] This affects the size, volume and draft of ships (see Chapter 5).

The ocean areas west of the Yamal Peninsula are fortunate in having a slightly deeper shelf and lighter ice conditions in average than the eastern sector. This is partly due to the circumstance that the Kara Sea is to the north surrounded by several archipelagos which usually prevent heavy multiyear ice from the Central Arctic Ocean from penetrating into these waters. Multiyear ice, which is extremely hard and consequently a serious obstacle to navigation (see below), has survived the summer melt season and is typically 1–5 meters thick.[7] The eastern sector lacks this kind of land protection and is more open to the influx of multiyear ice from the Central Arctic Basin. However, even in the east, ice conditions are changing due to global warming. Here, new extreme minima of summer ice extent have been established repeatedly ever since 1979.[8] In six of the last nine years, the Chukchi Sea was ice free with periods extending from 1 week to as much as 2.5 months. In contrast, there was always ice over the Chukchi Sea shelf in all of the previous 20 years (1979–98).[9] In 2010, the Norilsk-Nickel operated vessel "Monchegorsk" became the first cargo ship to sail the entire NSR without icebreaker assistance.

In light of global warming the question has been raised whether we are "approaching a break even point when the extra investment cost on ice capability of the merchant ship is more cost effective than providing an icebreaker escort."[10] This is a valid question, but still debatable, not least for the NWP and TPP, and even for certain portions of the NSR (see discussion on the *concept of ice free* below).

In cases where the convoys along the NSR enter the high sea, prominent Soviet ocean law experts claim that the navigation lanes used are national and under full Russian control and jurisdiction: "The integral nature of the Northern Sea Route as a transport route is not affected by the fact that individual portions of it, at one time or another, may pass outside the aforesaid boundaries (i.e. boundaries of internal waters, territorial waters and economic zone) where the USSR exercises its sovereign rights or sovereignty in full (i.e. it may pass into the high seas)".[11] Thus, as long as part of the voyage includes waters under Russian jurisdiction, the Russian Federation has, according to this reasoning, the right to define the NSR to include sea lanes running beyond its own economic zone in high latitudes, even close to the North Pole. In principle, this implies that all conceivable lanes south of the North Pole, and even across the Pole itself, might be part of the NSR as long as the voyage passes through North Russian coastal waters. In line with this reasoning, Russian scientists

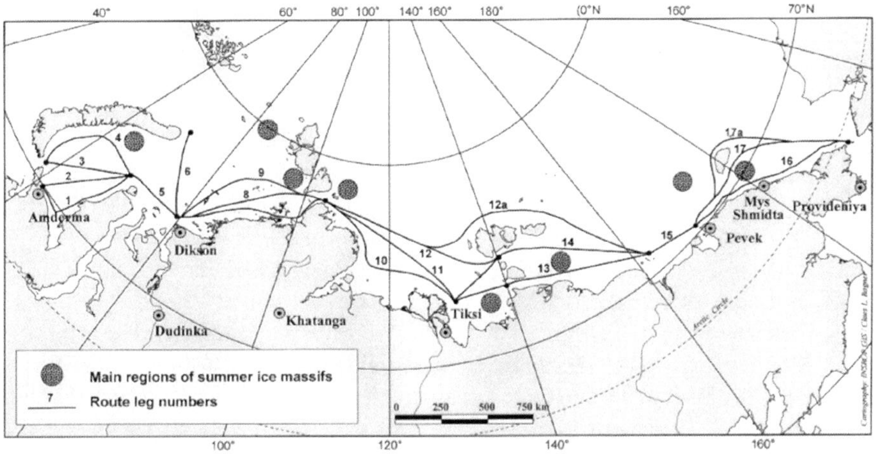

Source: INSROP Working Paper no.108 , 1998

Figure 1.3 The ice massifs along the NSR.

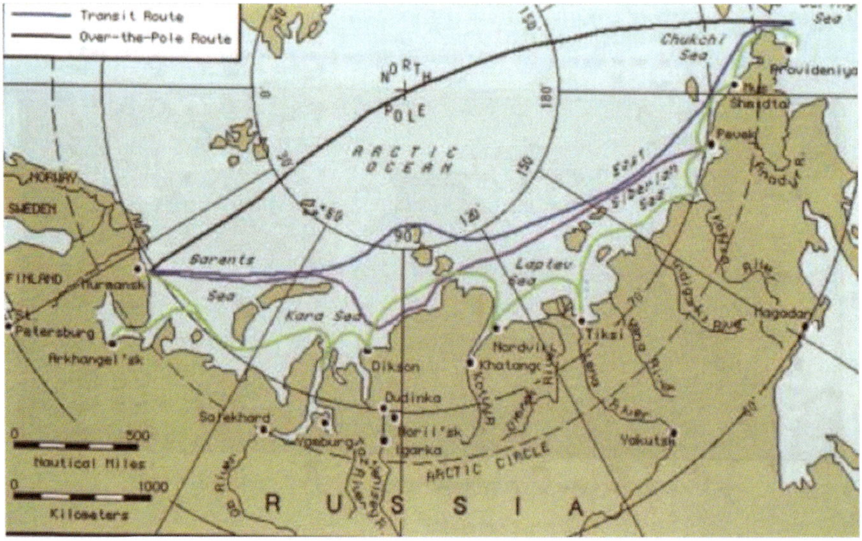

Source: http://www.arctic-search.com

Figure 1.4 The NSR defined as part of Russian coastal waters and the High Seas

employed by the Federation claim that, "Voyages along the NSR are carried out along coastal, marine, high-latitudinal and near-pole routes. Coastal routes are the most traditional"..., whereas "the fourth route, which is 700 miles shorter than the coastal route, passes the large circle across the geographical North Pole" (Figure 1.4). [12] In this interpretation, the NSR overlaps with the TPP, covering huge expanses of the high seas that, according to the *UN Convention of the Law of the Sea of 1982* (UNCLOS), is open to all nations and where ships are subject to flag state jurisdiction only (see Chapter 6).

The NSR is part of an interconnected rectangular transportation system for the Russian north. The legs of the rectangle consist of, in addition to the passage itself, the big Siberian rivers and the east–west running railways in the south connecting with the rivers thousands of miles from the coast. Ocean-going vessels sail from the port of Igarka, which is 670 km south of the estuary of Yenisei, to Yakutsk which is 1160 km south of Igarka. The rivers Ob, Yenisei, Lena and Kolyma are navigable to the Trans-Siberian railway which is 2270 km south of Siberian coast. The river Lena connects with the Baikal-Amur railway.[13] As has been stated, "The NSR and the river system is the primary mode of transportation in this remote part of the world apart from airborne transportation. Nearly all human activity in the Russian Arctic is in some way dependent on the NSR."[14] In this perspective and interpretation the NSR extends northward and southward from the coast, servicing huge ocean and land territories, covering thousands of kilometers from the North Pole to the railways of the south.

Unofficial functional definitions of the NSR

The official Russian definition operates with fixed geographical endpoints in the east–west direction: the Bering Strait in the east and the Novaya Zemlya in the west. Also functional definitions have geographical endpoints, but there are requirements as to what should characterize these ends. A sea route, in the functional tradition, is a trading link between towns and cities – between ports with loading, service and reception facilities, transport networks, sizeable populations, etc. Neither the Bering Strait nor Novaya Zemlja meet any of these criteria. These are desolate, environmentally hostile places with no needs, abilities or capabilities to take part in trading – not even small-scale trading. On this backdrop, it has been argued that the NSR should be defined functionally as connecting sizable ports on the Pacific side of the Russian far east with those in the European part of Russia, for instance Murmansk.[15] As a secluded part of the NEP, the NSR has no meaning in large-scale trading other than securing Russia gateway control over the main part of the NEP. As has been pointed out: "If Vladivostok is the functional Russian eastern end point of the NSR, then the neighboring countries of Japan, North Korea, South Korea and China can easily become functional end points as well."[16]

This functional reasoning has aptly been summarized: "In compliance with the definition of the term 'Arctic', adopted by the Russian marine fleet, the ports of Murmansk, Kandalaksha, Naryan-Mar, as well as the ports of the Far East, south of the Bering Strait, are not related to the Arctic, but of course this cannot lessen their role in the Arctic transport supply. The ports of the Barents and White Seas also play a significant role in the economic

Source: Barents Interactive Geography Viewer, www.metainfo.se/gitbarents/barents.html

Figure 1.5 The Barents region.

linkages between Russia and West Europe."[17] The same authors take full cogniscence of this point by discussing the role of the Barents Euro-Arctic Region (BEAR), which lies west of Novaya Zemlya in the *economic development of the regions along the NSR* (Figure 1.5).[18] In line with this analysis, it is interesting to note that Russian researchers use maps depicting the four principal transit routes of the NSR – the coastal, marine, high-latitudinal and near-pole routes – as starting out in Murmansk going through the Bering Strait to ports in the North Pacific (see Figure 1.4).[19] Given the fact that the BEAR, comprising the 11 northernmost counties of Russia, Norway, Sweden and Finland, host a Working Group for the NSR, the southernmost boundary of the route can be claimed to coincide with the Norwegian BEAR county of Nordland on the Atlantic.[20] In functional terms, the NSR stretches from the ice free portions of the North Pacific to Norway's Nordland County on the Atlantic. In this definition it is more appropriate and even accurate to use the term 'the Northeast Passage' rather than the geographically confined term of the 'Northern Sea Route'. To make the confusion complete, Russian researchers recently claimed that the Council of People's Commissars of the USSR " ... named the Northeast Passage the 'Northern Sea Route' ... to make the NSR a functioning, practicable shipping route from the White Sea to the Bering Strait."[21] Here the NSR is made the same as the NEP, placing the White Sea, which is west of Novaya Zemlja, the endpoint of both the NSR and the NEP.

The definition and legal sensitivity of the Northeast Passage

The Northeast Passage complies with different versions of the functional definition of the Northern Sea Route as long as this corridor connects the Atlantic and Pacific Oceans in a continuous stretch. A minimum definition of the NEP is that it is made up of all the marginal seas of the Eurasian Arctic, i.e., the Chukchi, the East Siberian, the Laptev, the Kara and the Barents Seas. As such, the NSR makes up approximately 90 percent of the NEP.

Formally, Russia opened the NSR to international shipping on 1 July 1991 on the premise that the users would comply with coastal state regulations. Since the archipelagos of the NSR are legislated to become internal waters, Russia claims the same sovereignty over these parts of the route as over her land territory (see Chapter 6). This stand provides Russian authorities with an unlimited regulatory power which is challenged by both the United States and the European Union (EU) (see Chapters 2 and 6). Their position is that the NSR is an *international strait* open to international shipping on the condition of transit passage as defined in *Law of the Sea Convention of 1982* (UNCLOS).[22] If , as indicated in the official definition of the NSR, Russia extends her jurisdiction also to the high seas of the Arctic Ocean, diplomatic protests will be heard from Washington, Brussels and capitals of smaller states (see Chapter 2). In the spirit of UNCLOS, the high-latitudinal and near-polar routes claimed to be part of the NSR regime may be interpreted as "creeping jurisdiction" and a blatant violation of the High Seas regime (see Chapter 6).

1.1.2 Intra-Arctic, destination-Arctic and transit routes of the Northern Sea Route

Immediately following the 1917 October revolution, the Soviet authorities authorized hydrographical surveys, primarily in the Kara Sea, for the purpose of improving navigation there. Geographical observatories were constructed on Novaya Zemlja, Franz Josef Land, Severnaya Zemlja and the New Siberian Islands. In 1930, plans for developing navigation possibilities in the Kara Sea were expanded drastically, and the entire NEP was to be opened to transport and transit sailings.[23] To attain this goal, research was stepped up. By 1937, Glavsevmorput – the Directorate of the Northern Sea Route, formally established in 1932 – had spent an equivalent of 1 billion US dollars on activities north of 60 degrees N latitude and had some 40,000 persons on its pay role.[24] From 1937 until 1956, the Soviet Union equipped scientific expeditions to a total of 524 different destinations in the Arctic. These research efforts did not pass unnoticed abroad. During the 1960s and 1970s it was conventional wisdom in western research circles that the Soviet "knowledge of the region (was) much more extensive than that of the sum of the other nations bordering the basin."[25]

After World War II, Soviet research efforts were followed up by an ambitious program for constructing a large fleet of powerful icebreakers. At its peak in the 1980s, this fleet counted 38 vessels operating along the route and southward on the big Siberian rivers. Six of these icebreakers were nuclear powered of which the biggest exerted 75 000 horsepower. In addition, a fleet of close to 700 ice-strengthen vessels was built to operate along the route on a year-round basis.[26] These efforts notwithstanding, on occasions convoys of ships had to overwinter in the NSR before they were freed by icebreakers in late spring the following year. Accidents happened and freighters were damaged and lost. According to Russian sources, in the period 1954–1990 the total number of ice damage to ships traversing the NSR averaged 800, i.e., 22 a year. The accidents were distributed as follows: the Kara Sea 49% (here the intensity of navigation is the highest, see below); the Laptev Sea 20%; the East Siberian Sea 2%; the Chukchi Sea 14% (here the density of ships is the lowest and ice conditions the worst).[27] Many ships' captains experienced the truth that "... being at sea is risky, being at sea in ice is twice as risky, and being at sea in a convoy with an icebreaker present is three times the risk."[28] In the period 1945–70, the sailing season of the eastern part of the NSR was restricted to about three months, whereas ice conditions in the western part allowed for an extended sailing season of up to 4.5 months.

The NSR never got its intended significance as a transit route between the two world oceans. Transit traffic reached its maximum cargo volume in 1993 with 208,600 tons brought in by 30 voyages of multipurpose ships of the Norilsk type (SA-15).[29] Mostly, the NSR served regional developmental purposes. In the 1980s, more than 400 Soviet ships were active in cargo shipments supporting several destinations at the estuaries of the great rivers of Siberia. Raw materials were transport out of the region and necessities of life were brought in. In addition, some 100 scientific, commercial and military outposts were supplied this way. The areas attracting intra-Arctic and destination-Arctic shipping on a regular and long-lasting basis were, and still are the Barents Sea, Dickson, Cape Chelyuskin, Tiksi, Kolyma river, Pevek, Cape Schmidt and the Bering Strait (see Figure 1.6).[30] Cargo shipping along the route to these ports increased steadily for some years. In 1987, it peaked at 6.58 million tons – the heyday of Soviet Arctic marine traffic in the Arctic.[31] Since then, and in particular following the dissolution of the Soviet Union in 1991, there has been a steep drop in the volume of cargo. In 1998, it dropped to an all-time minimum of 1.5 million tons, increasing to 2.13 million tons in 2007, which equals the volume transported in the late 1960s. Current indications are that this negative trend may continue for the immediate future.[32] But there are exceptions to the rule for certain sections of the NSR.

Since 1978 and up to the present, the Russian icebreaker fleet has succeeded in keeping the stretch from Murmansk to Dudinka on the banks of the Yenisei River (231 nautical

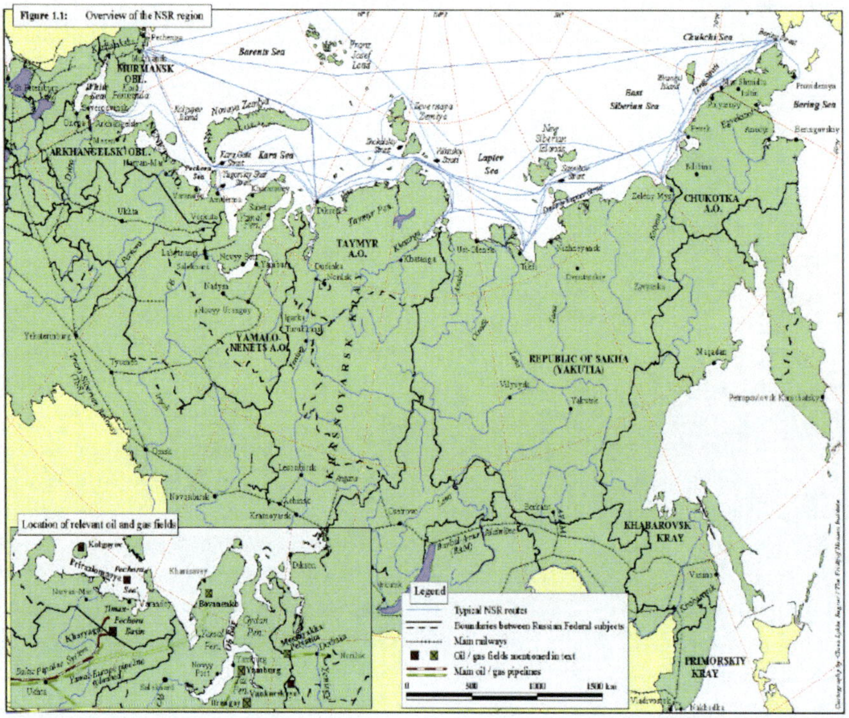

Source: Ragner FNI Report 13/2000b

Figure 1.6 Ports along the NSR.

miles from the river's mouth) open for sailings 12 months a year. This means that more than 1000 nautical miles, or some 10 percent of the NEP between Murmansk and the Bering Strait, is now kept open for shipping all year round. This stretch is what throughout history has been labelled the *Kara Sea Route*. The driving force behind this achievement was the prospect of increased revenues stemming from year-round shipments of nickel from Igarka. In 1980 this transportation provided revenues. The nickel industry then got an extra income of 71 million roubles, whereas the extra costs of the shipping activity amounted to 23 million roubles.[33]

In January 1988, the Ministry of the Fleet was instructed to run the NSR on a commercial basis, signalling that the state subsidies for the route would stop and the management of the route should be secured by the income it generated. Today, modern ice strengthened oil and gas tankers ply the Kara Sea Route along with the nickel industry. A clearly identifiable intra-Arctic route has been established across the politically defined geographical divide between the NSR and the NEP. In 2006, regional transportation of hydrocarbons within the Barents and White Seas alone amounted to 8.5 million tons,[34] which is four times more than the volume of cargo transported through the rest of the NEP/NSR. Profitability is one decisive key to increasing shipping. History provides a source of evidence. The most massive, sustained activity in terms of Arctic shipping was that of whaling. Between 1610 and 1915, just over 39,000 voyages were undertaken to the Arctic in pursuit of the bowhead whales. This activity took place in ice infested waters in four main areas: the Svalbard/ Greenland Sea areas, Davis Strait and Baffin Bay, Hudson Bay and the Bering, Chukchi and Beaufort Seas.[35] The prospects of profit will be the driving force in the future as well.

Russian authorities are planning to expand the volume of cargo transported along the NSR to between 7.8 and 11,4 million tonnes in 2015.[36] Of this total, oil products are expected to make up 4.6 –5.9 million tons, i.e., close to 60% of the total cargo flow.[37] Independent research estimates indicate an increase in transit cargo by 2020 of about 5–6 million tons per year in the eastern direction and 2–3 million tons in the western direction.[38] The dominant shipping activity is and will be destination-Arctic in character, involving among others indigenous communities.

Within the northern territorial frontiers of the Russian north there are 26 different indigenous groups or nationalities. All together, they number close to 180,000 individuals. The effects of increased shipping on Arctic communities – coastal as well as inland – are assumed to be many, positive as well as negative.[39] If, however, due consideration to indigenous livelihood, culture and social life is not an integral part of the economic planning process, these peoples may be victimized by the increasing developmental pressure exerted on the region from the south.[40]

The problem relating to the Arctic transportation corridors in general is that there "... is insufficient information to identify with any precision the likely effects of marine shipping for most Arctic coastal communities. No current database exists for indigenous use in local Arctic waterways that could be used to develop multiple use management measures and potential mitigation strategies."[41] Until such a database is established, the interests of indigenous peoples find protection only in various documents of Human Rights, among them the *ILO Convention Document no. 169*, and in soft law institutions like the Arctic Council (see Chapter 6). The protection offered by these instruments is totally dependent on the ability and willingness of the Executive to put their spirit and letter into practice. In this respect, Soviet history is no source of encouragement.[42] As such, the NSR is a "one state and multiple nation route", whereas the NEP is a "two state and multiple nation route". All voices claim a role in development planning.

In the Barents Sea there are two shipping routes between the Norwegian archipelago of Svalbard, which has no native population of its own. and cities on the Russian and Norwegian mainland. Cargo ships from Murmansk supply the Russian community of Barentsburg in Spitsbergen and bring coal back to the city of Murmansk, whereas Norwegian ships do the same between Longyearbyen in Spitsbergen and the city of Tromsø in north Norway. These routes have been in operation in the summer season ever since the early 1900s. In recent years, the Norwegian route has been operated on a commercially-viable basis. In addition, a rapidly increasing number of cruise ships call on different locations in the archipelago during the summer (see Chapter 3).

Huge areas of the Arctic currently have inadequate infrastructure to support marine shipping. This includes such infrastructure components as availability of ports and port facilities needed for different types of vessels operating in Arctic waters, the accuracy and availability of information needed for safe navigation, and availability of search and rescue assets.

Along the NSR/NEP there are several ports and port points. Even if there is adequate access to icebreaker assistance, only very few of the ports have the essential facilities such as adequate water depth, berths and mechanizations needed for increased shipping. Adequate marine communication systems exist in some parts of the NSR but not in others. Communications using VHF-radio, MF- and HF-systems and satellite are generally adequate for the lower parts of the NSR, but data transmission becomes problematic when the High Arctic is reached. Russia has currently several on-going projects developing systems to meet the demand of better communications for ships operating along the NSR. Several search and rescue centers are located along the NSR, but only a few can give the support needed for ships sailing along the route. Russia has made structural plans for implementation of search and rescue technology along the NSR up to 2020, but needs great financial support to complete the plans (see Chapter 5).

1.1.3 The Northwest Passage

The Northwest Passage is the name given to a set of marine routes between the Atlantic and Pacific Ocean, spanning the straits and sounds of the Canadian Archipelago, the Davis Strait and Baffin Bay in the east and the Beaufort Sea in the west. Like the NSR, it is a transportation corridor channelled through islands occupying broad expanses of water and land in the north–south direction. The base of the archipelago stretches some 3000 km along the mainland coast, covering about 80 degrees of ocean and land territories.[43] The tip of Ellesmere Island is less than 900 km from the geographic North Pole. The archipelago is one of the largest in the world and consists of a labyrinth of islands and headlands of various sizes and shapes. There are 73 major islands of more than 50 sq. miles in area, and some 18,114 smaller ones.[44] If islets and rocks are included, the archipelago comprises approximately 36,000 pieces of dry land above sea level, making it one of the most complex geographies on Earth.[45]

The Canadian Archipelago is divided into two main parts by the Parry Channel: the northern part consists of the Queen Elizabeth Islands, whereas the southern part comprises all islands located north of the Canadian mainland and south of the Parry Channel. Thus, the most troublesome part of the NWP, as seen from a mariner's point of view, runs through a continuous archipelago with narrow straits often jammed with impenetrable multiyear sea ice drifting in from the Central Arctic Ocean. The NWP consists of seven different routes, of which six run through the southern part of the archipelago (see Figure 1.7a).[46] Every route starts out in the eastern part of the Parry Channel heading westward, until six of them break off at different points along the channel and take a southerly direction to reach the coastal waters of the mainland before making a turn westwards towards the Beaufort Sea. The seventh route employs the Parry Channel in its entirety until it reaches the Beaufort Sea through the often ice clogged M'Clure Strait between Melville and Banks Islands. Of the six southerly routes, the most navigable to be employed by international shipping comprises the Parry Channel until it takes a southerly direction and goes through the Prince of Wales Strait between Banks and Victoria Islands, and then west again along the north coast of the Canadian mainland and Alaska to the Bering Strait (see Figure 1.7).[47] Like the NEP,

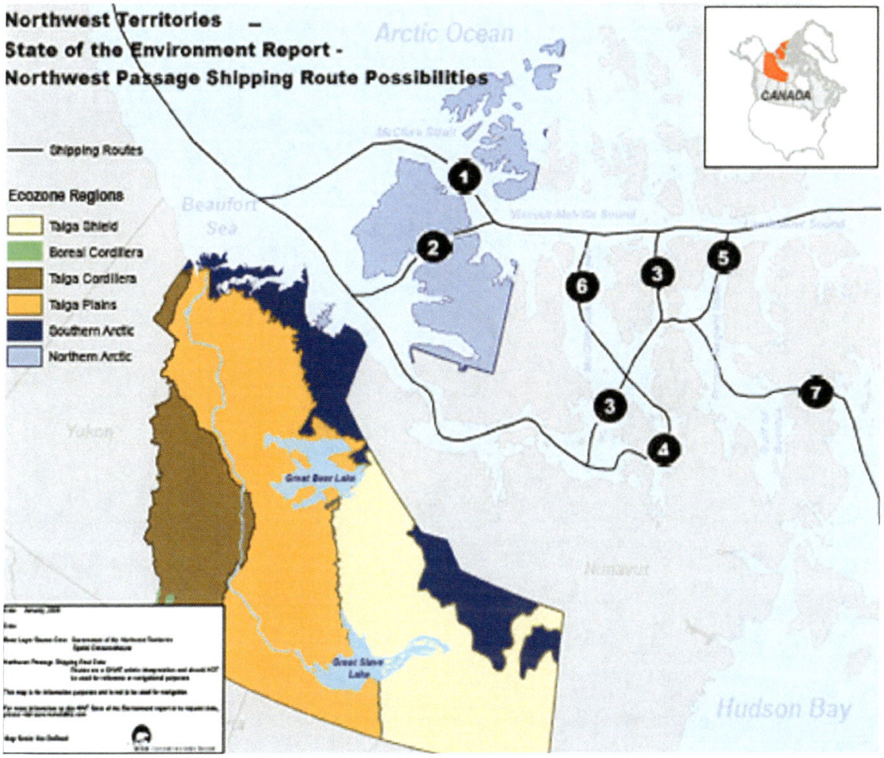

Source: Stuart et al. (2005), from http://www.enr.gov.nt.ca/_live/pages/wpPages/soe_human_activities.aspx

Figure 1.7 Seven possible routes through the NWP.

there is no single set channel for ships to follow. The channel used is based on which one offers the best sea ice conditions at any one time and place. Thus, the NWP, like the NEP/NSR, is a transportation corridor through one massive archipelago until it reaches open, but ice infested waters in the Baffin Bay (east) and Beaufort Sea (west). From Baffin Island (east side) to Banks Island (west side) it covers a distance of about 2400 kilometers, and the size of this whole archipelago is approximately 2.1 million sq. kilometers, i.e., about the size of Greenland.[48]

Sea ice condition within the archipelago varies dramatically from year to year, presenting unpredictability to any surface operation. There is mounting evidence that sea ice reduction will continue, although there is great uncertainty over the rate at which sea ice will continue to diminish. In the summers of 2007 and 2008 most of the archipelago was so-called ice free, promising to open the NWP to high volumes of intercontinental commercial shipping. This warrants a comment on the concept of *ice free*.

Most Arctic shipping experts view this term as meaning ice infested with icebergs, bergybits and growlers present, even in the summer period. In fact, some believe shipping

Source: blogs.chron.com/sciguy/archives/2009/08/

Figure 1.8 The sailing lanes of the NWP.

operations in this environment can be even more dangerous than in ice covered areas. From a mariner's point of view it has been assumed that "... with less ice, more icebreaking capacity will be needed".[49] The reasoning goes as follows: "Initially, as first-year ice weakens and/or disappears, its ability to keep multiyear ice out of shipping areas will be adversely affected. This will mean that, even if there is less ice overall, it will be much harder, pose more of a damage risk and be more difficult to break the passage through. I have rammed multiyear ice with a heavy icebreaker, stopped and when the icebreaker was reversed, was not able to see any evidence of the impact of ice. The same lack of first-year ice will also allow for much more freedom of movement of the multiyear ice pack which will then likely compact in chokepoints, thereby compounding the problem. In the future then, as the climate changes, we can look forward to standard ice deviations in coverage, thickness and movement that will continue to increase dramatically, giving shipping some of the best "ice" years yet, but potentially some of the worst as well."[50] Ice free does not necessarily mean problem-free or, for that matter, preclude icebreaker assistance.

The inter-annual variability in sea ice conditions within the Canadian Archipelago will continue to be extreme. According to the Canadian Ice Service, "It is quite likely that the latter half of this century will still experience occasional summers with ice conditions as severe as those witnessed in the 1980s. Multiyear ice, particularly in low concentrations, will present the major hazard to shipping ... Since the oldest and thickest ice in the Arctic Ocean is that which is driven against the western flank of the Canadian Archipelago, this will likely be the last multiyear ice to remain. As long as this remains a source of multiyear ice in the Arctic Ocean, it will continue to drift through the Canadian Archipelago."[51]

M'Clure Strait between Melville and Banks Island is one of the straits that have a fairly long history of being blocked with multiyear ice drifting in from the Central Arctic Ocean. In addition come shallow waters and draft restrictions, narrow straits acting like chock points, and the combination of the two making navigation a regular and punctual activity hard to achieve. In the AMSA study the conclusion is clear: "Even during the most recent periods of reduced ice, the location of the ice, its thickness from year to year and the variability of ice-free areas makes it nearly impossible to schedule transits with any degree of certainty of reaching the desired port on schedule."[52] Model experiments suggest that the maritime accessibility of Type A vessels, which are capable of limited icebreaking, to the NWP will at baseline increase from 63% (2014) to 82% by mid-century (2045–2059) in the period July to September, whereas the NSR and TPP will be fully accessible.[53]

In 1969, the ice strengthened American supertanker *Manhattan* transited the Canadian Archipelago from the east coast of the USA to Prudhoe Bay in Alaska. The next year it returned the same way. The purpose was to prove that supertankers are capable of transporting petroleum through ice-covered waters. The two voyages were completed as planned and, in that respect, they were deemed successful. But *Manhattan* also encountered challenges, unleashing a heated debate in several countries on what the environmental consequences would be if the NWP became an object of large-scale regular shipping activities – if oil leaked into ice infested fragile waters. During the first voyage, *Manhattan* was accompanied by the Canadian icebreaker *Macdonald* which at one point registered an unexpected "shoal" in the vicinity of the sailing route. One year later, the Canadian research vessel *Hudson* proved with the help of side-scanning sonars that the "shoal" was a seabed pingo. A pingo is old ice shaped like a cone extending like a knife from the seabed upwards towards the surface of the ocean. At the base, the biggest measure more than 300 meters, and may rise more than 60 meters above the seabed. The assumption is that seabed pingoes, which are often covered and strengthened with frozen clay and mud, are relics from the time when the seabed was above sea level. The Hudson expedition registered over 100 pingoes scattered around the continental shelf of the Beaufort Sea and within the shallow channels of the Canadian Archipelago.[54] At the time, it was not known how many pingoes the NWP hosted, but the number was assumed to be significant.

Seabed pingoes undoubtedly represent a danger to ships with deep drafts. If a ship collides with a pingo the hull may be ripped open and an environmental catastrophe could result. An example: *Manhattan* had a draft of 56 feet whereas the peaks of the pingo often are no more than 40 feet beneath the surface of the ocean. The semi-official journal, *Canada Today*, claimed in 1974 that: " The discovery meant that the long-sought passage around the top of North America was at the time a dead end for super tankers and that the *Manhattan*, which had pioneered the route less than a year before, could be the last as well as the first to make the run ... One tanker ripped open by a pingo ... could disrupt the fragile ecological balance of much of the Arctic."[55] Even if all pingoes are registered and mapped so that ships can steer away from the danger areas, the risk of collisions is not eliminated. The strength, thickness and drift direction of the surface sea ice will impact the freedom of navigation and decide what route to choose. Despite *Manhattan* being a "giant" for its time with a displacement of 115,000 dwt. and exercising 43,000 horsepower, it got stuck in the ice no less than 25 times during her first trip through the NWP. For *Manhattan* to get free from the ice she needed icebreaker assistance multiple times. In the meantime, *Manhattan* could have drifted helplessly over pingo areas without being able to avoid collisions. On her return voyage, the sea ice of Lancaster Sound knocked out a huge panel from *Manhattan*'s hull, spilling 15,000 barrels of ballast water.[56] Thus, to avoid areas of pingoes, pilots and sea captains may be forced to choose areas of heavy sea ice. The combination of ice on surface and seabed leaves restricted space for maneuvring and safe navigation.[57]

After the startling discovery of seabed pingoes in the early 1970s, the impression was that the Canadian government would take action to map them and eventually destroy them where needed. If this was ever done is not known to these authors; however, we note that pingoes as a navigational challenge have vanished from recent literature on the NWP, and also from the AMSA report. Actually, during our fact-finding trip to Ottawa in October 2010 (for this book), we were surprised to find that when we posed the question of pingoes, Canadian bureaucrats seemingly did not know what we were talking about. This can be explained either as an indication that the problem had been resolved (through mapping and destruction of pingoes) and then was forgotten about, or that the pingoes are still with us without anyone knowing or thinking about it. If the latter is true, someone responsible should readdress the challenge.

In the Beaufort Sea, the NWP passes through the coastal waters of Canada and the United States and in the Davis Strait between Greenland and Canada, and as such is a "three-state and multi-nation route". The indigenous peoples – of which the Inuit population is the biggest – makes up the multination characteristics of the route. They are exposed to the same possible negative and positive effects of increasing shipping as their counterparts along the NEP and NSR (see above and Chapters 4 and 6). But there are important differences between the natives of the two passages. In Greenland the Inuit have enjoyed home rule since 1979 and in Nunavut (meaning "our land") a fair amount of native autonomy was achieved in 1999. This notwithstanding, the indigenous peoples of the two countries do have an image of governance and institutions that differs from that of their former colonizers. As has been stated, "(t)here is a need for more flexibility and a rethinking of the boundaries of Danish-European and Anglo-Canadian legal imaginations that would adopt to Inuit realities and better serve their aspirations to self governance." [58] The assumption is that with increasing autonomy follows increasing political influence on regional affairs. That may provide the natives of the two countries with a place at the table and a more decisive role in national and international decision making (see Chapter 6).

Canada claims full and unlimited jurisdiction over the archipelagic section of the route which was declared part of Canadian internal waters, effective January 1986. The United States and later the EU have protested against this claim, which they find illegal. According to the US government, the NWP is an *international strait* open to international shipping on the basis of the principle of *transit passage* without any interferences from the coastal state. Thus, when it comes to the legal status of their respective passages, Canada and Russia are faced with the same legal and political opposition (see Chapters 2 and 6). This disagreement surfaced in 1970 when Canada enacted the *Arctic Water Pollution Prevention Act* (AWPPA), establishing a 100 nautical miles environmental zone north of the Archipelago as a precautionary step to prevent ship-based pollution. This was a unilateral reaction on the part of Canada to the roundtrip of the US tanker *Manhattan* through the NWP in 1969–70. At the time of its enactment the Act was not part of accepted international ocean law. Thus, the US government immediately declared the Canadian move a violation of international law and a threat to the status of the NWP as a strait open to international navigation. Diplomatic notes were exchanged and protests issued in both directions. In the 1980s, the two parties calmed the disagreement and agreed to disagree. However, in 2009 the USA restated her long-term opposition against the Canadian position, indicating that the question of freedom of navigation is high on her political agenda and that national interests are at stake if that freedom is curbed.[59] The EU Commission has followed suit and supported the US stand on the legal status of the NWP (see Chapter 6).[60]

1.1.4 Intra-Arctic, destination-Arctic and transit routes of the Northwest Passage

Despite attempts to find the NWP during the 19th century, the first transit passage was not achieved until Gjøa negotiated the southern coastal route in 1903–06. *Manhattan* completed the first round trip in 1969–70 and the cruise ship *Lindblad Explorer* made the first revenue-earning voyage in 1984. Between 1903 and 2004 there have been 94 single transit passages by larger ships, 30 round trips and 27 recreational small boat passages (from Atlantic to Pacific waters or vice versa). Thus, counting round trips as two passages, there have been 181 transits during the 101-year period and most of these have been through the southern coastal route. On average, 1.7 transits have been conducted per year in the course of 101 years. These voyages were undertaken by 67 vessels carrying 15 different flags. In the same period, 175 partial transits were recorded through the waters of the archipelago (not going through the whole of the NWP, including the Beaufort Sea).[61] A further brief examination of available data shows that transits of the passage remained a fairly sporadic affair until the 1970s, up to which point only nine complete transits had been made. In terms of flag activity, 63 percent of transits between 1903 and 2004 were Canadian flag, mainly Canadian Coast Guard icebreakers.[62] The last 34 years have seen the most transit sailings. From 1969 to the end of the 1980s, over 30 complete transits of the passage were undertaken by a variety of vessels. The bulk of these voyages were Canadian, of which most were involved in the search for hydrocarbon resources in the Canadian offshore area of the Beaufort Sea. Between 1984 and 2004, a total of 23 commercial cruise ships and 15 recreational yachts accomplished transits of the NWP. Cruise ships operating in these waters doubled in 2006, from 11 to 22. There were 26 transits in 2010, out of which only three were commercial.[63] The AMSA study concludes: "Modern history of Arctic marine transport indicates that despite the historical attempt to make the NWP a viable route between the East and the West, it is unlikely the passage will become the route it was originally intended."[64] Thus, destination-Arctic shipping is anticipated to increase incrementally in the years ahead mostly driven by the search for resources, in particular oil and gas.

The Hudson Bay Company started destination-Arctic shipping to the area as early as 1670. In terms of commercial vessel traffic, throughout the 19th century all commercial shipping in the Canadian Arctic was to the account of this company, ending in 1913. Resupply of commodities has been and still is the most stable element of shipping in the Canadian Arctic. Nearly all the voyages are destination-Arctic coming from the Atlantic to support the sealift of cargo to local communities.

The war years brought marine activity in support of the United States Air Force construction of the airfield of Frobisher Bay, and later the Cold War brought the mid-Canada line (1947) and the DEW line (1954). All these activities required active marine support with some seasons having as many as 50 vessels on charter. The DEW line shipping activity segued into a combination of defence and community resupply which, with changes to ships, focus and administration, continues today as a community lifeline.

Current shipping demand in the Eastern Arctic involves up to 22 seasonal trips and occurs during the 100-day navigation season that spans from mid-July to the end of October (most communities receive at most two resupply calls a year). Each voyage can include deliveries to 8–9 communities. All cargo is lightered from the anchored resupply ship to a beach landing and delivered at the high water mark. Recently, marine operations averaged 100 voyages by large ships in summer.

Churchill is a prime trading port in the east. In 2004, 14 out of 18 foreign voyages to the Canadian Arctic called at the port of Churchill, shipping wheat and grain to international markets. The first Russian shipment of fertilizer to the port of Churchill, coupled with a perception of an extended sailing season due to global climate change, has renewed planning for establishing an international "Arctic Bridge" between Murmansk and Churchill. In September 1992, the *Arctic Bridge Agreement* was signed between Canada and Russia. The major goods that might become the basis for significant trade between the two regions are bulk commodities, mainly the export of an estimated 315,000 tonnes of mineral products and lumber from Murmansk, and a minimum of 600,000 tonnes of export grain from Churchill.[65] The Bridge goes through the Barents and Norwegian Seas and proceeds south of Iceland and Greenland before making a northward turn through the Davis Strait to reach Churchill in Hudson Bay (see Figure 1.9). It will then partly overlap with the transoceanic branch of the NMC (see below). For this bridge to materialize, Churchill will require significant additional port and rail investments, as well as further study by both countries regarding costs, cargos and volumes. The port of Murmansk is in need of similar improvements before this intra-regional traffic scheme – or intra-international route – can be realized. Model experiments suggest, however, that global warming will continue to keep the Bridge fully open for maritime transport throughout the century.[66] It is the endpoints that need fixing, not the bridge itself.

In the western Canadian Arctic cargo is handled by tugs and barges, with most cargo shipped down the Mackenzie River to Tuktoyaktuk for transfer to ships with deeper ocean drafts. Current shipping demand involves 14–15 seasonal tug-barge trips. These operations take place in what has been labeled the *Mackenzie River route*[67] – a Canadian intra-Arctic route of some regional significance. The western Arctic sailing season is typically 60 days between mid-July and mid-September. These are some of the more important destination-Arctic shipping locations within the Canadian Arctic: the port of Churchill, the oil field at Cameron Island, the Nanisivik zinc-lead mine in the vicinity of Arctic Bay, Tuktoyaktuk, Cambridge Bay, Resolute, Pond Inlet, Inuvik and Whitehorse. It is anticipated that by 2020, annual resupply demand will require up to 30 ship trips, and destination-Arctic export shipping will probably include Mary River Iron ore, Port at Steensby Inlet, Roche Bay magnetite, Igloolik, Grays Bay and Bathurst Inlet (see Figure 1.9).[68]

The NWP has essentially no adequate ports with the necessary facilities to support the growing amount of commercial traffic along the northern slope of Alaska and throughout the Canadian Archipelago. Several ports have been proposed and are under development, but it is unclear if these ports will have the necessary facilities to meet the demand of increased shipping along the NWP. Currently there is a lack of adequate communication

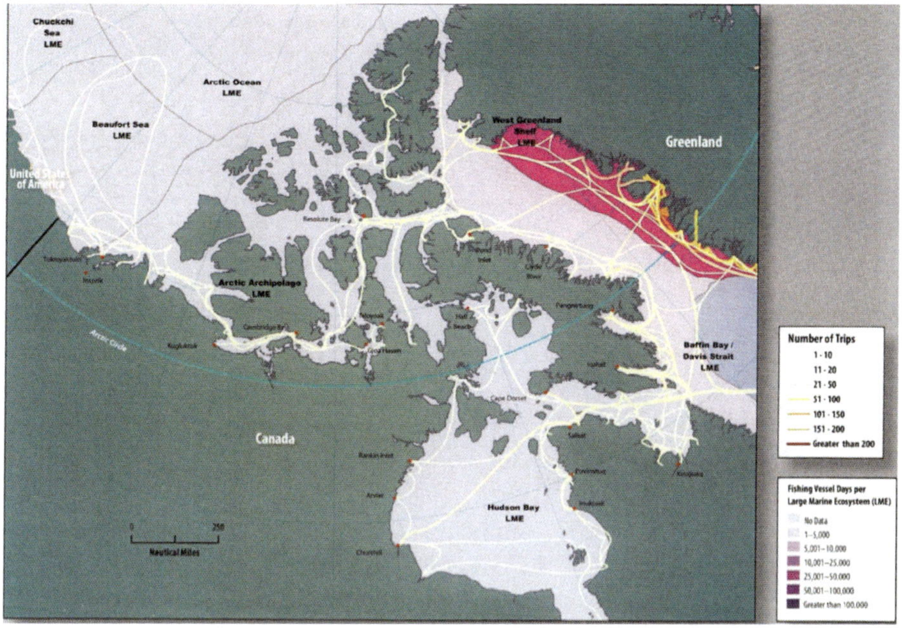

Source: AMSA (2009)

Figure 1.9 Transportation patterns of the NWP.

systems along the route. The United States has made few contributions in the development of adequate communication systems and services, while Canada operates with seasonal systems. The Canadian Coast Guard (CCG) is the primary agency along the NWP relative to rescue, safety and environmental response. Since the CCG icebreakers leave the Arctic at the end of October there is a lack of emergency response along the NWP on a year-round basis. Longer active shipping seasons along the NWP raise a number of service level issues for the governments of Canada and the United States.

The AMSA report makes three important conclusions when it comes to the future economic attraction of the NWP:

1. Despite climate change, the NWP will continue to be controlled by ice conditions at four choke points (see Chapter 4).

2. It may be years before the Canadian Arctic matches the resources extracted when compared with Alaska or the Russian Arctic (see Chapter 3).

3. Other transit routes are more attractive compared with the NWP (see Chapter 7).

A Canadian paper adds a fourth point: "...ice conditions are widely expected to improve more rapidly in Russia's Northern Sea Route than in the Canadian. In a survey of shipowners' motivations to engage in the Arctic, eight out of 15 businesses favor the Northeast

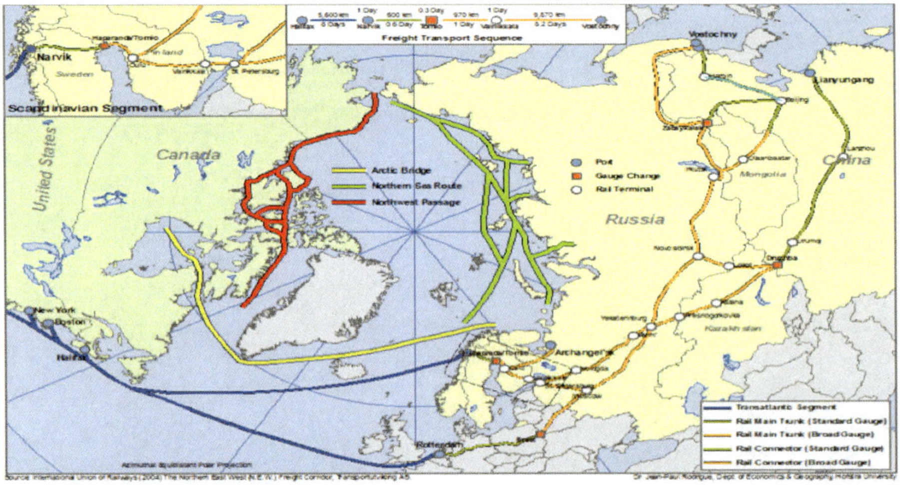

Source: International Union of Railways. The Northern East West (N.E.W) Freight Corridor, Transportutvikling AS (2004)

Figure 1.10 Arctic bridges, including the Murmansk-Churchill bridge.

Passage (Northern Sea Route), which has better infrastructures, more local ports to service and more mining and oil and gas operations (than the NWP).[69] In fact, this is probably one reason why most of the 'pioneer commercial transit'[70] to date has been through the NEP rather than the NWP. This far, passages of this kind are for the 'adventurously minded' .[71] On these grounds, it seems a reasonable conclusion that the NWP in a short and medium term will not be a serious competitor to the NSR or will open up to high volumes of international commercial shipping."[72] Most Arctic shipping scenarios indicate an increase in cargo shipped in the years ahead, but the bulk will be destination-Arctic and intra-Arctic and mostly resource driven.

1.1.5 The Transpolar passage

Transpolar routes outside of national jurisdiction in the Arctic Ocean cover all waters that are part of the high seas and where the freedom of navigation applies. This definition includes two sections of water expanses: The first is the Central Arctic Basin which is 4.7 million sq. km in area. Here, coastal states have no jurisdiction at all apart from the flag state jurisdiction they exercise over their own ships and crews. The second section includes all ocean areas beyond the territorial seas of 12 nautical miles and within the outer limits of the 200 nautical miles exclusive economic zones (EEZ). This belt is 188 nautical miles in extension measured outwards from the outer limits of the territorial sea. Here, coastal state rights and obligations mix with the rights and obligations of all other states. In this belt the coastal states have sovereign rights over certain issue areas, among them the exploration and exploitation, conservation and management of natural resources – living and nonliving – on and in the sea bed and in the water column above. The coastal states also exercise rights to adopt and enforce nondiscriminatory laws and regulations for the prevention, reduction and control of marine pollution from vessels in those areas that are ice covered within the limits of the EEZ. In the ice covered areas of these stretches, Article 234 of UNCLOS provides coastal states with some additional powers to apply pollution regulations. At the same time, Article 87 of the same Convention claims these waters to be part of the High Seas, guaranteeing the freedom of navigation to coastal and landlocked countries alike. Thus, this belt is where coastal state jurisdiction meets with the freedoms of the High Seas and where all parties shall have due regard to the rights and duties of other states and where all are obligated to act in a manner compatible with the spirit and letter of the Convention (see Chapter 6). The biggest chunk of Arctic water is that designated the High Seas, i.e., both the Central Basin and the 188 nautical miles belt of the EEZs. The direct distance across the Central Arctic Ocean from the Bering Strait across the North Pole to the Fram Strait is 2100 nautical miles. All other transit distances within the Arctic Basin are longer. The state of sea ice at any one time will decide if the distance advantage of using the TPP instead of the alternatives will attract the interest of international shipping.

Contrary to popular belief the ice cover of the Central Arctic Ocean is not a static unbroken surface. It is constantly in motion, breaking into pieces, and building up pressure ridges above and below the surface where floes grind together. Because of the cleavage of the sea ice canopy, leads and areas of open water (called polynyas) and thin ice (called skylights) are present all year round. As early as the 1960s (before global warming of the present was recorded) these open water areas constituted from 5 to 8 percent of the total area of the Arctic Ocean during the winter season, and approximately 15 percent during summer.[73] By way of example, the open space available to shipping in the Central Arctic Ocean during the summer season was about twice the total size of the Caspian Sea (approx. 700,000 sq. km) and about the size of the Caspian in winter (approx. 235,000–376,000 sq. km).[74] The distribution and frequency of polynyas and skylights are random, but they appear even in the vicinity of and at the very North Pole throughout the year.[75] Some of the leads and skylights in the vicinity of the Pole were assessed to be nearly half a mile in diameter.[76]

The sea ice varies in shape, thickness, age and hardness, presenting different challenges to navigation. These are some of the more frequent:[77]

- *New ice*: Recently formed ice composed of ice crystals that are only weakly frozen together (if at all) and have a definite form only while they are afloat.
- *Nilas*: A thin elastic crust of ice (up to 10 cm thick), easily bending on waves and swell and under pressure growing in a pattern of interlocking "fingers" (finger rafting).
- *Young ice*: Sea ice in the transition stage between nilas and first-year ice and 10–30 cm in thickness.
- *First-year ice*: Sea ice of not more than one winter's growth, developing from young ice having a thickness of 30 cm or more.
- *Old ice/multiyear ice*: Sea ice that has survived at least one summer's melt. Its topographic features generally are smoother than first-year ice and can be a few meters thick. Old ice is also much harder than first-year ice, and can be much more damaging to ships, if hit at a normal cruise speed.

- *Ice massifs*: Extensive accumulations of close or very close ice that are found in the same region every summer.[78]
- *Drift ice*: Ice that floats on the surface of the water in cold regions, as opposed to *fast ice*, which is attached "fastened" to a shore. Usually drift ice is carried along by winds and sea currents, hence its name, "drift ice".[79]
- *Pack ice*: When the drift ice is driven together into a large single mass, it is called pack ice. Wind and currents can pile up ice to form ridges 3–4 meters high, creating obstacles difficult for powerful icebreakers to penetrate. Typically, areas of pack ice are identified by high percentage of surface coverage by ice, e.g., 80–100%.[80]
- *Ice floe*: A large piece of drift ice that might range from tens of meters to several kilometers in diameter. Wider chunks of ice are called *ice fields*.[81]

Although reliable information about the proportion and distribution of different ice thicknesses is difficult to obtain, a Swedish study dating back to the 1970s estimated the area of the Arctic Ocean covered with ice thinner than 4–5 feet to be 30 percent or more.[82] Forty years later, global warming has increased the portion of navigable waters in the central part of the Arctic.

Expert opinion is that the present thawing is long-term and that the ice-edge will steadily migrate northward along with a further thinning and weakening of sea ice. Over the last 30 years, sea ice thickness in the Central Arctic Ocean has decreased by 42%, a decrease of 1.3 meters from 3.1 to 1.8 meters,[83] with an accompanying reduction of some 73% in the frequency of deep pressure ridges.[84] As a consequence, the influx of multiyear ice from the Central Arctic Ocean to the coastal areas has decreased by 14 percent from 1978 to 1998. This decrease greatly benefits economic activities in coastal waters (see Chapter 4).

On the basis of these and other scientific observations, model experiments suggest a further decrease in sea ice thickness of some 30%, and an ice volume decrease between 15 and 40% by 2050.[85] If this trend continues, one postulate is that summertime disappearance of the ice cap is possible in the course of this century and that significant areas of the Arctic Ocean may become permanently free of sea ice in summer.[86] One of the models simulates an ice free Arctic Ocean in summer by 2050 (see the discussion of the concept of ice free above). This scenario implies that the physical occurrence of multiyear ice can possibly disappear from these waters in the future, further improving the conditions of economic activity in the Central Arctic Basin. The accessibility of Type A vessels (see above) at baseline to the TPP by mid-century (2045–2059) will, according to model experiments, increase to 100% from 64% at present for the period July–September.[87]

This is not to say that the Arctic Ocean will become an ice free ocean in winter. As concluded in the AMSA study: "Even after the first ice-free summer is recorded, there will almost certainly be subsequent years when all of the ice does not melt in summer but survives to become 'old' ice the following year. It is ... generally agreed that the Arctic waters will continue to freeze over in winter."[88] Russian scientists go one step further, maintaining that the likelihood of an ice free Arctic Ocean in the future is small even if the air temperature continues to increase. Their doubt is founded on the argument that the sustainability of the composition and functioning of the structure of the upper layers of the Arctic Ocean will control and reduce the melting process.[89] Let us illustrate the point: In mid-September 2007, the Arctic Ocean reached its absolute sea ice minimum so far covering only 4.1 million square km. One year later – in September 2008 – the extent of sea ice was about 1 million square km bigger than at the same time the year before, covering 5.2 million square kilometers.[90] In March 2008 the ice extent rebounded rapidly to a winter maximum that was actually higher than in the previous four years. On these grounds, sea ice experts expect strong natural variability events in the future, causing both decreases and increases of the Arctic sea ice cover on seasonal and decadal time scales.[91] Thus, different sources assume sea ice to be a lasting characteristic of the Arctic Ocean, although still very different from the conditions 30 years back.

As already discussed, ever since1978 the sea ice extent has been declining. This trend is most pronounced at the end of September, but all months show a declining trend. However, the retreat introduces the dangerous *polar lows* to new areas of the Arctic Ocean. Polar lows are high latitude, maritime small-scale cyclones caused by cold air interacting with relatively warm open oceans. These cyclones, which are difficult to forecast, appear suddenly and unexpectedly, and are often associated with strong surface winds and heavy snowfall. They usually originate north of 70 degree N and practically all move in a southward direction. Polar lows are not only found in ice covered areas as they also appear in southern waters, such as near Japan.[92] Recent studies suggest, however, that the likelihood of polar lows occurring in open waters for various reasons will decrease over time, whereas "... the retreat of sea ice will expose large new ocean regions to the atmosphere. In these regions, where *polar lows* and related weather have been non-existent so far (because *polar lows* need energy from the oceans), *polar lows* will make their first appearances in the future. These are the same regions that have been proposed as tomorrow's shipping lanes and oil/gas exploration areas."[93] In this respect, improvement in one operational condition for ships – retreat of sea ice – is counteracted by the introduction of a new challenge – polar lows. Among the passages, the TPP will be the least affected by the newcomer, which gradually will become more characteristic of Arctic operational conditions (see Introduction).

Transpolar routes can serve both intra-Arctic, destination-Arctic and transit purposes. The two former imply that vessels can use international waters for parts of their voyage, entering the NEP and NWP from the north to unload their cargo. On such occasions, Transpolar routes become affected by the controversial legal regimes of the NSR and NWP. For transit voyages between the Pacific and Atlantic oceans, this can be avoided by using the high sea sections of the Arctic Ocean, accessing or exiting through the Fram and Bering Straits (see Chapter 2).

1.1.6 Intra-Arctic, destination-Arctic and transit routes in the High Seas of the Arctic Ocean

The first surface ship ever to reach the North Pole was the Soviet nuclear icebreaker *Arktika* on 17 August 1977. *Arktika* departed from Murmansk on 9 August and sailed eastbound through the Vilkitskii Strait to the ice-edge of the Laptev Sea, then turned northward and sailed along longitude 125 degrees East, reaching the North Pole eight days later. The ship arrived back in Murmansk on 23 August after having sailed 3852 nautical miles in 14 days with an average speed of 11.5 knots. Parts of the voyage took place in heavy ice. This trip unleashed several more voyages from several other countries. Between 1977 and 2008, ship access to the North Pole in summer has been attained in all regions of the Arctic Basin. Data show that 77 voyages have been made to the geographic North Pole by icebreakers from Russia (65), Sweden (5), USA (3), Germany (2), Canada (1) and Norway (1). Of all the visits, 85% have been undertaken by Soviet/Russian icebreakers. Nineteen of these trips were in support of scientific exploration and the remaining 58 were for the entertainment of tourists. Eight icebreakers reached the Pole in summer 2004, and during the four consecutive summer seasons, 33 ships reached the North Pole mainly for tourist and scientific purposes. Of the 76 icebreaker trips that have been to the Pole in summer, the earliest date of arrival has been 2 July 2007 and the latest 12 September 2005. This indicates that the navigation season has been restricted to about 10 weeks for highly capable icebreaking ships. The only voyage of the 77 not conducted in summer was that of the Soviet nuclear icebreaker *Sibir*, which supported scientific operations during the period from 8 May to 19 June 1987, reaching the North Pole on 25 May.[94] As has been stated, "Sibir navigated in near-maximum thickness of Arctic sea ice while removing the personnel from Soviet North Pole Drift Station 27 and establishing a new scientific drift station (number 29) in the northern Laptev Sea. This trip is the most demanding icebreaker operation to date."[95]

Eleven of the 77 voyages were conducted by diesel powered icebreakers, the rest had nuclear propulsion.[96] The fact that conventionally-powered icebreakers have conducted successful operations to high latitudes in all regions of the Central Arctic Ocean implies that such voyages are not entirely dependent on nuclear propulsion (see Chapters 4 and 5). New icebreaking technology may enhance the capabilities of diesel-powered ships to operate in the waters around the North Pole.

No commercial ship has ever conducted a voyage across the Central Arctic Ocean. The seven transArctic voyages – all taking place in summer – have been conducted by icebreakers, nuclear- as well as diesel-powered. The first transect of the Arctic Ocean was undertaken by the Canadian icebreaker *Louis S. St-Laurent* and the *Polar Sea* of the United States in July–August 2004.[97] Both ships started their voyage from the Bering Strait and sailed northward to the North Pole and exited through the Fram Strait between Greenland and Svalbard (see Chapter 2). One year later, the Swedish icebreaker *Oden* and the American icebreaker *Healy* also made a successful transArctic passage (see Figure 1.1).[98] It is easy to subscribe to the preliminary AMSA study conclusion: "One of the historic polar achievements of the 20th and early 21th centuries has been the successful operation of icebreakers at the North Pole and across the Central Arctic Ocean."[99]

The capabilities demonstrated by these voyages are interesting in light of the thinning and diminishing of sea ice in the Central Arctic Ocean. This combination invites for "... an extraordinary future with implications for ice navigation and Arctic ship constructions".[100] At the same time, one should bear in mind that there will always be an ice covered Arctic Ocean in winter, but the ice will be thinner and contain a smaller fraction of multiyear ice (see Chapter 4).[101] In other words: sea ice conditions are changing in all sections of the Arctic Ocean on a year round basis, if not in all respects improving (see above). In light of inter-seasonal sea ice variability, there is still a need for vessels with modern icebreaking capabilities to operate effectively for commercial or other purposes in all regions of the Arctic Ocean. But the dwindling sea ice cover has given some extra impetus and nourishment to the old idea that commercial ships will be able to operate in ice infested waters without icebreaker escort or convoy. This ability was actually demonstrated as early as

1932, when the icebreaking steamship *Alexander Sibiryakov* sailed from the west to the east of the NSR without icebreaking assistance during one navigation season. Today, the idea is coming to fruition and materializing on realistic technological terms. As discussed above, Norilsk Nickel already operates two such ships on the Kara Sea Route, and four more of the same class are under construction in Germany. Thus, Norilsk Nickel will in a short time have a fleet of six operational icebreaking carriers, "all highly capable of operating independently through the winter season to serve the port of Dudinka."[102] In 2006, these ships began shuttling between Dudinka and Murmansk along the Kara Sea Route independent of icebreaker assistance.[103] The experiences gained and the technology developed for these operations may in the long-term perspective be applied to the Central Arctic Ocean for route developments.

Actually, regular shipping operations in the high seas of the Arctic Ocean are up against multiple challenges of a nontechnological nature. Among the more obvious is the lack of governmental or commercial salvage response to support shipping in faraway waters, There is also the lack of communications and there are no routinely-produced ice information products at navigation scale for the high seas beyond coastal state waters. Although all Arctic states provide marine weather information for their coastal waters, none has as yet been assigned the responsibility to do so for the high seas regions – although an initiative seems to be under way in this respect. When considering possibilities for the Transpolar Passage, there are few adequate ports in the United States' or Russian waters near the Bering Strait, which is a crucial chokepoint for any transArctic shipping and for vessels transiting the region. The closest US harbor with deeper water is Dutch Harbor in the southern Bering Sea. On the Russian side, the nearest and largest deepwater port is Provideniya located in the northeastern part of Chukotka (see Chapter 5).[104] These infrastructural shortcomings may partly explain why a sample of 98 shipping companies in a recent study showed a lack of enthusiasm to engage in Arctic transportation. The conclusion being that "although marine traffic in the Russian and Canadian Arctic seems to be definitely on the rise, this is far from being an explosion". One of the companies, however, cited the possibility of the transpolar route becoming navigable over the long term, "which would make the route passing through Arctic straits obsolete".[105]

1.1.7 Connecting corridors in southern waters

On the Atlantic side of the Arctic Ocean there are three corridors: the *Northern Maritime Corridor*, connecting the NEP to Europe and North America; the *Fram Corridor* connecting the TPP to the North Atlantic and ultimately to the NMC and the Arctic Bridge; and the *Davis Corridor* connecting the NWP to the western branch of the NMC and the east coast of North America. On the Pacific side, the three Arctic corridors connect with one joint southern corridor: the *Northern Pacific Corridor* going through the Bering Strait connecting to the west coast of North America and northeast Asia and ultimately to the Great Circle Route. These corridors are two-way corridors made up by resupply and destination-Arctic shipping and possibly, in the long term, transit shipping.

1.1.8 The Northern Maritime Corridor

The Northern Maritime Corridor stretches from the Whites Sea in the north, with partners in Murmansk, Nenets and Archangel regions, to multiple ports in the North Sea.[106] This corridor – also called the "new motorway of the sea" was approved as an inter-regional project by the European Union (EU) in 2002, involving partners in 22 regions in eight countries. The NMC is by western analysts regarded as a most important linkage to north-west Russia, connecting "... the NMC to the ... Northern Sea Route which connects North West Russia to the Pacific Ocean" (see Chapter 2).[107] From a natural and geographical point of view using the NMC, NEP and NPC from London to Yokohama it is seldom necessary to negotiate more than 400 nautical miles of difficult ice conditions, i.e., 5–6 percent of the total freight distance (see Chapter 7). Well over half of the route goes through ice free waters and about 40 percent through navigable ice. The bulk of the route runs through international waters not legally disputed by anyone. Thus, in a geopolitical perspective the physical and political challenges of these interconnecting corridors apply only to a small part of the total distance in the Arctic proper (see Chapter 2).

In our definition of the NEP, the NMC overlaps with the latter in the White and Barents Seas. In this definition, the NMC overlaps with the traditional geographical conception of the NEP, making the Barents Sea a definitional venue of four overlapping routes: the NEP, the NMC, the Kara Sea Route and the functional extension of the NSR (see Figure 1.10).

Marine transport of Russian oil through the NMC has been going on for some years, but increased dramatically in 2002. The oil comes from production sites in Western Siberia. As the existing pipeline from Siberia to southern Russia was oversubscribed at the time, oil was instead shipped by train to the White Sea, transferred to tankers and shipped on to the European market through the NMC. Oil production in Russia's Arctic deposits is expected to increase – some suggest a production level by 2021 of 55–65 million tonnes (see Chapter 3).[108] Crude oil, bunker oil and refined products are shipped out on small ice strengthened tankers from different ports in the White Sea to Murmansk where it is transferred to large tankers for export. The transport capacity was originally about 5.4 million tons a year, but is expected to triple and quadruple over a short period of time (see Chapter 3).

Also the transAtlantic branch of the NMC, which connects the NEP to the east coast of North America has been activated for transport as the first load was delivered to the US east coast in February 2008. Since transport costs from Murmansk to North America are comparable to those from the Middle East, this trade is expected to increase rapidly.[109] By 2020 the estimate is that 20 million tonnes of LNG will be transported from Russian Arctic gas fields to North America.[110] In this perspective, Siberia is linked to Washington via two or three Arctic routes and the transoceanic blue water branch of the NMC.

Due to Iceland's geographical location en route to the North American east coast, Icelandic authorities and shipping companies have plans to service the transoceanic branch of the NMC by offering deep ocean ports, repair facilities, reloading of cargo from small to large tankers, etc. The idea is to establish a transhipment port at Isafjordur in northwest Iceland. Previously, the harbors at Reykjavik and Reydarfjordur in East Iceland have been suggested.[111] The government points out that the deep fjords in west Iceland, like Hvalfjordur, offer good natural conditions for ports for big ships and even "better than other options in the northern part of the Atlantic."[112] The Icelandic government does not only suggest Iceland to be a transhipment country for the east coast of North America, but also for Northern Europe. The geopolitics of this scheme is that Iceland can serve international trade"... as a transhipment hub ... between the continents of Europe ... North America and

Source: Ocean Futures Maps: The Northern Maritime Corridor (2006)

Figure 1.11 The NMC–NEP connection.

Asia across the Central Arctic Ocean through trans-arctic sea routes."[113] The reference to Asia has among other things to do with the close cooperation that has developed between Iceland and China in the course of the three last years on Arctic shipping (see Chapter 2).[114]

1.1.9 The Northern Pacific Corridor

The Northern Pacific Corridor on the Pacific has not yet been established or for that matter got an official name. For simplicity and for the purpose of this study we call it the *Northern Pacific Corridor* (NPC), which starts out in the Bering Strait, overlapping with the functional definition of the NSR on the Pacific.

The Bering Strait is a narrow international strait that connects the Arctic Ocean to the North Pacific Ocean. It is the geographical venue of the NWP, NEP, TPP and NPC – a choke point through which all vessels have to pass to exit or access the Arctic Ocean on the Pacific (see Figure 1.12). At the strait's narrowest point, the continents of North America and Asia are just 90 km apart. The biggest depth is 60 meters. In the midst of the strait are two small islands belonging respectively to the USA and Russia – Little and Big Diomede.

Seasonally, dynamic sea conditions are found in this natural bottleneck labeled by some the "navigator's nightmare" clogged as it is with first-year sea ice more than 4 feet thick. Multiyear sea ice is known to move through the strait at speeds approaching 27 nautical miles per day.[115] In the land territory surrounding the strait there are three US ports: Nome, Kotzebue and the Red Dog mine harbor. Also on the Russian side the ports are located to the south of the strait. These are: Providenija, Anadyr and Egvekinot. The water depth in all these ports is less than 10 meters.[116] The closest US harbor with deep water is Dutch Harbor at the Aleutians in the Southern Bering Sea. On the Russian side, the nearest deep water port is Providenija. Thus, the regional shortage of suitable and effective infrastructure is striking and in need of cost-intensive improvements (see Figure 1.12). Current shipping activity in the area is based on community resupply and destination-Arctic traffic.

When the Russian Navy undertook and planned the Great Northern Expedition to the Arctic in 1733–1743 the objective was to explore the American (Alaskan) coast and to reconnoitre a sea route from Kamchatka to Japan. Although the expedition resulted in 62 maps and charts of these uncharted waters,[117] a sea route was never formally set up. Unlike the NMC, the NPC is still only an expanse of water stretching from the Bering Strait southwards through the North Pacific ending up nowhere and everywhere. Or put differently: The NPC has not yet formally been established as an official link between ports in the Arctic and ports in southern waters by any countries or groups of countries. One important reason for this is that "Because of Russia's European identity, analysts often seems to overlook the Russian Far East (RFE) as an actor in the region when analysing North East Asia. Maps of East Asia list the names and major cities of all the other states in the region, but often let the Russian Far East remain a dark unnamed mass of territory in the north, often cutting the map north of Japan ... There are a host of structural and cultural obstacles to overcome before RFE and its Asian neighbors can in fact reach a level of mutual trust high enough to ensure dynamic cooperative development in the region."[118]

Thus, regional cooperation in the northern areas of the RFE is an issue on the absolute fringe of most of the political relations and perceptions in northeast Asia. "For all the stumbling blocks and hindrances ... to be overcome many mutual perceptions and not least realities have to change both in Russia and in Japan, China and the Koreas." [119] Here politics do not reflect commercial needs. To keep up the economy of northern regions like Magadan, Kamchatka, Sakhalin Oblast and the northern areas of the Republic of Sakha and Khabarovsky Krai, "sea transport is practically the only means of cargo haulage".[120] Therefore, the waters of the Bering Strait have been used in summertime by US, Russian and Canadian vessels servicing communities and industry in northern Alaska and ports along the NWP and NSR in both directions through the Bering Strait and the Aleutian islands. Overall, approximately 159 large commercial vessels pass through the Bering Strait every year during the open-water period from July to October. These estimates exclude fishing vessels and fuel barges serving coastal communities, in particular in Alaska.[121] The volumes of cargo taken through the Strait are small by any yardstick. In recent decades proposals have been put forward to change the state of affairs and to put the region on political and economic maps through the establishment of interconnecting transportation corridors.

In 1992, the State Advisor to the Russian Federation, Professor Alexander Granberg, suggested that it would be attractive "... to set up a system of food supply to the eastern sector of the Arctic (i.e., the Russian Far East) through regular deliveries from the US

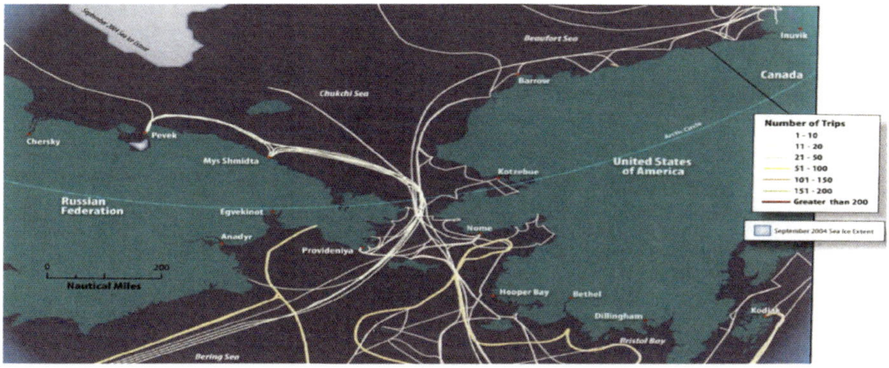

Source: AMSA (2009)

Figure 1.12 Transportation through the Bering Strait and the NPC.

Pacific coast and south-East Asia ... This system assumes particular relevance since many of the former food suppliers to the (Russian) Arctic are now far 'abroad' (Ukraine, Belarus, Central Asia)". [122] Granberg also suggested to deliver oil from the Russian Arctic to the west coast of the United Sates in exchange for American food supplies to the Russian Far East through a transoceanic sea lane across the Pacific, connecting with the NSR. In this scheme, the NSR would be used to distribute large portions of US supplies to the Arctic regions through connecting rivers. This suggestion has not yet materialized in interstate politics. However, in 2007 the Canadian firm Broe/Omni Trax announced its intention to collaborate with Russian companies in the international "Arctic Bridge project" for large-scale transport of transit cargoes via the NSR in the west–east direction through the Bering Strait. This company owns and operates the Canadian port of Churchill and is the largest private railway company in the United States (see above).[123]

According to Japanese experts, "... progress in international specialization and economic globalization has accelerated and broadened the interrelationship between the two regions (RFE and NEA) both socially and economically. It is widely known that oil and natural gas development off the coast of Sakhalin Island provides a wide range of multiplier effects in these areas. And in this way, the globalized face of the economy and industry will play an important role in the sustainable development of the Russian Far East and East Asia for many years to come ... (T)he abundant natural resources in the extreme north area of the Russian Far East will draw (the) ... attention of the international market" (see Chapter 3).[124]

South Korea, being the fourth largest oil importing country and the tenth largest oil consuming country in the world, is dependent on oil deliveries from the Middle East. From a logistical point of view, the security of marine transportation routes for oil between north-east Asia and the Middle East has been seriously threatened by piracy and conflicts among Asian countries, not least threats by Iran to close the Hermus Strait. Due to piracy, the cost of insurance for ships travelling via the Gulf of Aden towards the Suez Canal increased

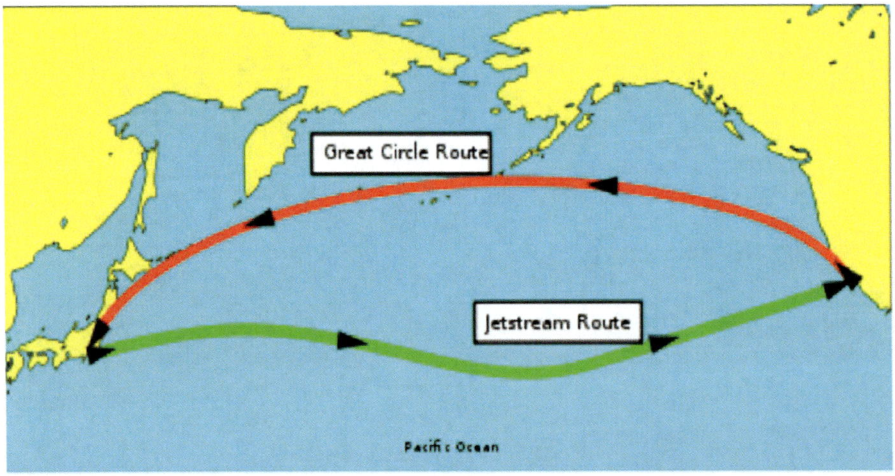

Source: http://en.wikipedia.org/wiki/Great_circle

Figure 1.13 The Great Circle Route.

more than tenfold between September 2008 and March 2009.[125] In response to the threat to these southern supply routes, "it is necessary to exploit various other transportation routes and modes for natural resources ... In addition to the exploitation of transportation routes, naval cooperation among the countries in East Asia should be reinforced to promote maritime safety" (see Chapter 2).[126] For this reason, Russia " is strategically important to Korea as a new alternative energy source in accordance with Korea's strategy of diversifying the countries it imports from."[127] For these reasons, cooperative measures should be taken into account in order to decrease and eventually remove problems among Asian countries.[128]

Chinese researchers claim that the opening of the Arctic routes "will advance the development of China's north-east region and eastern coastal area, (and)… it is of importance to East-Asian cooperation as well".[129] As has been pointed out, "… non-Arctic states, China, Japan, North Korea and South Korea are all in the same boat" when it comes to the prospects of an ice free Arctic. Each of these countries "... stands to benefit enormously from shorter commercial shipping routes and possible access to new fishing grounds and other natural resources. A unified Arctic strategy would be of their mutual interest. Finding ways to jointly use an ice-free Arctic has the potential to create a genuine win-win situation for both China and Japan, the two East-Asian powers which in so many other areas find it difficult to find common ground."[130] Arctic shipping could contribute to economic development in east and northeast China. "Known as the rust belt, China is actively promoting the economic and industrial revitalization of this region, which lags behind other major industrial and manufacturing centers."[131]

Thus, the countries bordering on the northeast Pacific may in due time formally establish a Northern Pacific Corridor connecting the northeast Asian countries to the NEP (NSR), NWP and TPP with a transoceanic branch to the North American west coast. In the meantime, and while waiting for regional cooperation to establish such a corridor, these waters are freely available to increasing international shipping as high seas servicing both transit and destination sailings along all the transportation corridors of the Arctic Ocean.

If and when this route is formally established as a cooperative project between regional states and provinces, the transoceanic branch will interact with the Great Circle Route (GCR) that connects the west coast of North America (including Alaskan waters) to ports in the South Korea, Japan and the Russian Federation through the Panama Canal. This route passes through the Aleutian Islands. In 2004, 2759 vessels passed through the GCR, which was more than all of the shipping traffic in the entire circumpolar Arctic.[132] For this branch of the NPC to be made into a fully integrated route, the Great Circle Route will have to be extended northward from the Aleutians through the Bering Sea and the Bering Strait (see Figure 1.13).

1.1.10 The "Fram Corridor"

The Fram Corridor has not been formally established and/or baptised with a name of public recognition. It is simply a label provided for the purpose of this study, borrowing its name from the Fram Strait which separates Svalbard and Greenland in the north with a minimum of 540 km and in the south with a maximum of 900 km. The Molly Deep provides the deepest point not only in the Fram Strait but also in the whole of the Arctic Ocean with a depth of some 5607 m. In the center of the Strait, depths in general are around 2000 m, with coastal depths ranging from 100 to 500 m.[133] The Fram Corridor in our definition includes the Fram Strait and the Greenland Sea connecting in the south with the transoceanic branch of the NMC north and/or south of Iceland.

Ninety percent of all sea ice that leaves the Arctic Ocean goes through the Fram Strait at high speeds – between 10 to 25 cm per second.[134] Previously, this strait was the outlet of thick multiyear ice extending down to the Denmark Strait between Jan Mayen, Iceland and Greenland. Thus, among the three connecting corridors, the FC is the only one containing Arctic waters – as defined in the introductory chapter – in its full stretch.

Parts of the Fram Corridor is what whalers and sealers for centuries have called the *West Ice* (Vestisen) outside the east coast of Greenland – a hostile sea ice area of many tragic ship losses. Today, the Fram Strait is mostly the outlet of young and first-year ice with a thickness of up to 1.5 meters. Only on rare occasions has multiyear ice been recorded going through the Strait in recent years. This is due to the rise of the air temperature in the area of 2–3 degrees C in the course of the last decades.[135] Thus, climate change has made the FC more accessible to surface shipping than before.

Both Denmark and Norway have established 200 nm zones in the Svalbard/Greenland area. Those zones overlapped with some 150,000 sq km. In 2006, the two countries reached agreement to delimit the disputed area on the basis of the median line principle.[136] The shelf area outside of 200 nm north of the Fram Strait has not yet been delimited between the two countries. In 2010, Norway was the first Arctic coastal state to receive the recommendations of the UN Commission on the Delimitation of the Outer Continental Shelf. Until

Denmark receives her recommendations, probably in 2014, the area will not be finally delimited. Politically and legally, however, the area attracts no disagreements between the two states.

Unlike the Northern Pacific Corridor, the Fram Corridor is not being used on a regular basis for shipping purposes, neither destination-Arctic nor transits. It is known that it has been used by a small number of submarines exiting or accessing the Arctic Ocean,[137] and as an exit area for the Canadian icebreaker *Louis S. St-Laurent* and the US *Polar Sea* in August 2004 (see above). In addition, a few research ships and even drifting ice stations[138] have been operating in the area for research purposes, but the volume of this traffic has been limited and is fairly recent. As a connecting corridor of active seasonal use, the Fram Corridor at best belongs to the future, but ice and navigation conditions are in steady improvement for more active shipping.

1.1.11 The "Davis Corridor"

The Davis Corridor (DC) is not a formal name depicting an established transport route between adjacent countries. It is used as a label for the purpose of this study, borrowing its name from the Davis Strait which separates Greenland and Baffin Island with depths varying between 350 and 3600 m. The Strait is known for its fierce tides ranging from 30 to 60 feet, which discouraged many early explorers. A cold ocean current of heavy ice runs southward along the banks of Baffin Island emptying itself into the Labrador Sea in the North Atlantic at speeds of 8 to 20 km a day. This makes the northern part of the Labrador Sea ice-infested and similar to the waters of the Fram Corridor. The DC includes the Davis Strait and the Labrador Sea and extends southward connecting to the western branch of the NMC. It includes the whole of the North American East coast and passes four national territories – Greenland, Canada, Iceland and the USA.

Shipping through the Corridor is modest, counting between 100 to 200 vessels a year. It is seasonal and mostly conducted by Danish, Greenlandic and Canadian vessels. Since 2002 the amount of sea ice in the Strait has decreased, and today there is open water all year round making commercial activities possible. The Strait has a fairly long history of large-scale commercial fisheries (trawling for scallops, Pollock and cod) and has also been subjected to petroleum prospecting.

There have not been any serious political attempts to establish a cooperative Corridor in these waters. The present level of shipping activity is handled sufficiently effective by the informal cooperation that already takes place between the bordering countries.

Notes

1 Okhuizen (1998), p.4

2 Regulations for Marine Operations Headquarters on the Seaways of the NSR of 1976, Regulations for Navigation of the Sea Ways of the NSR of 1991, Guide to Navigation through the NSR of 1996, Regulations for Icebreaker-Assisted Pilotage of Vessels on the NSR of 1996, Federal Law of Internal Sea Waters, Territorial Sea and Contiguous Zone of July 1998, no. 155-F3, and The Regulations for Navigation on the Sea Ways of the Northern Sea Route, Marine Doctrine of the Russian Federation for the Period 2000 –2020, and Tariffs for Icebreaking Fleet Services on the Seaways of the NSR of 2005.

3 Peresypkin and Yakovlev (2008), p. 1

4 Jørgensen (1991), pp. 77–89
5 Johannessen et al.(2007), pp. 283–285
6 Brubaker and Østreng (1999), p. 309
7 AMSA (2008), p. 178
8 Weller (2000), p. 43
9 USGS (2007)
10 Riska (2011), p. 63
11 Kolodkin and Kolosov,(1990), p. 164
12 Johannessen et al. (2007), pp. 21–23
13 Østreng (1991), pp. 14–15
14 Simonsen (1996), p.73
15 Simonsen (1996), p. 6
16 Simonsen (1996), p. 6
17 Tamvakis, Granberg and Gold (1999), p. 264
18 Tamvakis, Granberg and Gold (1999), p. 264.
19 Frolov, Borodachev and Alexandrov (2007), pp. 22 and 217
20 Østreng (1999b), pp. 169–174
21 Peresypkin and Yakolev (2008), p. 1
22 *US Presidential Directive* (2009) and *EU Communication* (2008)
23 Gordienko (1960), p. 17
24 Kassel (1959)
25 Sater (1969), pp. 12–13
26 Østreng (1991), pp. 9–12
27 Lensky
28 Arikainen & Chubakov (1987), p. 130
29 Peresypkin and Yakovlev (2008). p. 1
30 Frolov, Boradachev and Alexandrov (2007), pp. 21–22
31 Bulatov (1997), p. 82
32 Østreng (1999 II), pp. 18-22
33 Armstrong (1981), p. 452
34 Frolov and Krutskih (2008).
35 AMSA (2009), p.39
36 *Arctic Marine Transport Workshop* (2004), p. A-7
37 *Arctic Marine Transport Workshop* (2004), p. A-19
38 Ramsland (1999)
39 Golovnev and Osherenko (1999)
40 Dallmann (1997)
41 AMSA (2009), p.133
42 Osherenko, Schindler et al. (1997)
43 Riska (2011), 57
44 Pharand (2007), p. 15
45 AMSA (2009), p. 20
46 For a detailed description of all seven routes see *Arctic Marine Transport Workshop* (2004), p. A-20
47 Pharand (1973), pp. 50–54 and Brigham (2006), p.138
48 AMSA (2009), p.20
49 Marr (2001), p. 1
50 Marr (2001), p. 1
51 Falkingham (2004)

52 AMSA (2008), p. 139
53 Stephenson, Smith , Agnew (2011), p. 159
54 *Information Canada* (1972)
55 *Canada Today* (1974), p.4
56 Pharand (1973), p. 211
57 For more on this discussion see Østreng (1979), pp. 216–220
58 Loukacheva (2007), pp. 59–60
59 *US Presidential Directive* (2009)
60 *EU Communication* (2008)
61 *Arctic Marine Transport Workshop* (2004), pp.A-20–A-21
62 AMSA (2008), p. 38
63 Lasserre and Pelletier (2011), p. 1470
64 AMSA (2008), p. 39
65 Arctic Bridge (1994)
66 Stephenson, Smith, Agnew (2011), p. 159
67 AMSA (2008), p. 79
68 AMSA (2008), pp.148-149
69 Lasserre and Pelletier (2011), p. 1469
70 Concept used by Chircop (2011), p. 11
71 Riska (2011), p. 60
72 Griffiths (2009)
73 Molloy (1969)
74 Østreng (1982), p. 128
75 Lyon (1963)
76 Calvert (1960) and Dyson (1963)
77 Definitions derived from: http://ice.ec.gc.ca/WsvPageDsp.cfm?ID=181&Lang=eng
78 Johannessen et al. (2007), p. 257
79 Description taken from: http://en.wikipedia.org/wiki/Drift_ice
80 Description taken from: http://en.wikipedia.org/wiki/Drift_ice
81 Description taken from: http://en.wikipedia.org/wiki/Drift_ice
82 Samuelson (1970)
83 Weller (2000), p. 40
84 Wadhams (2004)
85 *Naval Operations in an Ice-free Arctic* (2001), p. 3
86 Ibid.
87 Stephenson, Smith, Agnew (2011), p. 159
88 AMSA (2008), p. 178
89 Ugryumov and Korovin (2005), pp.110–111.
90 Doyle (2007)
91 Johannessen (2008), p. 52
92 Mauritzen and Kolstad (2011), pp. 25–36
93 Mauritzen and Kolstad (2011), p. 34
94 This section is based on the AMSA (2009)
95 AMSA (2008), p. 86
96 *Arctic Marine Transport Workshop* (2004), p. A-26
97 Brigham (1995)
98 AMSA (2008), p. 86
99 AMSA (2008), p. 85
100 AMSA (2008), p. 74

101 AMSA (2008), p. 11
102 AMSA (2008), p. 83
103 AMSA (2008), p. 139
104 AMSA (2008), p. 182
105 Lasserre and Pelletier (2011), p. 1468
106 Solheim, Hauge and Leknes (2004), p. 70
107 Solheim, Hauge and Leknes (2004), p. 71
108 Peresypkin andYakovlev (2008), p. 4
109 Focus North 4-2007, p. 2, and Østreng (2005), pp. 86–90
110 Lassere and Pelletier (2011), p. 1469
111 Jakobsson (2004)
112 *North Meets North* (2006), p. 39
113 Heininen (2011), pp. 29–35
114 www.Barentsobserver.com: Iceland invites China to Arctic Shipping, 2010-09-22
115 AMSA (2008), pp. 141–142
116 AMSA (2008), p. 144
117 AMSA (2008), p. 37
118 Simonsen (1996), pp. 4–5
119 For a thorough and insightful discussion of these aspects see Simonsen (1996)
120 Otsuka (2006), p. 74
121 AMSA (2008), p. 144
122 Granberg (1992), p. 13
123 Peresypkin and Yakovlev (2008), p. 5
124 Otsuka (2006), p. 71
125 Jacobson (2010), p. 5
126 Yeong-Seok Ha (2006), p. 106
127 Yeong-Seok Ha (2006), p. 99
128 Yeong-Seok Ha (2006), p. 103
129 Jacobson (2010), p. 7
130 Jacobson (2010), p. 13
131 Chircop (2011), p. 12
132 AMSA (2008), p.170
133 Lysaker (2009)
134 Lysaker (2009)
135 WWW.npweb.npolar.no
136 Overenskomst av 20. februar 2006
137 Østreng (1979), pp.70-132
138 Althoff (2007)

2 Geopolitics and power constellations in the Arctic*

Geopolitics concerns the relationship between *geographical space* and *international relations*.[1] This relationship is by some assumed to be extremely tight. E. W. Said reasons: "Since no State is outside or beyond geography, no State is completely free from the struggle over geography."[2] Or in J. Painter's imagery: "There can be no politics which is not geographical."[3] These statements are geographical determinism in the extreme. A more relaxed version of the role of geography in politics claims that: the world image of States is conditioned by their own geographical location and horizon; technological changes transform the strategic significance of an area; and supply lanes of energy and mineral resources tie regions together and show their vulnerability and interdependency.[4] This version is applied here.

The geographical area addressed in this book is the *space of the Arctic* and, in particular, the *Arctic Ocean* as it relates to *national and international shipping* (see the Introduction). The Arctic Ocean attracts political interest from an increasing number of states, European as well as Asian, big as well as small, polar as well as tropic. This interest relates in varying degrees to six geopolitical features of the Arctic:

1. Its geographical location in between three continents – America, Europe and Asia – offering short trade distances, destination as well as transit.

2. Its assumed abundance of strategically important industrial resources and mineral deposits, in particular oil and gas, offering degrees of increasing economic and energy security to the parties taking part in resource extractions (discussed in Chapter 3).

3. Its sea lanes – inside and outside of the region – and its natural and man-induced operational conditions (discussed in Chapters 1, 5 and 7).

4. Its dwindling sea ice regime resulting from global warming and climate change, offering easier access to resources and better exploitation conditions in the region (discussed in Chapters 1 and 4).

*This chapter has been written by Willy Østreng, Ocean Futures.

5. Its unique environmental fragility, vulnerability and eco-systemic interconnections with ecosystems in southern latitudes (discussed in Chapter 4).

6. Its regulatory affinity to existing global ocean conventions, in particular the third *Law of the Sea Convention of 1982* (UNCLOS) (discussed in Chapter 6).

In this chapter, these six features will be related to the national interests of Arctic-interested states – inside and outside – of the region aiming at providing an imagery of what the geopolitical future of the Arctic may look like. The assumption or hypothesis is that the common denominator of big power interests (as defined in policy documents) will forge an "*informal operational regime*" for the region for other states to accept and/or comply with – willingly or unwillingly. Four categories of states are included: (1) the *big Arctic insiders*[5] – USA and Russia; (2) the *Arctic insider and outsider*[6] – the EU; and (3) the *Arctic outsiders*[7] – Japan, China, South Korea and India. As indicated by the insider–outsider dichotomy, these states are differently positioned in Arctic collective decision-making, but all possess – directly and/or indirectly – an ability to influence the long-term direction of the "informal operational regime", not least if their interests coalesce.

The big power emphasis is not to say that small developed states or alliances of such states might not impact regional operational conditions through an active political posture and diplomacy, but the combined interests of big powers are assumed to be even more important in defining the intrinsic conditions, values and features of such a "regime": Votes counts, power decides. The role of smaller Arctic states is category (4) and will be discussed under the heading: The *small Arctic insiders*[8]: Norway, Denmark, Canada, Sweden, Finland and Iceland (see Section 2.3).

2.1 The state of the Arctic: The six features combined

The interaction of features 1, 2, 3 and 4 (see above) has unleashed a fresh interest in exploiting Arctic resources for industrial and societal purposes, whereas features 5 and 6 basically have been articulated as responses to and "warning" of possible negative consequences of increasing exploitation. Let us start with the significance of location and distance.

None of the major industrial areas in Russia, North America, Europe or Japan are located more than 3,860 nm from the North Pole. That is to say: some 80% of world industrial production takes place north of 30° degree N latitude, and some 70% of all metropolises lie north of the Tropic of Cancer. The Arctic Ocean is thus an industrial "Mediterranean Sea" – a shortcut – between the world's most advanced and productive regions. Historically, this fact has attracted more attention from the military–industrial complex than from the shipowners and industrialists of the world. As early as 1935, General Billy Mitchell maintained in a speech to the US Congress: "Alaska is the most central place in the world for aircrafts. He who holds Alaska holds the world."[9] A scant decade later, in 1944, US Air Force General Henry A. Arnold was to echo this, stating that the North Pole would become the strategic center point if a third world war should break out.[10] Some 200 years before them, the famous Russian scientist Mikhail Lomonosov had stated: "The power of Russia shall be increased by Siberia and the Arctic Ocean."[11] These and similar pronouncements, controversial as they were at their respective times, pointed to an undisputable geopolitical fact*: the shortest distance between three continents is above and through the Arctic Ocean*. The first to put this feature into operational effect was the military establishments of the superpowers.

Developments in military technology during World War II, combined with the geostrategic location of the Arctic Ocean between the superpowers, made the Arctic a suitable deployment area for strategic, high-tech weapon systems. In the 1950s and 1960s, Arctic airspace served as a deployment area and as an attack route for strategic bombers. This deployment pattern was further accelerated in the 1970s with the deployment of new generations of intercontinental ballistic missiles. In the course of the 1980s, the Soviet Northern Fleet based at the Kola Peninsula moved its strategic submarines from the fringes of the Arctic Ocean to the water columns beneath the ice cover of the Central Arctic Basin.[12] Thus, in a short time, the Arctic transformed from being a *military vacuum* prior to world war II to a *military flank* in the period from 1950 to 1970 to a *military front* in the 1980s. Due to its location, the Arctic became an integral part of conflicts originating in southern latitudes. This rather quick transformation of the region into a military arena points to the link between geography and technology on the one hand, and military and political might on the other.[13] In theory, the same geographical features can be put to work for the economic benefit of civil societies.

There is an obvious, and at times considerable, distance advantage involved in using the three Arctic transportation corridors between ports in the Pacific and those in the Atlantic, as compared to the Suez and Panama Canals. The distance between Yokohama in Japan and Hamburg in Germany, for example, is only 6,600 nm by way of the NSR, as against 11,400 nm through the Suez Canal. This implies a 42% reduction in freight distance. Another example: Between the town of Tromsø in northern Norway and Vancouver on the Canadian west coast, 3,350 nm can be saved by using the NSR instead of the Panama Canal – a distance reduction of about 37%. If one uses the TPP across the North Pole, the distance is shortened by yet another 700 nm miles (see Table 2.1).

As is evident from Figure 2.1, the distance from London to all ports north of Hong Kong is shorter via the NSR than through Suez. It is just as far from London to San Francisco through the Panama Canal as by the NSR. Most of the North American west coast, the Russian east coast, Japan, China, Korea and Taiwan are all closer to the European Union/ European Economic Area in freight distance through the Arctic than by way of the Atlantic and the Mediterranean. Thus viewed, the Arctic Ocean is indeed an "industrial Mediterranean" in the true geographical sense of the word.

Several examples from the 1960s and 1970s – before the beginning of the ice melt triggered by global warming – illustrate the potential of the NEP/NSR/TPP as alternative transportation corridors to existing trading routes in southern waters. Voyages previously undertaken by Russian freighters confirm the time-saving benefit of using the NSR. In September 1989, a Soviet vessel carried cargo from Hamburg to Osaka in 22 days by the NSR. Previously, such sailings had taken less than 20 days in summer. By the Suez Canal, voyages from Continental Europe to the same destination would normally take between 30 and 33 days. Thus, in summertime, 10 to 15 days could be saved by using the NSR instead of the Suez Canal between Japanese and North European ports.[14] Transit time between the

US northwest coast and Hamburg through the Suez Canal averages some 28 days. Via the Arctic Great Circle Route, passing north of the large island masses in the Arctic Ocean (the TPP), calculations indicate an 18-day voyage.[15]

In the summer of 1995, under the auspices of INSROP, Japan's Ship and Ocean Foundation organized an experimental voyage through the NSR, starting 1 August from Yokohama and ending up in the township of Kirkenes in North Norway (Figure 2.2). The whole trip lasted 28 days, including three days of scientific work. Due to lack of ice in the Chukchi Sea, the multipurpose cargo vessel *Kandalaksha* used for the voyage had to pass north of Severnaya Zemlya in order to encounter sea ice and conduct the ice studies planned for INSROP. *Kandalaksha* did the trip without any assistance from icebreakers and took what the Russians refer to as the "the high latitudinal-route" of the NSR (see Chapter 1). It took only 13 days for the ship to sail the 3,140 nm between the Bering Strait and Kirkenes, including the time it required to conduct the scientific work. According to the expedition leader, this part of the voyage might have taken "...less than 10 days without the time taken for special mission tasks".[16]

Route	Panama Canal	Northwest Passage	Northeast Passage	Suez and Malacca
London–Yokohama	23,300	15,930	13,841	21,200
Marseilles–Yokohama	24,030	16,720	17,954	17,800
Marseilles–Singapore	29,484	21,600	23,672	12,420
Marseilles–Shanghai	26,038	19,160	19,718	16,460
Rotterdam–Singapore	28,994	19,900	19,641	15,750
Rotterdam–Shanghai	25,588	17,570	15,793	19,550
Hamburg–Seattle	17,110	15,270	13,459	29,780
Rotterdam–Vancouver	16,350	14,330	13,445	28,400
Rotterdam–Los Angeles	14,490	15,790	15,252	29,750
Gioia Tauro–Hong Kong	25,934	24,071	21,556	14,093
Barcelona–Hong Kong	25,044	23,179	20,686	14,693
New York–Shanghai	20,880	17,030	19,893	22,930
New York–Hong Kong	21,260	18,140	20,982	21,570
New York–Singapore	23,580	20,310	23,121	18,770

Source: http://www.diis.dk/graphics/Publications/Briefs2009/sac_northern_searoutes.pdf

Table 2.1 Distance in km between harbors using various southern and northern routes.

Source: http://www.fni.no/INSROP/image1.gif

Figure 2.1 Sea routes between the Atlantic and Pacific Oceans.

The *Kandalaksha* voyage is interesting also in the light of global warming and ice melt. Due to a notable temperature rise at the time, the Chukchi Sea was to a large extent ice free. Thus, the vessel took to 81 degrees and 23 minutes N before it encountered sea ice. In comparison, when Fridtjof Nansen made his quest for the North Pole in 1893/96, his vessel *Fram* was stuck in "fairly heavy ice in all directions" at approximately 79 degrees N. [17] This happened on 24 September – the best summer month for sailings in the Arctic, whereas the crew on *Kandalaksha* had an easy ride under blue water conditions much further north in the month of August, which usually offers more difficult sea ice conditions than late September.

Since the time of the *Kandalaksha* voyage, sea ice conditions have become much more negotiable. In the summers 2009–2011, multiple cargo transits through the NEP have repeated the time-saving experiences of the late 20th century (see Chapter 5). These savings have paid off. In August and September 2009 the Beluga Shipping Group sent two ships, *MV Foresight* and *MV Fraternity*, through the NEP starting out in Ulsan in South Korea

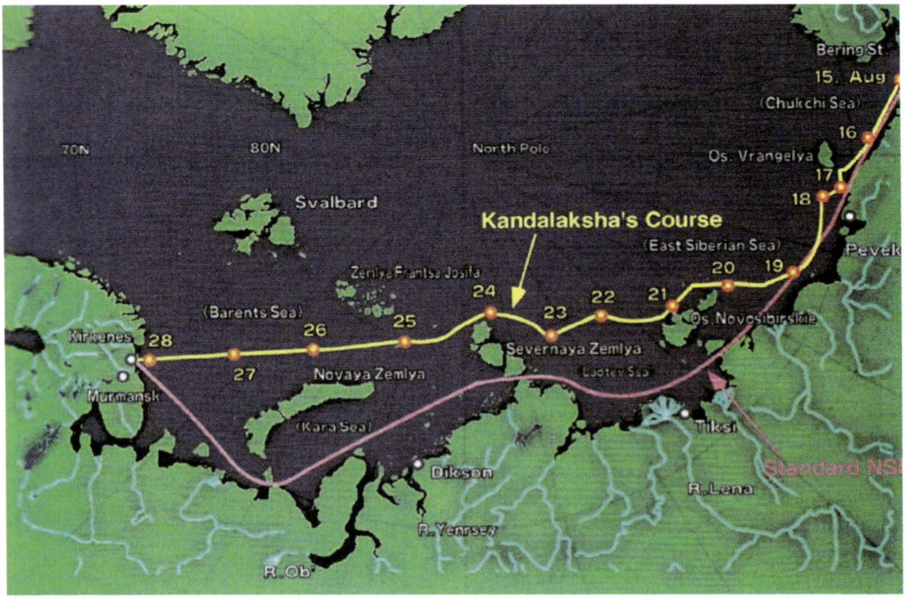

Source: *Northern Sea Route; Future & Perspective*. The Proceedings of INSROP Symposium Tokyo (1995)

Figure 2.2 The *Kandalaksh*a voyage.

and picking up steel pipes in Arkhangelsk for delivery in Nigeria. According to the company, by using this route instead of the Suez Canal they saved some 3,000 nm and reduced fuel consumption by roughly 200 tons in total per vessel. "This resulted in financial savings of about 100,000 US dollars alone for bunker costs with Beluga F-class vessels plus 20,000 US dollars daily for each day travelling the Northeast-Passage which shortens the usual voyage time ... All in all, about 300,000 US-Dollars per vessel were saved by these transits ... instead of taking the long route round."[18] At the time, the President and CEO of Beluga Shipping GmbH, Niels Stolberg, indicated that he expected the new generation vessels of Beluga P-class would result in financial savings of about 600,000 US dollars per vessel and transit.[19] This experience is not unique. In September 2010, the Hong Kong-flagged *MV Nordic Barents* (Ice-class 1A) transported iron ore from Kirkenes in Norway to Shanghai. This voyage was one-third shorter than the Suez route, saving time, fuel and carbon dioxide emissions. Estimates show that about 180,000 US dollars worth of fuel was saved (see Chapter 5).[20]

The Arctic Ocean does not offer shorter and less expensive transport for all maritime trade between Europe and Asia. Mediterranean ports, for example, are closer to Shanghai and Hong Kong using the Suez Canal than northern routes. Thus, trade passages through the Arctic Ocean will not displace the current routes through the Suez and Panama Canals, especially not with the latter's enlargement to be completed in 2014. The expectations are, rather, "that the Arctic will be increasingly attractive for certain kinds of trade, in particular related to hydrocarbons and other minerals produced in the region and exported to China and other Asian countries, and for some seasonal transcontinental trade."[21]

Despite the savings – in time, carbon dioxide emissions and capital – offered by using the Arctic instead of southern routes, Lasserre and Pelletier conclude that those routes "... will not become the new Panama of the 21st century."[22] In their survey of 98 shipping companies, only six answered "Yes" or "Maybe" to the question whether they were planning for transit sailings in the Arctic Ocean. On this outcome the authors conclude that "... many marine carriers are far from being won over by the prospects of significantly less expensive transit opportunities."[23] Why is this? Three explanations can be applied: (1) attitudes to and lack of Arctic-specific knowledge; (2) lack of adequate applied shipping technology; and (3) subordination of civil needs to military interests.

2.1.1 Western attitudes and lack of Arctic-specific knowledge

Throughout history, the Arctic Ocean has remained off the beaten track of large-scale world shipping operations. It has been the operational backyard of a few Soviet/Russian ships. In the West, the general tendency has been to play down the geopolitical aspects of the Arctic Ocean and emphasize the climatic, hydrological and bathymetric restrictions of the NSR and NWP. On the whole, the conclusion has been that these routes are of no interest for market-based economies. Lack of regular sailing schedules, limited length of sailing seasons, costs of icebreaker assistance, high insurance premiums, limited sailing speeds and cargo capacity, and cost of building ice-reinforced freighters are just a few of the numerous cost factors invoked when disregarding the Arctic Ocean as an alternative transportation medium between the Atlantic and Pacific. Even the NSR – the most developed of these transportation corridors (see Chapter 7) – has been regarded as nothing but a misbegotten product of communist command economy rather than as a legitimate offspring of market forces. In most Western quarters, the Arctic routes have never been seen as viable and realistic alternatives to the Suez and Panama Canals. Their fleets are mostly fitted and designed for blue water operation, and have never been built with Arctic operational capabilities in mind. To a large extent, this is still the case. Lasserre and Pelletier show that, in spite of the increasing focus on Arctic transit shipping in media and elsewhere in recent years, "there is no explosion (in the orders to shipyards) of Arctic-plying ships."[24] As of February 2011, most of the ice class ships ordered or built worldwide – 120 in total – were directed at the Baltic and Northwest Atlantic, and only 24% were built for the Arctic Ocean, in particular the Barents Sea and the NSR. The building of ice class ships peaked in 2005 with 234 vessels. Since then it has decreased and, again, most of those ships were not designed for deployment in the Arctic Ocean, but in the Baltic Sea and northern Pacific.[25] By and large, Western shipowners still lack the capacity to undertake shipping assignments in the Arctic Ocean. No wonder the majority of them signal a lack of interest to take on the challenge. Their lack of capability directs their interest towards blue waters.

The blanket rejection by Western shipping companies to engage in Arctic operations and their lack of capability to undertake them have failed to take into consideration the fact that there are several feasible uses apart from all-year transit sailings, among them destination

Arctic-applied transport.[26] The Kara Sea Route is the most prominent example of this (see Chapter 1). As stated by the former President of the Norwegian Shipping Association, Rolf Saether, in 1999: "To study the possibilities facing sea transportation [in these waters] was a task too complex for individual companies. This [called] for international cooperation. When the INSROP programme[27] was presented, we decided to follow it closely. The programme seemed well organized and the most outstanding experts worldwide were invited to cooperate in INSROP. Our purpose for joining in was that the research programme could reveal the potential of the northern sea route to commerce and shipping both economically and technically. We expected the programme to answer the questions we had about the ice and weather conditions along the route. Further, we expected to learn about the environmental challenges along the Northern Sea Route, and how these challenges could be met. We knew that the natural resources in the northern regions were very rich, but we needed more accurate information about oil and gas, minerals, forests and other resources. Finally, we hoped the INSROP would indicate the type of know-how a shipping company would need to be a quality operator in the area."[28]

INSROP was the first international program to address the complexity and multiplicity of questions involved in Arctic navigation. Several multilateral projects have since been launched to follow up on INSROP: ARCOP, JANSROP, AMSA, etc. The knowledge base is building and expanding and, as also demonstrated by Lasserre and Pelletier, a slow but gradual change in the attitudes of the shipping industry to the opportunities of the North is emerging, not least in destination shipping.[29]

2.1.2 Lack of adequate applied shipping technology

One argument put forward for not building ships suited for ice-infested waters was that the shallowness of the shelf areas of the coastal corridors did not allow for building ships of sufficient size and cargo capacity (see Chapters 1, 3 and 5). The draft of ice class ships operating along the NSR excluded vessels of more than 20–25,000 dwt, whereas the routes through the Suez and Panama Canals can accommodate ships many times that tonnage. Case studies undertaken by INSROP suggested that if a cargo vessel is required to call on NSR ports, draft and breadth will be limited to 9 m and 30 m, respectively, with cargo capacity probably restricted to 20,000 dwt. For more northerly routes (without)port calls a maximum draft of 12.5 m and a breadth of 30 m can yield a maximum cargo capacity of 50,000 dwt. Such cargo ships would be approximately one-third the size of those sailing the Suez Canal route.[30] The restrictions underlying these assumptions do not necessarily reflect the current situation.

The Ship Research Institute in Tokyo has been working on the construction of an icebreaking tanker of 200,000 dwt for transport of oil from the Arctic to Japan on a year-round basis. Model trials made as early as in the early 1980s show that this vessel is capable of forcing its way through 2 meter-thick ice by own propulsion at a speed of 5 knots, and in ice free waters, at 15 knots. Cargo capacity amounts to 246,000 cubic meters and keel draft 11 meters, corresponding to the keel depth of the larger nuclear icebreakers, which is 12–13 meters.[31] Thus, according to this study, existing technology in the 1980s allowed for the building of icebreaking supertankers that not only satisfy the requirements of special draft conditions along the NSR, but are also capable of progressing without icebreaker assistance, a point which could increase the profitability of major freighter transportation, destination-Arctic as well as transit. These ships can be built even bigger if transpolar passages over the Central Arctic Basin are being employed. Does this study imply that avail able ship technology as of 2012 suffices the needs of sufficiently safe Arctic operations and that fresh development research related to ships in ice is not in high demand?

Kaj Riska claims that the AMSA study downplays the needs for more research in this respect, referring to a list – "Areas and Issues of Research" – containing 27 items of which the majority relate "...to infrastructure, regulations and ship emissions, but only one fully and one partly relates to ship technology". He finds this remarkable "...as the two new year-round operations that have started recently are related to improved design – and not at all to infrastructural related matters".[32] The examples put forward by Riska are the delivery of high Arctic container carriers to the Norilsk Nickel company and the opening of the offloading terminal at Varanday in the Pechora Sea in northwest Russia (see Chapters 1 and 3). This depiction is not accurate. The AMSA report contains multiple lists on *Research opportunities* scattered around the text. All together, those lists contain 33 items, of which four are directly relevant to ship construction developments. One item asks for a comprehensive study of the history, *design evolution* and use of icebreakers; another urges for the establishment of a database of *damage to ships for use in risk assessments*; the third suggests a comprehensive review of *changes in Arctic marine technology especially for Arctic commercial ships* and how those changes might influence the future of Arctic marine transport systems,[33] whereas the fourth calls for a comparative investigation of *national construction and equipment standards for ships*.[34] Notwithstanding this, Riska may still have a valid point: In the public debate, the emphasis is more on infrastructure than on ship design.

In Riska's mind, the traditional *linear innovation chain* usually applied for ship design improvements should be developed into what he labels an *innovation ring* allowing basic and applied research to play concurrently into a more dynamic and holistic design process of polar ship technology.[35] This is probably true, but it reflects more on the methodology applied among ship engineers than on priorities among items. We find no problem in subscribing to his overall conclusion that "the analysis of means to reduce risks in the Arctic operations suggests that technological innovations are quite important in enhancing shipping and other maritime operations in the Arctic."[36] However, the pioneer transits that have been conducted in recent years show that part of the "conventional wisdom" – the doubts – about Arctic sailings is being challenged.

In 2010, the *SCF Baltica* of Sovcomflot, which displaced 117,000 dwt, transported 70,000 tonnes of gas condensate from the ports of Vitino/Murmansk to Ningbo in China. *Baltica* had a draft of 15 meters, being the first ship ever to ply these waters with a draft this deep. The vessel was 44 m wide – much wider than the assisting icebreakers. The same season, the Norilsk-Nickel operated vessel *Monchegorsk* became the first cargo ship to sail the entire NEP without icebreaker assistance (for more, see Chapter 5). New vessel designs will conquer more territory and accommodate more to the operational conditions of the Arctic transportation passages than what has been accomplished so far. Costs will accrue.

The cost of building double-hulled ice-class vessels is a good deal higher than for conventional vessels. Capital investments will therefore increase the costs of using the three Arctic transportation corridors. Some of these extra costs have already been covered in the building of modern conventional tankers. The international trend is to fit blue water tankers with a double hull to reduce risk of environmental pollution. This is reflected in the legislation of more and more countries – among them the USA – that requires every tanker calling on their ports to have a double hull. To operate in ice-infested waters, a double hull is a requirement. Thus, the price differential between conventional and ice-class ships is getting smaller. There are, of course, more cost-intensive requirements for achieving ice-class status than double hulls, but this is a step in the direction of gradually reducing the difference in technology and price (see Chapters 5 and 7). As pointed out by Kaj Riska, "The use of icebreakers in Russian waters has been the rule but the experience from the advanced Norilsk-Nickel ship series of high Arctic ships has shown that icebreakers are not necessarily needed, if the ship owner is prepared to accept the higher cost of the ships (some 30% as compared to just ice strengthened ships)."[37] The relevant question for economics "is how low the ice class – and how close the ship can be to open water ship – to make the voyage without undue risks" (see Chapter 5).

Attitudes and technology are not the only reasons for the low profile of Western shipping in Arctic waters. Politics, naturally, play a major role.

2.1.3 Subordination of civil needs to military interests

The gradual inclusion of the High North into Cold War strategic planning made most governments conceive of Arctic security solely in military terms. Due to the hegemonic features of the East–West conflict, areas of civilian concern such as circumpolar transportation, resource extraction, scientific research, environmental protection and trade were not accorded autonomous significance of their own. The two sets of issues – civil and military – were, for strategic reasons, tightly interlinked. Not only were few distinctions made between them, but civil affairs were subordinated to military requirements. As a rule, military interests and security considerations gained the upper hand in national priorities for the North. Whenever the military establishment perceived a conflict between the two types of interests, the obligation to yield rested with the civil sector. According to Russian analysts, the Arctic was "...viewed in the USSR in terms of strategic interests, and this ... produced a situation in which all issue areas concerning the region – including economic, social and in some cases even cultural ones – (was) shrouded in secrecy to a degree that (was) uncommon even by Soviet standards. Even non-military matters concerning the Arctic were discussed by a limited group of people within a special interdepartmental body (State Commission on Arctic Affairs) and all publications about Arctic-related issues were subject to special censorship procedures."[38] Use of the NSR fell prey to the security-based hegemonialism of the Cold War and became integrated into it.

The first public attempt to break out of the Cold War security thinking came from the party that most rigorously insisted on it in the past. On 1 October 1987, Secretary General Mikhail Gorbachev gave a speech in Murmansk in which he signalled willingness to initiate international cooperation in four civil issue areas: *energy planning*, *environmental protection, scientific cooperation*, and *transportation*.[39] In identifying these areas, Gorbachev also introduced a distinction between *military* and *civil* security. Both were regarded as vital for safeguarding national security, but the civil component was to be given priority from then on.[40] Actually, the new approach held that security lay in the political rather than military sphere and that national security was a comprehensive and complex matter based on two principles: First, national security is an integral part of the security of others, implying that no country can be more secure than others and that one country's insecurity equals the insecurity of the rest. Thus, military imbalances and asymmetries should in the long term be eliminated. Second, common problems of a transboundary nature – ecological, economic or whatever – can only be resolved through international cooperation. The increased complexity and interdependence between states had, according to the new thinking, created a need to develop a comprehensive system of international security based on a mechanism capable of discussing common problems in a responsible way and at a representative level. The need was to extend the concept to comprise, in addition to military matters, economy, ecology and human rights. Thus, international cooperation in civil issue areas was defined as a measure to bolster national security.[41] The purpose was to create *extended security* through international cooperation by *decoupling* military and civil issue areas. Coexistence between rather than exclusion of interests was the prescription suggested to transform the region into a cooperative place for civil activities to take place on their own preconditions and on an equal footing with military activities.[42] The separation of Arctic security into interrelated parts was an acknowledgement of the complexity of national security, the military component being only one. Whereas, the Cold War concept was one of *military partiality*, regarding civil activities as a potential obstacle, or even threat, to military security, the new thinking was one of *comprehensiveness*, regarding civil cooperation in many fields as one of two sets of measures to prepare nations to meet all kinds of threat to national security, military as well as civil. The new security concept was one of *comprehensive complexity, extended to comprise and counteract all possible threats to the wellbeing of states, civil as well as military. In achieving this, some civil issue areas should be the object of international cooperation*. Thus, the Soviet Union was on a brand new track in regional security thinking – a track that the western arctic states had been on for some years already.[43]

In accordance with the new thinking, on 1 July 1991 the Soviet Union opened the NSR for international shipping. In the course of the 1990s a whole new set of cooperative forums –among them the Barents Euro-Arctic Region, the Arctic Council, the Northern Forum etc. – were established for scientific and environmental cooperation at different levels of government.[44] Thus, a new era of low politics and civil involvement in regional affairs has been put in the post-Cold War melting pot of Arctic affairs. Low politics (civil issues) have become part of high politics (national security issues). Military security has become extended security, inviting the civil sector to take part in securing national interests in the region. A new page is being turned in Arctic history.

The incentives to utilize this fresh political foundation for civil purposes are being strengthened by global climate warming and changes in the sea ice cover of the Arctic Ocean. The Arctic is in a slow and gradual process of being tied in to the economic securities of South East Asian, European and North American countries, relieving them of some of their dependencies of energy imports from the Middle East and the political turmoil of southern waters. Here global climate is forcing a change in the global policy focus (see Chapter 4).

2.1.4 The rise of new economic interests in Arctic energy

The annual reductions in sea ice extent in the Arctic Ocean amounts to 45,000 sq km, i.e., more than the size of Denmark. Since 1978, the overall reduction of sea ice extent has been over 10%.[45] In the period 1976–1990, the extent of sea ice was reduced by 1 million sq km, i.e., an area bigger than Norway, Denmark and Sweden combined (see Chapter 4). This reduction is the most pronounced in the marginal seas of the Arctic Oceans. This ice melt is accelerating and opening up sizeable chunks of previously "ice-closed" continental shelf areas for exploration and exploitation. In its search for oil and gas, expectations are that the petroleum industry will follow the ice edge northward until it reaches the edge of the continental slope bordering on, but not overlapping with, the operational area of strategic submarines – the Central Arctic Basin (see below). The access to Arctic resources is continuously improving.

According to the US Geological Survey's (USGS) most recent estimates, the Arctic could hold up to 22% of the world's undiscovered hydrocarbons, i.e., 50 billion tons of oil equivalents.[46] Of these resources, the shelf is supposed to contain a reasonable share – up to 84% – not least the Russian, which takes up about 40% of the seabed of the Arctic Ocean.

In the view of the USGS, most of the undiscovered oil and gas resources is concentrated between the shoreline and the 500 m contour and within the 200 nm limit (see Chapter 3). The interest in exploiting these resources is fuelled by two extraterritorial and geopolitical circumstances.

First, the global rate of oil finds has dropped drastically since the late 1960s, indicating that world energy production may be on a steep downhill track in the years ahead. At the same time, the demand for oil is expected to increase by some 60 percent over the next 30 years. Here, find rates, supply and demand are on a course of fatal collision. New energy forms, sources and provinces are in high demand. The assumed role of petroleum in this rather bleak futuristic scenario is that oil and gas will remain the dominant form of energy supply for at least 40 more years.[47] As stated in the *US National Security Presidential Directive* and the *Homeland Security Presidential Directive* of 9 January 2009: "Energy development in the Arctic region will play an important role in meeting growing global energy demand as the area is thought to contain a substantial portion of the world's undiscovered energy resources."[48]

Second, to take energy resources from the Arctic complies with the policy of most oil- and gas-importing countries to reduce their vulnerability to being subjected to energy blackmail from governments in politically unstable and volatile areas. The attraction for Arctic oil and gas is fed by the war against terrorism, piracy in southern waters and the enduring political dramas of the Middle East and Central Asia providing the bulk of fossil energy to import-dependent countries in the Western world. The Suez and Panama Canals run through some of the most politically turbulent and volatile areas of the world. At the same time, the effectiveness and commerciality of international ship-based trade has long been founded in and dependent on the enduring and continuous functioning of these waterways. The Suez Canal has been closed two times for extended periods of time in the post-World War II period, forcing international ship-based trade to go around the Cape of Good Hope to markets in Western Europe and North America. In addition, there have been six Israeli/Arab wars, the Gulf war, the Afghan War, and the 11-year war between Iraq and Iran, all introducing severe uncertainties in the supply of oil to energy-dependent countries in the West. These events caused energy prices to soar to unprecedented levels, seriously hurting the world economy, in particular the least developed countries (LDC). Piracy outside of the east coast of Somalia and in the East China Sea and conflicts among Asian countries make supply routes eastward and northward from the Middle East an uncertain business. As has been pointed out, "From a logistical point of view, the security of marine transportation routes for crude oil between Northeast Asia and the Middle East has been seriously threatened by piracy, conflicts among Asian countries and military affairs."[49] These countries are looking for alternative routes of supply through politically stable waters. This again draws attention to the formal establishment of a Northern Pacific Corridor, which in due course might increase the role of Arctic transportation corridors in international trade in relation to the sand-swept tropics and the pirate-infested waters of the Indian Ocean and East China Sea (see Chapter 1). As has been suggested, "The prospects that ships with western cargoes will one day be sailing through the Arctic, instead of the Suez Canal, would seem to be one good reason for opening a dialogue with the (Russians) on ice navigation technology."[50] The former President of the Norwegian Shipowners' Association, Rolf Saether, follows suit, stating that "... we should all join forces to promote the sustainable development in the (NSR), a development that might stimulate the economy and improve the living conditions ... (which) might be a stimulus to international trade and shipping. If so, it might turn out that the Northern Sea Route will eventually emerge as a new and important trade route in world commerce. We would all like that to happen."[51]

Will this civil wish – shared by more and more – cause competition between the civil and military sectors over access to the same ocean expanses, and introduce fresh conflicts to the region?

2.1.5 Availability of ocean space and extended security

The Arctic Ocean is an ocean suited for submarines, not for surface warships. Surface warships are constructed with thin hulls because of speed requirements, whereas cargo vessels are ice-strengthened to cope with ice at the expense of speed. According to Admiral Anatoliy Yakovlev, this is why the NSR has never been used for the operation and battle training of Soviet/Russian surface warships, and why "... transference of Russian warships (between the Pacific and Northern Fleets) along the NSR has nearly stopped".[52] What is more, the coastal shelf areas are not suitable for strategic submarines due to the combination of their size, shallow shelves and sea ice. These vessels experience limitations in maneuvring ability through the restricted underwater spaces between deep extending ice and the extremely shallow waters covering the shelves off Siberia and North America. In general, strategic submarines (SSBNs) are simply too large to operate in most of these areas. As an example, the Russian Delta class submarine is much longer (150 meters) than a soccer field, wider than a handball field and as tall as a 10-storey building (approximately 25 meters). It goes without saying that such a vessel, as big as the largest battleships of World War II, need considerable space to maneuvre, horizontally as well as vertically. As stated by US submarine experts: "It is axiomatic that a short submarine is more maneuvrable than a long one. Any submarine longer than 350 feet (approximately 105 meters) probably is incapable of meeting maneuvrability requirements under ice in shallow waters."[53] The operational space needed between surface (ice) and sea bottom has been estimated to between 180 to 200 meters.[54] The Kara, Laptev and Chukchi Seas all have average depths of less than 100 meters. In the East Siberian Sea, depths are generally less than 40 meters and 53% of the Laptev seabed has a depth of less than 50 meters.[55] In fact, in extensive parts of these areas, a Delta class submarine sitting on the sea floor would have its tower protruding above the surface of the ocean. The depth of the Canadian Archipelago is on average somewhat deeper than the offshore areas of Siberia, but several of the NWP channels are both too narrow and shallow to allow the access and effective operations of big strategic submarines. With or without the presence of ice, the depths of these seas do not suffice to fulfil optimal operational requirements of large strategic submarines.[56] No admirality in its right senses would gamble with the survivability of its strategic submarines in such waters as long as better alternatives are immediately at hand. The deep Central Arctic Basin is such an alternative, offering depths of thousands of meters and a moving ice sheet producing more than enough ambient noise to provide adequate noise protection to strategic submarines. The strategic interest for the area is picking up. In 2007 and 2008, 14 longer patrols were conducted by Russian submarines in the Central Arctic Ocean, and a US attack-submarine surfaced in the vicinity of the North Pole in mid-October 2009. The increased Russian submarine activities in the Arctic Ocean come after nearly two decades of little Russian naval activity in the region.[57] The only area where clashes of interests may occur is in the Barents Sea, with an average depth of 229 meters, and where submarines in a future scenario may have their thoroughfare restricted and narrowed in by oil, gas and other civil installations.[58]

Thus, by and large, nature to a certain extent sees to it that the civil and military sectors can fulfil their objectives and activities in separate parts of the Arctic Ocean, thus avoiding interference in each other's activities. The shelf areas seem to be "reserved" by nature for the civil sector – shipping and resource mining – whereas the Central Arctic Basin is the prime operational space of strategic submarines – a strategic "sanctuary" of submarines.[59] This natural separation is in no way absolute, but it seems to meet the needs of the two sectors to conduct most of their activities without continuous and undue interference from the other.

If in the future the TPP is used for regular freighting purposes, the shipping industry will share ocean space with strategic submarines. This, however, does not provide an insurmountable risk for clashes of interests in that freighter traffic across the North Pole area will only create additional noise to the noise created by moving sea ice, providing added protection for submarines against detection from enemy sonar. If collisions of interests should occur in this part of the Arctic Ocean – which cannot be ruled out completely – the shipping industry still has the option of using the coastal corridors, and even high-latitudinal routes above the shelf north of the archipelagos. To an astonishing extent, natural parameters provide for cohabitation between military and civil actors in the Arctic Ocean, and for both sectors to fulfil their obligations on their own terms without necessarily interfering in the activities of the other. Thus, military-strategic interests are no longer an absolute obstacle to the civil sector as they were assumed to be during the Cold War. Here nature (ice melting and shallow shelves) and politics (Murmansk speech) coalesce.

Recently, the USA, Russia and the EU have released policy documents defining their national interests – military as well as civil – in the Arctic for the 21st century. To see how these fresh documents relate to the state of Arctic geopolitics, we will address them systematically along six policy dimensions: *national security*, *resource extraction*, *international governance*, *environmental concerns*, *international cooperation* and *sea transport*.

2.2 The big Arctic insiders: The USA and Russia

Both countries are littoral to the Arctic Ocean and full members of the Arctic Council. They belong to the inner circle of Arctic affairs – to the Arctic Eight – and are part of the regional cooperative structures and decisive players in regional deliberations.

2.2.1 US Arctic policy

On 9 January 2009, the White House issued two Presidential Directives defining US policy in the Arctic: one for *National Security* and one for *Homeland Security*. The Directives state that the United States has varied and compelling interests in the region relating to national and homeland security, resources extraction, governance, environmental protection, scientific exploration, international cooperation and shipping. Due to its bipartisan flavor, the two documents, which were issued by President Bush just weeks before his presidency ended, are still considered relevant for current and future administrations.

Eight months later – in October 2009 - the US Chief of Naval Operations issued the *U.S. Navy Arctic Road Map* in support of the Presidential Directives, addressing among other things the changing Arctic environment, the potential increase in natural resource extraction, inter- and intra-Arctic shipping, the activities of other Arctic nations and current Naval Fleet capabilities and limitations for operations in the High North.[60]

In summer 2010, President Barack Obama issued a *Presidential Memorandum* assigning responsibility for Arctic research to the White House National Science and Technology Council.[61] This move was meant to secure a US leadership in Arctic research. The Memorandum did not in any serious way alter the content of the two Presidential Directives.

National security and homeland security

National security issues in the region are defined as both broad and fundamental, and to be safeguarded either independently by US action only or in conjunction with other states. These interests include missile defence and early warning, strategic deterrence, maritime presence to project sea power throughout the region, maritime security operations, ensuring freedom of navigation and over flight, and the prevention of terrorist attacks. It is stated that freedom of navigation is a "top national priority", underscoring that the NWP and the NSR are international straits subjected to transit passage. This stand is, according to the Directives, important to support the United States' "...ability to exercise these rights (freedom of navigation) throughout the world, including through strategic straits". Here US Arctic policy is defined in the context of her global interests.[62] The regional situation is part of the global situation.

To keep up her readiness, the US Navy, while being committed to conducting military activities in an environmentally sound manner, "... is opposed to any framework which unreasonably restricts or prevents (her) ability to train and operate effectively."[63] The Road Map is to be reviewed and revised every four year to provide fresh five-year actions plans specifying Navy objectives and "desired effects regarding the Arctic".[64] Among the areas to be addressed are undersea warfare, expeditionary warfare, strike warfare, etc.[65]

International governance

During World War II, the US Vice-President, Henry A. Wallace, proposed that his country should lead the way in establishing an *Arctic Treaty* for, *inter alia*, scientific exploration and cooperation among the Arctic States.[66] The drafting and implementation of such a treaty never materialized, partly due to the hegemonic features of the Cold War, and partly due to lack of pressing contemporary needs for international cooperation in most civil fields. In light of this episode, the Directive is quite firm in denouncing the drafting of "an Arctic Treaty of broad scope – along the lines of the Antarctic Treaty".[67] Such a move is regarded both as inappropriate and unnecessary due to "geopolitical circumstances". Instead, the President urges the Senate to act favorably on US accession to UNCLOS to protect and advance national interests both globally and with respect to the Arctic. Among the Arctic states, the US is alone in not having ratified the Law of the Sea Convention. Until that happens, the Directive shall be implemented in a manner "consistent with customary international law as recognized by the United States, including with respect to the law of the sea".[68] This implies that the United States, for the time being, is in a position to pick and choose among the articles of this and other ocean-related conventions that best suit the fulfilment of her national interests. This comes to expression in the Directive stating that the "most effective way to achieve international recognition and legal certainty for our extended continental shelf is through the procedures available to State Parties to the U.N. Convention of the Law of the Sea". This flexibility of action is also evident in the willingness of the White House to consider new and enhanced international arrangements to address issues like shipping and development of energy and other resources. Such arrangements are seemingly not to be part of the work of the Arctic Council, which is regarded as a useful instrument within its limited mandate of environmental protection and sustainable development. The Council should not, however, "... be transformed into a formal international organization, but should remain a high-level forum devoted to issues within its current mandate. The United States are willing to update the Council's structure to improve its work, as long as the update is consistent with its limited mandate". This is just another way of saying that the AC shall not extend its agenda, for example to cover "... matters related to military security"[69], which has been consistently rejected by the USA ever since the establishment of the Council in 1996.

Transportation/shipping

The United States defines three priorities for maritime transportation in the Arctic:

1. *To facilitate safe, secure, and reliable navigation*, which depends on reliable infrastructure to support shipping activities in cooperation with other nations.
2. *To protect maritime commerce*.

3. *To protect the Arctic environment* through a risk-based capability to address environmental hazards.

In light of these priorities the United States aims to develop Arctic waterway management regimes in accordance with accepted international standards (not Arctic-specific standards), including: vessel traffic monitoring and routing; safe navigation standards; accurate and standardized charts; and accurate and timely environmental and navigational information (see Chapter 3). The Roadmap explicitly states that regional developments in terms of sea ice melting and resource extraction may "reshape the global transportation system".[70]

The Directive is positive to evaluate the feasibility of using access through the Arctic for strategic sea lift and humanitarian aid and disaster relief, as suggested by former State Advisor to the Russian Federation, Alexander Granberg, in 1992 (see Chapter 1).

Resource extraction

According to the Directive, resource extraction shall be conducted in a sustainable way. That is to say, the United States will seek to balance access to, and development of, energy and natural resources with the protection of the Arctic environment by ensuring that continental shelf resources are managed in a responsible manner through working closely with other Arctic nations. Best practices and accepted international standards will be applied to ensure that hydrocarbon and other developments in the Arctic are conducted in a sustainable way. Consultation with other Arctic nations and international cooperation is defined as means to discuss issues related to exploration, production, environmental and socioeconomic impacts of economic utilization of resources. The new policy also states that the United States will protect its interests with respect to hydrocarbon reservoirs that may overlap boundaries to mitigate adverse environmental and economic consequences related to their development.

Environmental protection

An interesting aspect of the Directive is that it uses the concept of climate change consistently to describe the changing Arctic environment, despite the Bush administration's long-term resistance to accepting that the present global warming is indicative of climate change. It still claims that high levels of uncertainty exist concerning the effects of climate change and human activities in the Arctic region and that much more research is needed to make sound political decisions based on scientific evidence. This being said, and taking into account the limitation of existing knowledge and data, the efforts to protect the Arctic environment and to conserve its national resources "... must be risk-based and proceed on the basis of the best available information".[71] In other words, the United States is not inclined to postpone regional economic development until more scientific certainty is available about the potential effects of exploitation. Uncertainty will always be part of political decisions and should therefore not stay in the way of responsible exploitation. To enhance certainty in this respect, marine *eco-system management* should be applied in the region.

The two Presidentail Directives, the Arctic Road Map and the Presidential Memorandum indicate that in the early 21st century the Arctic region is steadily emerging as a new important area in US foreign policy.[72]

2.2.2 Russian Arctic policy

On 27 March 2009, the Security Council of the Russian Federation released an Arctic policy document: *Fundamentals of the Russian Federation's Policy in the Arctic for the Period up to 2020 and Beyond*. In this document, the concept of the *Arctic Zone of the Russian Federation* is launched. In terms of ocean space, this zone comprises the internal waters, land and islands, territorial sea, the exclusive economic zone and the continental shelf of Russia, within which "Russia possesses sovereign rights and jurisdiction in accordance with international law".[73] The Russian government states her national interests, or fundamentals, in the Arctic as follows:

1. To use the Russian Arctic Zone as a strategic resource base for the socio-economic development of the country.

2. To conserve the Arctic as a zone of peace and cooperation.

3. To preserve the Arctic's unique ecosystems.

4. to use of the NSR as a unified national transport link.

On the basis of these interests, the following objectives and priorities have been formulated.

National security

In the sphere of military security, the defence and protection of Russia's boundaries of the Arctic Zone are regarded as vital to the interests of the Federation. This involves effective maintenance of the required combat potential of conventional troops of general purpose and "military formations and bodies in the region (primarily border-guard authorities)". Such measures are required to ensure "the military security in preview of various military and political situations".[74] There is also a need to create an active functional system of a Coast Guard fighting terrorism at sea and in precluding smuggling and illegal migration. The overall objective is to create an integrated security system to protect the territory, population and sites critical for the national security of Russia within the Arctic Zone against both natural and "technogenic emergency risks".[75] Basic and applied science will be used to develop tools for solving problems of defence.

It is interesting to note that the policy document is tacit about the use of Arctic waters for strategic deterrence purposes. Neither is there a mention of the Northern Fleet based at the Kola Peninsula which is well within the Russian definition of what territories constitute her Arctic sector.[76] Until recently, it was Soviet/Russian strategic submarines that had a need to use the water columns of the Central Arctic Ocean for strategic operations, not US submarines or, for that matter, French and/or British, all of which have easy access to the High Seas from home bases.[77] Although not explicitly mentioned in the Document, there is every reason to assume that strategic deterrence will continue to be a fundamental Russian interest in the Central Arctic Basin in the medium and long term (see above).[78]

Resource extraction

In the sphere of socioeconomic development, the object is to ensure a significant increase in the exploitable mineral reserves of Arctic offshore deposits, among other methods through the implementation of a State program to study and develop the Russian continental shelf and to initiate the development of oil and gas fields within her own Arctic Zone. The government will be active in supporting entities operating in the Zone, primarily in the development of hydrocarbon resources. A main goal is to expand the resource basis of the Arctic Zone "that can largely satisfy Russia's needs for oil and gas and other types of strategic raw materials". The economic development of the Arctic Zone is a primary Russian goal through the creation of a favorable operational regime to be applied within its boundaries. In the period 2016–2020 the Russian Federation will transform her Arctic Zone into a leading strategic resource base in Russia. The future objective is to expand the Zone in a comprehensive manner to consolidate Russia's position "as a leading Arctic power". The Document is not so much focused on the Arctic as a whole as on the Russian Arctic Zone. As such, it is more inward than outward looking in orientation.. The only area outside of the Zone mentioned in the Document is the Norwegian Archipelago of Svalbard where the government will ensure Russia a "mutually beneficial presence".[79]

International governance

In line with the policy of the other big powers, Russia also opts for the region to be cooperative and peaceful. In particular, cooperation is required in defining the unresolved maritime boundaries, to come up with an all-Arctic system of search and rescue, strengthening cooperation within AC and BEAR, facilitating the efficient use of transit and cross-polar air routes and international navigation along the NSR. The legal instruments to be applied to regulate developments are domestic Russian legislation and the norms of international treaties and convention to which Russia is a party, among them UNCLOS (see Chapter 6).

Environmental concern

The Document states that it is in the national interest to preserve the unique Arctic ecosystems. This is a political emphasis that the former Soviet system did not have. There is explicit mention of the term *environmental security,* aiming at preserving and protecting the Arctic environment against the "ecological consequences of ... increasing economic development and global climate change".[80] Sustainability is never used as a concept in its own right to describe how to implement environmental security, which to a large extent is restricted to be applied to the Arctic Zone. Here the objective is to establish special environmental regimes for harvesting natural resources and for monitoring pollutants, primarily "in areas of compact settlements". The Document has less of an pan-Arctic orientation and does not reflect clearly the transboundary character of international pollution.

International cooperation

It is stated as one of four national interests that the Arctic should be a zone of peace and cooperation. Thus, there is an interest in developing a cooperative regime of mutually beneficial utility between the Arctic states on the basis of agreements and treaties to which Russia is a party. The Document is concrete and specific in mentioning the need for Russia to interact with its neighbors in defining maritime borders, as well as in achieving international justification for the boundaries of the Arctic Zone. Since Russia has unresolved boundaries with the United States in the east and Denmark and Canada in the north, the Document goes on to state that the boundaries of the Arctic Zone will be more precise in the future in accordance with "Russian domestic legislation and international conventions to which Russia is a party". The Document is quiet about the relative weight and significance to be paid to domestic legislation versus international conventions in defining these boundaries. Russia sees it in her national interest to strengthen the cooperation within the Arctic Council and the Barents Euro Arctic Region.

Transportation/shipping

As mentioned already, the NSR is singled out as one of four prime areas of national interests in the Arctic. The Document is very specific about this priority. No other means of transportation, for instance pipelines, are mentioned in the Document, and the objectives are grand. First, to facilitate the organization and efficient use of the NSR for international navigation under the jurisdiction of the Russian Federation and in accordance with international law. Second, to restructure the volume of freight traffic along the route, among other methods through state support to the construction of ice-breaking, rescue and support fleets, as well as coastal infrastructure. Third, to establish a system of monitoring the safety of navigation and transport flow management in areas of intense traffic, among others, by implementing measures of hydro methodological and navigational support in the Russian Arctic Zone. Fourth, to optimize the instrumental technological monitoring of areas adjacent to straits (58 along the NSR), river estuaries and coastal lakes along the route.

2.3 The small Arctic insiders

It stems from the assumption/premise of this chapter that the small Arctic insiders will have less of an influence on the content of the future *informal operational regime* than the big Arctic insiders. Although the small states are in the majority, they lack, even combined, the power potential and projection ability of the big insiders. To a certain extent: Votes count, power decides. This statement is, however, partly contradicted by history showing that big powers under certain conditions are being challenged by smaller states in matters of national urgency. For instance Canada is up against the USA and the EU when it comes to freedom of navigation through the NWP, and Iceland unleashed the Cod War with Great Britain in the 1970s when extending her fishing zone. Thus, exceptions to the rules seem to be the only true rule of international relations. But this rule does not dispose of all differences between states. All states are inclined to seek the support of others in safeguarding national interests – small ones more so than big ones. This measure is for all categories of states the less costly way of success. Therefore, the big Arctic insiders are all for a *cooperative Arctic* – as a first line of action. If success cannot be achieved through international cooperation all states are in principle able and reluctantly willing to apply power – small states less so than big powers. The United States has already declared her readiness to act unilaterally in the Arctic if need be – as a second line of action. Thus, in the willingness to cooperate and/or readiness to project power there is a difference of degree rather than kind between big and small states. They differ in their ability to project power effectively. This difference is what provides big powers with the extra ability to shape Arctic international relations to the best of their own national interest.

As the least costly alternative – politically as well as materially – international cooperation in the region will create a win–win situation for all parties. This is why cooperation is at the top of all political agendas. Nevertheless, the small insiders are those that will benefit the most from a collaborative atmosphere. In this perspective, they are the most likely and prone to seek political *adjustment*, *compromise*, *alliance building* and *mitigation* within the framework of *international law*.

In the period 2008–2011 all Arctic states produced policy documents on the Arctic. Although there are differences between the small Arctic states in certain fields – for instance, when it comes to freedom of navigation (see Chapter 6) – there is overall a remarkable political agreement between them on most issue areas. They are all for sustainable development of resources, the application and governance of international law, the regulation of international shipping, the improving of sailing conditions, the preservation of the Arctic environment from pollution, seeking international cooperation when need be and the utilization of opportunities of melting sea ice for economic purposes.[81] On these policy items the small agree with the big. The disagreements between the Arctic states is not so much between big and small as between the *Arctic Five* (coastal states) and the *Arctic Three* (Sweden, Finland and Iceland) and indigenous peoples.

2.3.1 The Arctic Five and the Arctic Three

On 28 May 2009, the coastal states of the Arctic Ocean met in Greenland to negotiate and agree on the so-called *Ilulissat-Declaration* on the legal foundation of governance in the Arctic Ocean (see Chapter 6). After the establishment of the Arctic Council in 1996 this meeting was the first exclusively for the Arctic Five. This prompted an immediate protest from the Inuit Circumpolar Conference (ICC) which underscored its legal right as a permanent participant of the AC to be present at the negotiating table when issues concerning the region are discussed. Iceland expressed concern at being excluded from the Greenland meeting as a permanent member of the AC. A year later a new meeting for the Arctic Five was organized in Ottawa on the invitation of Canada to discuss new ways of thinking about economic development and environmental protection. This time the reaction of the Arctic Three became more vocal, explicit and public. The Foreign Minister of Iceland expressed dissatisfaction at not being invited, and claimed that a better strategy on the part of Canada would be to try to build consensus among the Arctic Eight, rather than within an exclusive group of coastal states. Finish and Swedish officials made similar remarks, as did ICC and the Arctic Athabaskan Council (AAC).[82] These meetings prompted a policy process in Iceland.

In May 2010, the Minister of Foreign Affairs, Øssur Skarphedinsson, presented a Report to the Icelandic Parliament in which he defined Iceland's national interests in the Arctic. He wanted to secure his country's position as a coastal state (thus achieving the same status as the Arctic Five) by developing "legal and geographical arguments for Iceland's role in international decision-making regarding the High North", and further "to promote and strengthen the Arctic Council as the most important forum for circumpolar cooperation", opposing the exclusive meetings of the five littoral states.[83] The Foreign Minister got the support of the Parliament who urged the government to promote an interpretation of the Arctic that should not be limited to a narrow geographical definition. Here, an aerial definition has become international politics in which the aim is to extend the group of the Arctic Five to become a group of *Arctic Six* (see Introduction). If Iceland should succeed in her endeavors and get support for a redefinition of the Arctic (in line with what AMAP and AMSA reports have suggested) Finland and Sweden will lose out and form a group of *Arctic Two*. Such an outcome is not likely to be to the liking of Helsinki and Stockholm.

To the surprise of many, at the meeting in Ottawa US Foreign Secretary Hillary Clinton warned against the practice of inviting just a few of the Arctic states to these meetings and advised future discussions on regional issues to include all states with legitimate interests in the region. In principle, the Norwegian Foreign Minister agrees with his US counterpart,[84] but finds these meetings to be practical – as a means of resolving common challenges between parties sharing a problem. In his mind, participation of non-affected parties might only complicate negotiations and delay resolution.[85] Indigenous peoples join in with the critique articulated by the Arctic Three of being excluded from these meetings. They demand a seat at the negotiating table whenever Arctic questions are being discussed, and wonder whether, by this practice, the Arctic Five are trying to assume leadership in the governance of regional affairs at the expense of the Arctic Three and indigenous peoples – at the expense of the Arctic Council.[86] There are reasons to believe that these skirmishes will be sorted out in due time, not least because a big power wants *all* small states to be present at the table of the Arctic Council. A redefinition of the Arctic Ocean for the purpose of strengthening individual states' position in the decision-making process may nurture emerging cleavages within the Council, and weaken its position as a collective policy body for the Arctic Eight. Such an outcome will run counter to the declared collective interests of the Arctic states, which is to strengthen the role of the Council.

2.4 The Arctic insider and outsider: The EU

The EU does not belong to the inner circle of the Arctic Eight. It is half in and half out of the Arctic Council. It is half in by three of its member states (Denmark, Sweden and Finland) being full members of the Council, and partly out by the EU Commission not being a permanent observer to the Arctic Council. The EU also connects to the inner circle of Arctic States through Iceland and Norway's full membership in the AC and their status as members of the European Economic Area. The EU's Northern Dimension initiative, which is a shared policy among the EU, Iceland, Norway and Russia, also connects the EU to the core players of Arctic affairs as does the EU-Greenland Partnership. Several EU countries –France, Germany, the Netherlands, Poland, Spain and the United Kingdom – are permanent observers of the Council, which means that they participate in the meetings without seeking permission to participate. The political significance of the EU comes to expression in a recent White Paper on Norwegian Foreign Policy stating that "The EU increasingly stands out as Europe's spokesman vis-à-vis Russia, also in questions of significance to us (Norway) in the North".[87] Having a leg in both camps shows how the EU Commission can represent the interests of both insiders and outsiders to the Arctic Council.

2.4.1 EU Arctic policy

On 20 November 2008 the Commission issued a policy document to the European Parliament and the Council on EU interests and priorities in the Arctic. Three main policy objectives were stated:

5. To protect and preserve the Arctic in unison with its population.

6. To promoting sustainable use of resources.

7. To contribute to enhanced Arctic multilateral governance.

One year later – on 8 December 2009 – the Council approved the three policy objectives proposed by the Commission in the *Council Conclusions on Arctic Issues*. Ten months later, the Council document was followed by a *Report on Sustainable EU Policy for the High North* issued on 22 October 2010 by the Committee of Foreign Affairs of the European Parliament. By and large, the latter document confirms the policy recommendations of the two former.

National security

The Commission takes climate change as a "threat multiplier" altering the geo-strategic dynamics of the Arctic with "potential consequences for international stability and European security interests". The main cause of conflict relates to climate change and the quest and competition for Arctic resources, in particular hydrocarbons. The EU does not in this document discuss military issues per se. The security threats are civil connected to climate change and consequently the security measures to be applied should be civil related to the environment and to the sustainability of human activities both in and outside of the region. To be successful in these efforts, international cooperation is a prerequisite in all aspects of policy formulation (see below). The EU meets threats to international stability and European security with civil countermeasures. This mirrors the interconnectedness of civil and military issue areas in the concept of extended security also as defined in the Murmansk program.

Resource extraction

When it comes to resource extraction, the main focus is on hydrocarbons. Due to the assumed large untapped hydrocarbon offshore oil and gas reserves of the region, the Commission believes these resources can contribute "... to enhancing the EU's security of supply concerning energy and raw materials in general". In particular, Russia and Norway are singled out as long-term cooperative partners to facilitate a sustainable and environmentally friendly exploration, extraction and transportation of hydrocarbons in Arctic waters and through the Northern Maritime Corridor.

Transportation/shipping

The transportation focus of the EU is solely on shipping, reflecting the fact that the member states combined have the world's largest merchant fleet of which many ships use trans-oceanic routes. The melting of sea ice in the Arctic is progressively opening opportunities to navigate the transportation corridors of Arctic waters. As observed by the Commission:

"This could considerably shorten trips from Europe to the Pacific, save energy, reduce emissions, promote trade and diminish the pressure on main trans-continental navigation channels."[88] Here the main focus is on transit sailings to increase trade between Pacific and Atlantic countries through an Arctic shortcut, and to find supplementary transportation routes to main transcontinental navigation routes of troubled southern waters – the Suez and Panama Canals, and the East China and Indian Seas. The Commission therefore regard it an important asset for the future that European shipyards maintain their competitive lead in developing technology required for Arctic conditions, i.e., specially-designed and environmentally-friendly ships, including ice-breakers. The International Maritime Organization (IMO) should, according to the Commission, have a special role concerning the development of navigation rules, maritime safety, route systems and environmental standards in ice covered waters.

International governance

The Communication states that "No country or groups of countries have sovereignty over the North Pole or the Arctic Ocean around it", and there is no specific Treaty regime for the Arctic. The freedom of navigation and the freedoms of the High Seas shall rule these waters (see Chapter 6). In this regard, the EU Council goes one step further, reiterating "... the rights and obligations for flag, port and coastal states provided for in international law, including UNCLOS, in relation to freedom of navigation, the right of innocent passage and transit passage, and will monitor their observance".[89] In diplomatic terms, this wording is surprisingly clear relating directly to the controversies over the legal status of the NEP and NWP (see Chapters 1 and 6).

To mitigate conflicts, the EU Commission will work to uphold and further develop a "cooperative Arctic governance system" to secure security and stability, strict environmental management and sustainable use of resources. The UNCLOS provides the basis for the regulation of these waters. Thus, the full implementation of already existing obligations for settlements of disputes rather than proposing new legal instruments is being advocated. This, however, "should not preclude work on further developing some of the frameworks, adopting them to new conditions or Arctic specificities".[90] This flexibility is in accordance with the Union's Integrated Maritime Policy that each sea region is unique and needs individual attention in order to balance its uses in a sustainable manner. Only one article, article 234, out of the 320 articles of UNCLOS is specifically developed to reflect the specificities of ice-covered areas. The rest are basically developed to regulate blue ocean affairs.

The flexibility displayed by the EU is not mirrored in the US position, which claims that UNCLOS and relevant global treaties suffice to meet the regulatory needs of the Arctic Ocean. Here the Commission takes one step towards meeting the former position of the European Parliament and the WWF,[91] both of which have proposed to adopt a simple framework convention – an Arctic Treaty – to the best of arctic governance, but without accepting a fully fledged regional treaty (see Chapter 6).[92] This position has now been reversed in the last Report of the European Parliament underscoring that "the Arctic is not a legal and political vacuum as assumed by some observers, ... an Arctic Treaty is not only not promoted by the peoples and states in the Arctic, but also wouldn't be an appropriate way to deal with the challenges in the Arctic".[93] Here the Parliament refuses to engage in a decade-long UN process with the uncertain prospects of getting an international agreement for the Arctic, which would steal diplomatic energy from dealing with the practical and pressing issues of the region. This is a breakthrough in the effort to harmonize policy between the Commission, the Council and the Parliament which have now become one voice in stressing that the existing framework of international law, including UNCLOS, is the only basis of legal regulations in the region. The wording of the Parliamentary Report is direct, and even surprisingly frank: "Having a look at the map one will get the impression that most of the Arctic 'belongs' to the littoral states", and further: "The Arctic states want to have a privileged role in managing the region, which they interpret as being consistent with UNCLOS, based on their location, sovereign rights and economic and political interests."[94]

The Commission plans to enhance its input to the dealings of the Arctic Council, and as a first step will apply again for permanent observer status in the Council. Such a request was declined at the Ministerial meeting of the Council, both in Tromsø in 2009 and Nuuk in 2011. But, as has been rightly pointed out: "Permanent observer status is not just symbolic. It betters the positions of non-Arctic states to participate in the governance of the region."[95]

On this ground, the EU Council continues to support the applications of both Italy and the Commission to become permanent observers of the AC.[96]

Environmental concern

The main goal is to prevent and mitigate negative impacts of climate change as well as to support adaptation to inevitable change. In this respect, the Commission relates to global and trans-regional processes to battle the negative impacts of, among other things, long-range transport of pollutants. The Arctic is the recipient of pollutants from faraway places which should be involved in solving the problem (see Chapter 4). To counteract this, a holistic, ecosystem-based management system should be developed to ensure sustainability of the environment and by integrating environmental considerations at all levels of human activities. Broad international cooperation is called for to mitigate climate change, to promote high environmental standards, to screen and monitor chemicals in the region, to increase energy efficiency and to assess the impact on marine mammals of increased acoustic noise generated by human activity (see Chapter 4). One measure of achieving this is, according to the Commission, to promote a permanent dialogue with NGOs on the state of the Arctic environment. This policy item is unique to the EU. Another measure suggested is to conclude an agreement on emergency prevention and response with the Barents Euro-Arctic Council. The rights of indigenous peoples have become a thematic priority under the European Initiative for Democracy and Human Rights and in the Communication on the Arctic.

In connection with the International Year of Biodiversity in 2010, the Council invited Member States and the Commission to support efforts to protect Arctic ecosystems and their biodiversity, particularly in the high seas, and by encouraging Arctic states to develop marine protected areas on an individual and cooperative basis.[97] Here the Council goes one step further than the Commission.

International cooperation

As already stated, international cooperation permeates all aspects of the EU's Arctic policy. International cooperation is a prime measure to mitigate and cope with a malfunctioning environment that threatens to rock the boat of international stability and European security. The expressed hope of the Commission is that the present policy document "... will open up new cooperation perspectives with the Arctic states, helping all ... to increase stability and to establish the right balance between the priority goal of preserving the Arctic environment and the need for sustainable resources".[98] The Parliamentary Report states that the EU acknowledges that the Union has no Arctic coastline, but claims its interest in the region to be legitimate as a stakeholder by virtue of its rights and obligations under international law.[99] During the debate of the Report in the Foreign Relations Committee in October 2010, one Parliamentarian expressed the EU's position in the Arctic succinctly: "The EU is not a *shareholder*, but a *stakeholder* in the region."[100]

2.5 The Arctic outsiders: Asian states

Two issues seem to attract the collective attention of non-Arctic states: the oil and gas potential of the region and the transportation corridors through Arctic and southern waters. Or, in the words of the Norwegian Foreign Minister: "The energy sucking glances of the world are now being insistently directed towards the North."[101]

Energy prognoses suggest that Asian countries will be responsible for the biggest increase in the demand for petroleum products in the future. The International Energy Agency (IEA) predicts that India and China will account for just over half of the increase in world primary energy demand between 2006 and 2030, and fossil fuels will account for 80% of the world's primary energy mix in 2030 (see Chapter 3). For logistical reasons, most of these resources will in the short term be sold on the European and American markets but, as has been pointed out, "due to ice melting and shortened transportation routes, Asia may turn out to be an attractive market as well ...".[102]

About 90% of world trade and two-thirds of the trade with petroleum is transported on keel. Prognoses are that this will continue to be the case. Thus, transportation routes – old, new and not yet established (NPC) – are of paramount importance to the economic and energy security of petroleum-importing countries in Asia as elsewhere. The political focus of Asian states is now increasingly on the North Pacific and the Arctic (see Chapter 1). The renewed interest in the Arctic has made Singapore – one of the major shipping hubs of the world – examine what possible effects an opening of polar routes might have on her own shipping business.

2.5.1 China

In the fall of 2010, China's economy surpassed that of Japan, making China the second largest economy in the world. By some estimates, the Chinese economy will surpass that of the USA some time after 2020.[103] China is also a leading blue water shipping nation and nearly half of her gross domestic product (GDP) is dependent on shipping.[104] In the period

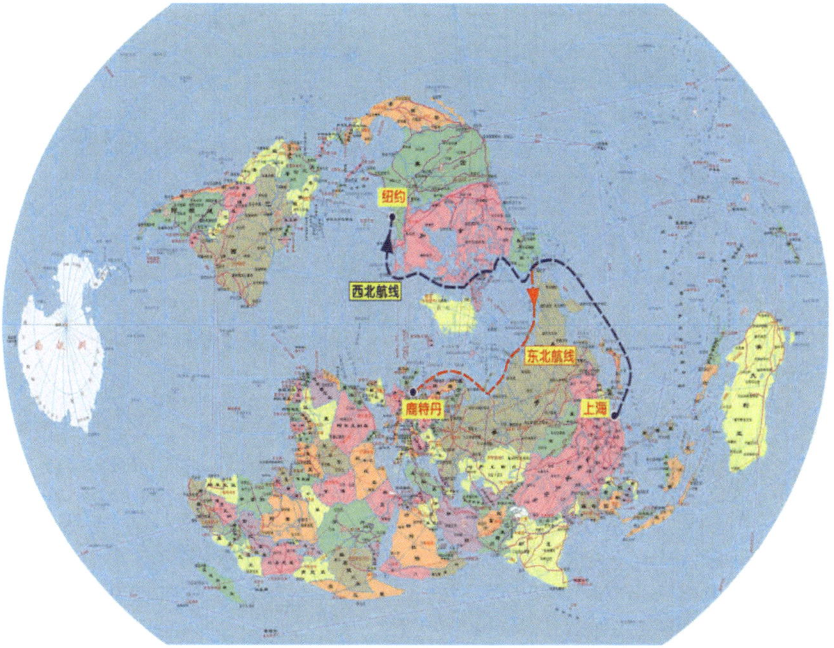

Source: Linda Jakobson: "China prepares for an Ice-free Arctic", *Sipri Insights on Peace and Security*, no. 2010/2

Figure 2.3 A Chinese view of the geopolitics of Arctic sea routes.

1993–2006, Chinese oil companies invested in 120 projects all over the world to a value of more than 27 billion dollars. A big surplus on the trade balance still makes it possible for the Chinese government to further its aggressive foreign investment policy in spite of worldwide financial crises. Recently, China gave a loan to Russia of 25 billion dollars as an impetus for Russian companies to increase their petroleum production, and in 2008 a Norwegian rig company was bought at the net sum of 12.5 billion NOK.[105]

Although the Assistant Minister of Foreign Affairs, Hu Zhengyue, claims that China does not have an *Arctic strategy*, it appears that she has an *Arctic agenda*. Expert opinion is that "China can be expected to seek a role in determining the political framework and legal foundation of future Arctic activities".[106] The prospect of the Arctic being navigable during more months of the year, leading to both shorter shipping routes and access to untapped energy resources, has moved observers to assume that China's large shipping companies can be expected to avail themselves of Arctic routes, even though those routes will be open only on a summer season basis.[107] At the same time, it seems as though China has acknowledged that she has to partner with foreign companies to get access to regional resources, not least because her own companies, which may have the investment capital, lag far behind in advanced offshore technology and suitable operational know-how in cold regions.

China – as a non-Arctic country – has one of the world's strongest polar scientific capabilities to be applied to the Arctic. Since 1984, the State has organized 26 expeditions and established three research stations in Antarctica.[108] China took an interest in the Arctic in the mid-1990s. Her first Arctic research expedition by sea took place in 1999, and since then she has carried out three more expeditions, in 2003, 2008 and 2010. The last expedition lasted for 85 days and involved over 120 scientists, logistical staff and media persons from China and eight scientists all together from Taiwan, Estonia, Finland, France, South Korea and the United States.[109] In October 2009, the State Council decided to replace the vessel they used for these expeditions because it no longer met the demands of the country's expanding polar research. The decision is to build a new high-tech polar expedition icebreaker at the cost of 2 billion yuan (US$300 million). The new vessel, which is to be built in China, is expected to be operational in 2013.

China's first Arctic research station – the Yellow River Station – was founded at Ny Ålesund in Svalbard in July 2004, accommodating up to 18 research personnel.[110] As observed by the Norwegian Foreign Minister, Asian countries like China express their interest in the Arctic "by establishing themselves with scientists in Svalbard".[111] To secure her national interests in the High North, the Chinese government in 2007 launched a national research program covering 10 Arctic projects of geopolitical interest: *the Arctic and human society*, *Arctic resources and their exploitation*, *Arctic scientific research*, *Arctic transportation*, *Arctic law*, *Arctic politics and diplomacy*, *military factors in the Arctic*, *China's Arctic activities*, *Arctic's strategic position*, and *China's Arctic policy and recommendations*. The program lasted for two years and was completed in 2009, but the reports have not been made public.[112] Vice Premier Li Keqiang in a recent speech urged Chinese scientists to continue to push forward in polar and oceanic exploration to serve the county's modernization drive because the oceans has become an important source of natural resources.[113]

Besides her own scientific endeavors, China has collaborated with international partners to monitor the environmental changes in the region. In 1997, China joined the International Arctic Science Committee (IASC) that aims to facilitate multidisciplinary research on the Arctic region and its role in the earth system. From March 2007 to March 2009 a group of Chinese polar experts worked in the International Polar Year Program.[114] China has also participated as an ad hoc observer in two Arctic Council ministerial meetings, in 2007 and 2009, and hopes that its application to be granted permanent observer status will be decided as soon as possible.[115] Moreover, for more than a decade Chinese researchers and officials have participated in international seminars on multiple topics of Arctic geopolitical relevance. Chinese officials seem to vacuum clean the international seminar market on Arctic issues to secure participation of their own polar peoples even in small faraway places.

As has been observed, to date "China has adopted a wait-to-see approach to Arctic developments, wary that active overtures would cause alarm in other countries due to China's size and status as a rising global power. Chinese officials are therefore very cautious when formulating their views on China's interests in the region".[116] Among other things, the Chinese government seems to prefer not to express views on the region's disputes, including the status of the NSR and the NWP, "possibly because in turn it does not welcome pronouncements by non-regional states on its sovereignty and maritime boundary disputes in the South and East China Seas".[117]

In June 2009, the Assistant Minister of Foreign Affairs, Hu Zhengyue, made a speech at Svalbard in which he expressed China's support for the Arctic countries' rights on the continental shelf, endowed by international law, including UNCLOS. He pointed out: "When determining the delimitation of the outer continental shelves, the Arctic states need to not only properly handle relationships among themselves, but must also consider the relationship between the outer continental shelf and the international submarine area that is the common human heritage, to ensure a balance of coastal countries interests and the common interests of the international community."[118] Here China speaks "... for the global commons in ways that no other major state has done in recent time."[119] In so doing she has given herself a voice for developing countries, and considering China's substantial development assistance and involvement in all developing regions, "this is a role which many developing countries are likely to endorse".[120]

Canada and Norway are the only countries to have engaged with China in formal bilateral dialogue on Arctic issues. China and Norway had a meeting in 2010, which also saw the first ever Sino-Canadian Workshop on Arctic issues.[121] In recent years, contacts have also been established between China and Iceland on Arctic shipping issues. In anticipation of Iceland becoming a major shipping hub for the western branch of the Northern Maritime Corridor, China has set up one of the largest foreign embassies in Reykjavik (see Chapter 1). From China's viewpoint, an ice-free Arctic will increase the value of strong ties and broader cooperation with the Nordic countries.

In light of the resumption of long-distance bomber flights in the Arctic in recent years and the planting of the Russian flag on the North Pole in 2007, China appears to be particularly wary of Russia's intentions in the region. [122]

2.5.2 Japan

Japan is the country in Asia where the NSR has received the most attention both from central governments and private foundations. The Japanese Ministry of Transport was actively involved in organizing the *International Northern Sea Route Programme* (INSROP) together with the Ship and Ocean Foundation (SOF) in the early 1990s. SOF, which was one of three principal partners in the implementation of INSROP lasting for six years, also organized and sponsored the scientific trial test, transiting the NSR with the Russian ship Kandalaksha from Yokohama to Kirkenes in August 1995. She also organized the international INSROP Symposium in Tokyo[123] in 1995 and a follow-up research program – JANSROP – which focused attention on the establishment of a transportation system to bring energy resources from Far East Russia to international markets.[124] Last but not least, the northern Japanese prefecture of Hokkaido knows the Arctic and the NSR from its participation in and membership of the Northern Forum.

2.5.3 South Korea

Being the fourth largest oil importing and the tenth largest oil consuming country in the world South Korea is dependent on oil deliveries from the Middle East. From a logistical point of view, the security of marine transportation routes for oil between Northeast Asia and the Middle East has been seriously threatened by piracy and conflicts among Asian countries. In response to the threat to these southern supply routes, it is necessary to exploit various other transportation routes and modes for natural resources.[125] For this reason, Russia has been identified as strategically important to Korea as a new alternative energy source in accordance with Korea's strategy of diversifying the countries she imports from.[126] In line with this, South Korea is planning to build an icebreaker to be launched in late 2012, primarily intended for scientific research in the Arctic Ocean and to develop Arctic transportation routes.[127] The South Korean government has recently restated her interest in involving herself in the development of Arctic sea routes.[128] In 2009, South Korea requested permanent observer status in the Arctic Council but was turned down together with requests from China, Italy and the EU.

2.5.4 India

India – the second most populated country on the face of the Earth – has expressed an interest in deliveries of oil and gas from the High North to feed her rapidly growing economy. For a state with no prior history of interest in the Arctic such expressions are signs of increasing and serious interest in northern affairs. It should be remembered that India, like China, already has a strong "polar connection" to Antarctica as a Consultative Party to the *Antarctic Treaty of 1959*.[129] Her Antarctic experience can be applied to secure her national interests in the Arctic.

As has been noted: "The Asian dimension will result in a possible change of the constellation of actors and the geopolitical dynamic of the North."[130] It is time to prepare politically and diplomatically for meeting "the strategic consequences of three continents growing together in the North".[131]

2.6 Geopolitics and the regional informal operational regime

In this section, the question addressed is whether the interests of the major powers will match to such an extent that they will form an informal operational regime for future utilization of regional resources. The question can be answered in the affirmative and the content of the regime can be summarized in nine points:

1. *All powers underscore their need for the hydrocarbon resources of the region*. The USA and the EU define their needs in terms of economic and energy security, whereas Russia needs these resources for socioeconomic development of their country. Japan, China, India and South Korea follow suit.

- *Prediction*: More and more parts of the Arctic will be opened up for oil and gas prospecting and production. In the ice-free parts of the Barents Sea production has started (the Snøhvit field) and two more fields – Skrugard and Havis – have recently been discovered and declared commercially interesting (see Chapter3).

2. *All Arctic insider states and the EU agree that the exploitation of resources should be conducted with due respect to a fragile and vulnerable environment*. The words vary somewhat between the respective policy documents but they all relate to the concept of sustain-

able development. The EU and the United States opt for ecosystem management and the application of best practices, whereas Russia uses the concept of environmental security. Asian states have not yet publicly expressed their policy on the Arctic environment, but it is reasonable to assume that they will comply with the principles of the environmental regime to get access to regional resources.

- *Prediction*: Environmental concerns will have a big say in oil, gas and mineral developments. Certain areas may even be closed to petroleum prospecting in order to meet environmental goals.

3. *All are in favor of international cooperation to resolve multiple challenges connected in particular to resource exploitation and shipping*. Russia is more inclined to focus her interest and attention on questions related to the working of her Arctic Zone, whereas the United States and the EU are somewhat more pan-Arctic in orientation. The EU seems the most cooperative, opting for a “cooperative governance system”. Asian countries have no option but to comply with the regime established by the big Arctic insiders.

- *Prediction*: All states are in need of peaceful relationships to fulfil their material needs from the High North. To achieve this they are likely to make every effort to sort out their differences by peaceful means and mitigation. If need be, they are likely to agree to disagree when resolution cannot be reached, for instance when it comes to the international status of the NWP and NSR. Failure to do this is failure to secure their own individual needs to achieve a higher degree of economic and energy security.

4. *All insiders are in favor of a peaceful resolution of conflicts and disputes*. This policy goal is a prerequisite for undisturbed and uninterrupted deliveries of petroleum to Asian countries.

- *Prediction*: See prediction in point 3.

5. *All seem to agree that the Arctic Council is a useful forum, but disagree when it comes to its mandate*. The United States is in no hurry to extend its present mandate, whereas Russia would like to strengthen it and the EU (Commission) would like to become a permanent observer in the short-term perspective to influence its working and decisions. China is an applicant for becoming a permanent observer and other countries, like Italy, would like to achieve the same status.

6. *All are in favor of shipping in Arctic waters*. The US seems the more occupied with destination-Arctic shipping in relation to resource transportation, whereas the EU also focuses on transit sailings. The likelihood is that destination-Arctic shipping is the most attractive in a short- and medium-term perspective.

- *Prediction*: Shipping activities will increase in all parts of the Arctic Ocean – more so along the NSR than through the NWP and TPP. Shipping is likely to become the most used mode of resource transportation in the Arctic.

7. *All parties are in favor of applying international ocean law, including UNCLOS, to regulate regional ocean territories and affairs*. The United States underscores the freedom of the high seas and accepts no special Treaty for the region, whereas the EU bases its policy

on UNCLOS, but is more inclined to accept deviations and additions of legal instruments to meet the specificities of the Arctic Ocean. Russia on her part accepts to bow to the Ocean Conventions to which she is a party, but underscores that domestic legislation is yet another instrument of regional regulation. At the moment, the United States invokes customary international law as a basis of regional regulations. This put her in a position to pick and choose what regulatory articles to apply in different outstanding questions, leaving her with some kind of flexibility to adopt to the varying needs of national interests.

- *Prediction*: Ocean Law Conventions – in particular UNCLOS – are the main instruments of regulation. A comprehensive Arctic Treaty is not a realistic option, but special amendments to accepted international law to accommodate regional needs might occur, for instance to preserve a unique and fragile environment, biodiversity, rescue and safety, etc (see Chapter 6).

8. *All are in favor of treating indigenous people and cultures with due respect and to have them reap some of the economic benefit of the new era.* Here, Asian countries have not expressed publicly any policy.

- *Prediction:* The voice of indigenous people in industrial development processes will be part of national and international decision-making. The time has long passed when governments and industries can plan at the expense of local communities.

9. *All seem to apply an extended security concept in the region, accepting that military and civil needs are taken care of in parallel in different parts of the Arctic Ocean.* Here politics and natural features play in concert. It follows from this that Asian countries may be in favor of the civil component of extended security, without expressing any viewpoints on the matter.

- *Prediction*: Military as well as economic activities in the region will increase in the years ahead – gradually. Conflicts between the sectors may occur (for instance in the Barents Sea and TPP), but "platonic cohabitation" seems to be the scenario that the parties are likely to plan with.

This "nine point operational regime" is to a large and compelling extent also supported by the small Arctic states. Small and big are in policy agreement to *extract and transport regional resources in a sustainable and cooperative manner within the framework of international law to the benefit of all peoples of the region*. What is more: the core elements of this regime seem to comply with the national interests of big non-Arctic states in the region, in particular the EU and China. The policy phrasings of all – insiders and outsiders – are non-confrontational.

This is not to say that there are no potential or real conflicts that could turn developments around. Let us see which elements in regional politics could reduce or overturn the overall agreement on the operational regime.

Both the EU and the US insist on the freedom of navigation and that the NWP and NEP are international straits allowing for transit passage, whereas Russia and Canada insist these passages are internal waters and open to international navigation only with the consent of the coastal states. The White House regards the freedom of navigation as a national security interest because it secures the freedom of navigation through strategically-important straits in southern latitudes. This stand reflects global superpower interest, and is hard to compromise. If Russia officially claims sovereignty over the "NSR-routes" in the Central Arctic Basin, the conflict most likely will escalate (see Chapters 1 and 6). Thus, the most potent and explicit conflict potential of the region is related to shipping and navigation. What is more: If need be, and if peaceful resolution is not possible, the United States is ready to resolve conflicts alone on a unilateral basis. The lines of conflict in these respects are mixed. The NSR controversy is between two big states, whereas the NWP is between a small and a big. If Russia officially claims sovereignty over the TPP, all categories of states are likely to react. The same pattern relates to the skirmishes over the exclusive meetings of the Arctic Five, in which the USA supports the criticism of the Arctic Three. Thus, the dichotomy between big and small as applied in this chapter did not reveal any significant power and policy differences. However, the agreement between them will provide small state legitimacy to big power actions if in the future the regime should be challenged or threatened.

Notes

1 Stokke (1999), p. 35
2 Said (1993), p. 6
3 Painter (1995), p. 20
4 Østerud (1996), p. 325
5 Members of the Arctic Council
6 The EU is an insider through the membership of its member states Sweden, Denmark and Finland in the Arctic Council and an outsider by not having its own separate seat in the Council
7 Extraterritorial states are outsiders in that they are not littoral states to the Arctic Ocean nor members of the Arctic Council, but still exert a market influence in international resource dealings that could affect the future direction of Arctic affairs
8 Members of the Arctic Council
9 Cited from Swartztrauber (1965), p.1
10 Gould (1958), p.10
11 Cited from Østreng and Simonsen (1992), p. 99
12 Østreng (1982), pp.59-69.
13 Østreng (1977), pp. 41–63
14 Wergeland (1992), pp. 43–57
15 Fuhs (1992), pp. 73–82
16 Yamaguchi (1995), p.62
17 Vogt (2011), p. 142 (the citation is translated from Norwegian by the author)
18 Beluga Group (2009), p. 1
19 Beluga Group (2009), p. 5
20 Chircop (2011), p. 11
21 Chircop (2011), p. 10
22 Lassere and Pelletier (2011), p. 1472
23 Lassere and Pelletier (2011), p. 1468
24 Lassere and Pelletier (2011), p. 1471
25 Lassere and Pelletier (2011), p. 1471
26 Appendix: Shipping in Arctic Waters
27 For a presentation of INSROP, see Østreng (2010), pp. 191–245

28 Saether (2000), p. 36
29 Lasserre and Pelletier (2011), pp. 1465–1473
30 Tamvakis, Granberg and Gold (1999), pp. 221–280
31 Fujita, Narita and Kitagawa (1986)
32 Riska (2011), p. 65
33 AMSA (2009), p. 47
34 AMSA (2009), p. 67
35 See Riska (2011), pp. 65–71
36 Riska (2011), p. 71
37 Riska (2011), p. 62
38 Vartanov, Roginko and Kolosov (1999), p. 60
39 Scrivener (1989)
40 Parrott (1988)
41 Granovsky (1989), pp. 215–222
42 Østreng (1992), pp. 26–51
43 Østreng (1999a), pp. 34–39
44 Østreng (1999a), pp. 29–48
45 *Impacts of Global Climatic Change in the Arctic Region* (1999), p. 12
46 Focus North 5 (2007).
47 *BP Statistical Review* (2005)
48 *US Presidential Directive* (2009), p. 6
49 Yeong-Seok Ha (2006), p. 99
50 Watson (1991), p. 175
51 Saether (2000), p. 38
52 Yakovlev, Kossov and Ushakov (1994), pp. 10–11 and 43
53 Boyle and Lyon (1998), p. 33
54 Bergesen, Moe and Østreng (1987), pp. 68–85
55 Hønneland (1997).
56 For a more extensive discussion of the operational conditions in these waters, see Østreng, Kolodkin, Brubaker and Jernsletten (1999), pp. 290– 297
57 *Barentsobserver* (2009), 11–16
58 Bergesen, Moe, Østreng (1987), 57–88
59 Østreng (1982), pp.106–113
60 *U.S. Navy Road Map* (2009), p. 5
61 Heininen (2011), p. 58
62 For a discussion of this long-term US interest see, Brubaker and Østreng (1999), pp. 299–331
63 *U.S. Navy Road Map* (2009), p. 12
64 *U.S. Navy Road Map* (2009), p. 27
65 *U.S. Navy Road Map* (2009), p. 15
66 Pollack and Anderson (1973)
67 *US Presidential Directive* (2009), p. 3
68 Ibid
69 "Declaration on the Establishment of the Arctic Council", footnotes 1 and 5 (1996)
70 *U.S Navy Roadmap* (2009), p. 6
71 *US Presidential Directive* (2009), p. 6
72 Heininen (2011), p. 58
73 *Fundamentals of Russian Arctic Policy* (2009), p. 1
74 Ibid. pp. 2 and 4

75 Ibid. p. 3
76 Ibid. p. 1
77 Østreng (1982)
78 Granberg (1992), p. 9
79 *Fundamentals of Russian Arctic Policy* (2009), p. 3
80 *Fundamentals of Russian Arctic Policy* (2009), p. 2
81 For a summary of this interstate agreement, see Heininen (2011), pp. 67–81
82 Nystø (2010), p. 22
83 *Report, Minister of Foreign Affairs*, Iceland (2010)
84 Støre (2008), p. 41
85 Hustadnes (2010), p. 14
86 Nystø (2010), p.22
87 *St. Meld.nr.* 15 (2009)
88 *The European Union and the Arctic Region* (2008), p. 8
89 *EU Council conclusions on Arctic issues* (2009), item 16
90 *The European Union and the Arctic Region* (2008), p. 10
91 Saksina (2008), p.19
92 Østreng (2009)
93 *EU Parliament Report* (2010), pp. 12–14
94 *EU Parliament Report* (2010), pp.13–14
95 Chircop (2011), p. 13
96 *EU Council conclusions on Arctic issues* (2009), item 17
97 *EU Council conclusions on Arctic issues* (2009), item 5
98 *The European Union and the Arctic Region* (2008), p. 12
99 *EU Parliament Report* (2010), pp. 5–14
100 Personal notes from the debate (2010)
101 Støre (2008), p. 37. The translations provided are by this author
102 *Dagsavisen* (2009)
103 Chircop (2011), p.10
104 Jakobson (2010), p. 5
105 Ibid.
106 Jakobson (2010), p. 1
107 Chircop (2011), p.11
108 Jakobson (2010), p. 3
109 Chircop (2011), p.13
110 Jakobson (2010), p. 2
111 Støre (2008), p. 40
112 Jakobson (2010), p. 5
113 www.xinhuanet.com (2009)
114 Jakobson (2010), p. 4
115 Jakobson (2010), p. 9
116 Jakobson (2010), p. 2
117 Chircop (2011), p. 14
118 Jakobson (2010), p. 10
119 Chircop (2011), p.14
120 Chircop (2011), p.14
121 Chircop (2011), p. 9
122 Jakobson (2010), p. 13
123 Kitagawa (1995)

124 Kitagawa (2006)
125 Yeong-Seok Ha (2006), p. 106
126 Yeong-Seok Ha (2006), p. 99
127 AMSA (2008), p. 177
128 Digital Chosunilbo (2008)
129 Jørgensen-Dahl and Østreng (1991)
130 *Dagsavisen* (2009)
131 Ibid.

3 Shipping, resources, economic trends and alternative means of transport*

3.1 Introduction

Shipping of any significance cannot exist without cargo, and without natural resources and their processed or manufactured derivatives, there will be neither cargo nor much shipping.

The Arctic possesses great quantities of natural resources without which little attention would be paid to this frigid part of the world except by scientists of various hues, and governments with national security and military strategic purposes in mind. Add to these the odd adventurers and aficionados who are looking for the unusual and exceptional.

But shipping is just one among several ways of transporting resources within and in and out of the Arctic. Inland waterways, roads, railways, pipelines and airplanes are part of the complicated Arctic transportation network.

The purpose of this chapter is to try to assess the abundance of the most important resources and their location in the Arctic landscape; their importance and competitiveness in the world economy and the global marketplace; and how exploitation of these resources could affect Arctic shipping as compared to other alternative modes of transportation in the Arctic.

The resources in question are petroleum, minerals, forests, and fish.

* This chapter is written by Dr Arnfinn Jørgensen-Dahl, Ocean Futures. Tor Wergeland, Ocean Futures (affiliated), has contributed material on the issues of pipelines and international shipping as well as the entire section on Arctic cruise activity.

3.2 A note on data, information, terms and definitions

When size and location of natural resources is the subject matter at hand, concerns about the veracity and source of the data and information will almost inevitably arise. Natural resources of the size and importance we are concerned with in this chapter carry significance and value well beyond the borders of the countries in which they are found and to which they belong. They become a source of international attention and demand

Estimates dealing with resources are subject to conjecture and are often and even routinely revised. As oil fields are developed or mature, initial estimates of reserves often turn out to have been too conservative. Improvements in extraction and production technology also lead to revisions of estimates, sometimes radical. Apart from the fact that different methods exist of calculating oil reserves, other factors may also play a role. In Russia, official estimates are kept secret for reasons not entirely clear and perhaps therefore all the more open to wayward and inaccurate speculation. There are also doubts about the reliability of official OPEC reserves estimates. Since the system of production quotas was introduced in the early 1980s, some OPEC members have reported dramatic increases in reserves. Some may have resulted from a change in ownership of the reserves away from international oil companies which are required to report reserves according to very strict standards set by the US Securities and Exchange Commission. These are said to lead to conservative estimates. A more important reason, however, is likely to be the introduction of quotas that related the size of the reserves to the allotment of production quotas – in other words, the greater the reserves, the greater the production quota. The increases in OPEC reserves over the last 20 years or so would in any case seem to have had little to do with discovery of new reserves.

At times, somewhat more unusual reasons for inaccurate estimates and reporting reach the public arena. In January 2004, Shell shocked the financial world and business rivals by revealing that it had overstated its proved reserves of oil and gas by 20%, or 3.9 billion barrels. Including unusual practices in the realm of corporate governance, the company admitted that in the late 1990s it had classified the oil and gas in an Australian project as proven even though the fields had yet to be developed. The same year, the El Paso Corporation of the US was compelled to cut its proven natural gas reserves by 41 percent.[1]

Corporate "mistakes" of this magnitude are a rare occurrence in the petroleum industry. But there are other reasons for uncertain and inaccurate estimates. Not every country with natural resources has a bureaucratic structure skilful and competent enough to meet the highest standards of reporting and statistics production. Some may even lack the inclination to do so. As mentioned above, some OPEC members might have inflated their reserves quite considerably without convincing evidence being brought forward to support the increases. Accusations to this effect have been levelled against countries such as Kuwait, Saudi Arabia, Iran, Iraq and the United Arab Emirates.[2]

But the petroleum industry is not the only one suffering from doubtful and inaccurate data. A reference is made below to the preference of the Food and Agricultural Organization of the United Nations (FAO) to deal separately with the data on fisheries submitted by China in some contexts because of uncertainties attached to data going as far back as the early 1990s.

There are also significantly different methods of reporting. The system of reporting mineral resources and reserves followed by Russia has made it almost impossible to compare Russian data with data on reserves in market-based economies. The Russian method is based on establishing drilling parameters to ascertain the certainty of reserves and does not include market-based economic criteria to establish the feasibility of producing these resources using current technology and prevailing market conditions. Like data on oil and gas reserves, data on reserves of many mineral resources is also either kept secret or is very difficult to obtain.[3]

In all reporting on natural resources the professional vocabulary adopted by those whose main preoccupation is to deal with a particular resource is, of course, crucial. It may well be that some of the uncertainties and discrepancies found in the data of the petroleum industry can be traced back to less than careful use of or unfamiliarity with prevalent and important terms and definitions adopted by the industry.

All estimates of oil and gas reserves involve uncertainty which in part is a function of the reliability of geologic and engineering data and the interpretation of these data. There are always *degrees* of uncertainty and this is reflected in the categories **proved** and **unproved** reserves.

Proved reserves are "... generally taken to be those quantities that geological and engineering information indicates with reasonable certainty can be recovered in the future from known reservoirs under existing economic and operating conditions.'[4] Proved reserves may be further subdivided into *Proved Developed* (PD) and *Proved Undeveloped* (PUD). Proved reserves are the only type the U.S. Securities and Exchange Commission allows companies to report to investors. Whereas companies listed on the U.S. Stock Exchange and other bourses must verify their claims, many governments and national oil companies are under no such obligation and therefore do not always disclose data to support their claims.

Unproved reserves are divided into **probable** and **possible,** the two terms reflecting the unequal degree to which they can be recovered. Unproved reserves are used by oil companies and government agencies for future planning purposes. Many countries also possess so-called *strategic petroleum reserves* for economic and national security reasons. These reserves are not included in the statistics for national oil reserves.

Two frequently-encountered categories should be noted. These are contingent and prospective resources. **Contingent resources** refer to petroleum estimated, as of a given date, to be potentially recoverable from *known* accumulations but are not yet considered ready for commercial development due to one or more contingencies such as no currently viable market, dependence on technology under development, and so on. **Prospective resources** refer to petroleum that, of a given date, is thought to be recoverable from *undiscovered* accumulations in future development projects. A last category to be included is **unconventional resources.** These are petroleum accumulations found over large areas. Examples are heavy oil, natural bitumen, and oil and gas shale deposits. These are resources that require specialized extraction technology to produce and may require significant processing before sale. The unconventional resources considerably exceed conventional oil reserves but are difficult and expensive to develop.[5]

When discrepancies in estimates occur it could well be that insufficient care is taken to avoid adding up estimates of different kinds of reserves. What is said to be *proved reserves* may be a cocktail of different types of reserves rather than the single type it purports to be. As reported in a recent study, the Russian Ministry of Nature Resources and Ecology estimated in 2006 that Russia possessed 12 percent of the world's oil resources and 45 percent of the world's gas resources.[6] What was included in the terms "oil resources" and "gas resources" was not specified but these estimates were nearly twice as high as the BP estimates of *proved reserves*.

Although no one in the petroleum sector claims that the industry and their own recording and reporting activities are infallible, data from some are used and referred to more often than others. Publications like the *BP Statistical Review of World Energy* and data published by journals such as *Oil and Gas Journal* and *World Oil* are generally regarded as reliable and they often refer to each other's data. In addition, a number of national governmental and intergovernmental bodies such as the *U.S. Department of Energy,* the *Norwegian Petroleum Directorate*, the *International Energy Agency* and several others are high on the ranking list of "reliable" sources. But almost as a matter of course they take care to stress the uncertainty of the accuracy of all important resource data, including those they themselves generate.

3.3 World economic and shipping trends

The Arctic economic landscape is part of the world economic landscape. The search for and exploitation of natural resources in the Arctic have in the past been and will continue to be influenced by similar activities elsewhere in the world. Developments in the global economy over the last 2–3 years have made this abundantly clear, very important among which are changes in world market conditions brought about by the global financial crisis.

But the world economic landscape is undergoing more fundamental change. For centuries the Mediterranean area was the centere of European economic, commercial, and trading activities. In the Far East, the Chinese Empire and other commercial and trading centers, ruled by feudal lords of varied magnificence and power, took part as best they could. But neither the East nor the West were islands unto themselves, and trade between the two was continuous and mutually advantageous. As late as around the year 1700 China and India together accounted for almost half the world's Gross Domestic Product (GDP), each with just under a quarter of the total. Two centuries later, the world economy was completely dominated by countries around the North Atlantic.

Spearheaded by the Spanish, the main causes of this drastic change were the onset of European imperialism, aided by superior military technology, and the industrial revolution. It made the North Atlantic area – Western Europe and eventually North America – the center of gravity of the world economic system. It has remained so for the best part of three centuries.

But the center of gravity of the world economy is again on the move and the North Atlantic area is now under serious challenge as the undisputed center of world economic power and influence. Led by countries such as China, India and Brazil, regions outside the North Atlantic area have accumulated an increasingly larger share of the world Gross Domestic Product (GDP). In the years 1995–2004 the North Atlantic area (the US, Canada, and Western Europe) accounted for 60–64% of global economic output. In 2009 the share is expected to drop to 49%, and it will continue to decline, predicted to fall to 45% by 2012 (Table 3.1).

Behind much of this massive 15 percent shift away from the North Atlantic area in a mere 15 years are the two most populous states of the world – China and India. Table 3.2 shows that both countries grew at impressive rates during the 1990s. Following the economic reforms introduced in the 1980s under Deng Chiao Ping, China in particular surged ahead, and with her in the lead both countries continued to grow strongly in the first decade of this century.

The year 2008 was a highly unusual one for the world economy and trade. Growth in the world GDP slowed dramatically during 2008 as the effects of the financial crisis that began in the US in 2007 were felt everywhere. The global economy began a serious downward slide in the second half of 2008. As a group, the advanced or developed economies achieved a growth of only 0.5 percent in 2008 and shrank by 3.2 percent in 2009. For the first time since the 1930s, the global economy showed negative growth.

Rest of the World		**North Atlantic Area**	
1995–2004	2009	1995–2004	2009
40–36%	51%	60–64%	49%

Source: Mark Deen Bloomberg, *The Financial Express*, "The West's share of global GDP set to fall", June 03.2009. www.financialexpress.com/printer

Table 3.1 Share of world Gross Domestic Product.

	Share of Global GDP 1990 %	Share of Global GDP 2003 %	GDP Growth Rate 1980–1990 %	GDP Growth Rate 1990–2003 %
China	1.6	3.89	10.3	9.6
India	1.5	1.64	5.7	5.9

Source: T.N. Srinivasan, "China, India and the World Economy", Table 2, p. 4

Table 3.2 China and India share of world GDP and growth of GDP.

Region/Country	2010	2011	2012	2013
WORLD	5.2	3.8	3.3	3.9
Advanced economies	3.2	1.6	1.2	1.9
of which				
United States	3.0	1.8	1.8	2.2
Euro area	1.9	1.6	-0.5	0.8
Germany	3.6	3.0	0.3	1.5
France	1.4	1.6	0.2	1.0
Italy	1.5	0.4	–2.2	–0.6
United Kingdom	2.1	0.9	0.6	2.0
Japan	4.4	-0.9	1.7	1.6
Developing countries	6.1	2.1	6.0	6.3
of which				
Brazil	7.5	2.9	3.0	4.0
China	10.4	9.2	8,2	8,8
India	9.9	7.4	7.0	7.3
Transition economies				
of which				
Russia	4.0	4.1	3.3	3.5

Source: International Monetary Fund. *World Economic Outlook Update*, 24 January 2012. www.imf.org

Table 3.3 Projected rates of economic growth.

	US	Japan	Russia	China	India	Brazil	Canada	Norway
Energy use (kg of o.e. per cap)	7,924	4,135	4,586	1,319	492	1,164	8,469	7,151
Electric power consumption (kwh per cap)	13,701	8,201	5,785	1,783	476	2,016	17,314	25,083

Source: *The World Bank Group 2009*. These figures are not strictly comparable because in some cases they refer to different years and should therefore be regarded as only indicative.

Table 3.4 Energy consumption in important states.

Table 3.3 tells the story of real and projected world economic growth for the years 2010–2013 (annual percentage change). The items showing *energy use* and *electric power consumption per capita* in Table 3.4 suggest that rapidly-growing economies such as China, India and Brazil will experience a hefty increase in demand for energy and other commodities in the years ahead if the economic development history of today's advanced economies is anything to go by. Although individually the latter have followed somewhat different paths to reach their present state of economic affluence, they have all relied on more or less the same mixture of energy resources. Economic development in the future will everywhere have to rely on a different mixture of energy sources. To develop economically in the future, countries like China and India are burdened with the colossal task of achieving a better life for their huge populations by relying on a different, less climatically and environmentally damaging mixture of energy than today's advanced economies have relied on, especially less reliance on coal. This may have a significant impact on patterns of investment and exploration for oil and gas in the Arctic.

3.3.1 Shifting patterns of trade

Economic activities and growth give rise to trade. From a shipping point of view, the most important factors are the volume of merchandise trade that these activities give rise to, their destinations, and the product composition. Table 3.5 shows some trade data that illustrate the rapidly growing importance of China in particular. The table also gives an indication of the current international trade deficit of the US which in considerable measure is a result of trade with China.

World trade in practically all types of merchandise products experienced rapid growth until 2008. Energy was at the center of this surge in demand. The decline that occurred during 2008 and 2009 will be followed by renewed demand in the years ahead. As the fastest growing non-OECD economies, China and India will be the key energy consumers of the future. From 1990 until 2006 China's and India's energy consumption together as a share of the world's total energy consumption increased from 10 percent to 19 percent. Based on predictions of strong economic growth in the future, the two countries are expected to increase their combined energy consumption to 28 percent of world energy use in 2030. By contrast, the US share of the total is expected to decrease from 21 percent in 2006 to 17 percent in 2030.[7]

	Exports 1948 %	**Exports 1993 %**	**Exports 2003 %**	**Exports 2008 %**	**Imports 1948 %**	**Imports 1993 %**	**Imports 2003 %**	**Imports 2008**
US	21.7	12.6	9.8	8.2	13.0	NA	16.9	13.5%
China	0.9	2.5	5.9	9.1	0.6	2.7	5.4	7.0%

Source: The World Bank Group 2009

Table 3.5 Share of world exports and imports.

Exports			Countries/Regions		Imports	
2008	2009	2010		2008	2009	2010
2.6	−13.6	16.2	**WORLD**	2.9	−13.6	15.2
5.5	−14.9	15.3	**US**	−3.7	−16.4	14.7
2.9	−14.7	18.2	**European Union (27)**	1.4	−14.8	14.1
16.8	−6.6	22.4	**India**	29.7	−0.8	11.5
10.5	−13.6	28	**China**	2.3	−1.7	27.1

Source: UNCTAD, *Review of Maritime Transport 2011*. Data from table on p. 6

Table 3.6 Exports and imports of world, important regions and China.

The non-OECD world consists of more than China and India. In non-OECD Asia, energy use is expected to rise by 104 percent between 2006 and 2030. Other non-OECD regions are also expected to be the source of a substantial increase in energy demand. The consumption of hydrocarbons is expected to increase at an annual rate of 0.9 percent from 2006 to 2030. It is also expected to remain the single largest source of energy, declining from 36 percent of world consumption of energy in 2006 to 32 percent in 2030. World coal consumption is expected to increase by 1.7 percent a year from 2006 to 2030 when it will account for 28 percent of total energy consumption. The US, China and India together are expected to account for 88 percent of the net increase in coal consumption between 2006 and 2030.

Natural gas and coal together account for the largest share of total world electricity generation. These two energy sources are expected to produce more than 60 percent of all electricity. In 2030, they will remain the most important sources of supply with a 64 percent share.

3.3.2 Trade by destination

Appendix 1 provides information about the export and import trade by destination of Canada, Russia, and the US, and important non-Arctic states such as Germany, Japan, China, and India.

These tables show, firstly, that the economically advanced countries trade mainly with each other and, secondly, that much of the trade of the developing economies is directed towards the advanced economies.

Table 3.6 shows the percentage growth in volume of merchandise trade by geographical region in the period 2008–2010. US export growth slowed down to a mere 5.5 percent in 2008 and the European Union recorded even less growth. Lacking demand for imported goods, especially consumer goods, led to a sharp drop in import volumes in the two regions. The exports and imports of the developing economies continued to drive world merchandise trade but at a reduced pace. China's exports in 2008 were less than half her exports a year earlier.

As mentioned earlier, China and India increased their share of world total energy consumption from 10 percent to 19 percent between 1990 and 2006. In 2030, they are expected to account for 28 percent. In 2008, primary energy consumption slowed, including growth in the consumption of fossil fuels. For the first time, however, non-OECD consumption surpassed OECD consumption with China accounting for no less than three-quarters and the Asia-Pacific region for 87 percent of world energy consumption growth.[8] The term "primary energy consumption", however, embraces a wide variety of energy sources.

Oil and gas are critical parts of this energy mix. Oil and gas are also the most valuable of Arctic natural resources. Table 3.7 shows the absolute and relative growth between 2000 and 2010 in consumption of oil and gas by countries important to this study. Canada apart, the developed economies in this table would appear to be reaching a consumption plateau as far as oil is concerned. Japan is the exception, having reduced her consumption by a significant 20.9 percent.

As far as international trade in oil is concerned, the year 2009 was highly significant in another way. For the first time, Saudi Arabia exported more oil to China than to the US. The global economic recession caused an economic slowdown in both the US and China. In the US it caused a 10 percent drop in oil consumption. China's economy also slowed down but has been quicker to begin the process of recovery. "While the exports of the United States might rebound this year, in the long run the decline in American demand and the growing importance of China represent a fundamental shift in the geopolitics of oil." Saudi exports to the United States fell to 989,000 barrels a day, the lowest level in 22 years, from 1.5 million barrels in 2008. China's imports of Saudi oil nearly doubled from 2008 to exceed a million barrels a day.[9]

When we turn to gas consumption, however, Table 3.7 conveys a different picture. Except for a rather modest US increase, the developed economies increased their consumption a great deal. The increase in the consumption of gas in Japan, set against the drop in oil consumption, suggests a policy of substitution on the part of Japan. This may also apply to the EU. Environmental considerations as well as technological developments are factors likely to be behind these changes. The outstanding features of Table 3.7 are, of course, the increase in the consumption of China and India. Both countries produce oil and gas, and Table 3.8 shows the relationship between consumption and production in China and India in 1998 and 2008. The gap between consumption and production of oil and gas is likely to rapidly grow in the years ahead.Given their enormous populations, both China and India (and other developing countries) will have to consume energy in quantities much greater than today to reach anywhere near the same level of economic development of today's advanced economies.

These observations concern mainly the demand side of oil and gas. The supply side was summed up by the BP Group Chief Executive Tony Hayward in the *BP Statistical Review 2009*: "Our data confirm that the world has enough proved reserves of oil, natural gas and coal to meet the world's needs for decades to come. The challenges the world faces in growing supplies to meet future demand are not below ground, they are above ground. They are human, not geological".[10]

At the end of 2010, world proved reserves of oil stood at 1,383 billion barrels and proved reserves of gas at 187.1 trillion cubic meters.[11] At the end of 2010, the US had 30.9 billion barrels of proved oil reserves. The National Petroleum Council of the US estimates, however, that 1,124 billion barrels are still underground, "of which 374 billion barrels would be recoverable with current technology". The USGS "estimates that the earth's remaining conventional oil (petroleum) deposits to be around seven trillion to eight trillion barrels". Even if only part of that oil can be recovered economically with today's technology, know-how and prices and be classified as proven reserves, the numbers are

Oil (Million tonnes)				**Gas (Million tonnes o.e.)**		
	2000	**2010**	**Change 2000–2010 %**	**2000**	**2010**	**Change 2000–2010 %**
US	884.1	850.0	–3.8	660.7	683.4	2.4
Canada	88.1	102.3	16.1	92.7	93.8	1.2
Japan	255.0	201.6	–20.9	72.3	94.5	30.7
EU	699.3	662.6	–5.2	372.9	441.1	18.3
China	224.2	428.6	**91.2**	24.5	109.0	**344.9**
India	106.1	155.5	46.6	26.4	61.9	161.7
Russia	129.7	147.6	13.8	354.0	414.1	17.0

Source: *BP Statistical Review of Energy 2011*. Data from various tables in the Review

Table 3.7 Oil and gas consumption trends.

	India		**China**	
	1998		**1998**	
	Oil	**Gas**	**Oil**	**Gas**
Consumption	92.5	22.0	197.0	18.2
– Production	–34.7	–22.0	–160.2	– 21.0
Deficit	57.8	0.0	37.2	Surplus +3.7
	2008		**2008**	
Consumption	135.0	37.2	375.7	72.6
– Production	–36.1	–27.5	–189.7	– 68.5
Deficit	98.9	6.7	186.0	4.1

Source: *BP Statistical Review of World Energy 2009*. Data from various tables in the Review

Table 3.8 Oil and gas consumption and production trends.

still staggering. Furthermore, only about a third of the sedimentary basins of the earth – the geological formations that may contain oil – have been thoroughly explored with modern technologies.[12]

An important element in any assessment of possible reserves is the role of technology. The recovery of oil is a three stage process as depicted in Table 3.9. Only an average of 35 percent of the oil known to be underground has been recovered, that is to say, oil recovery is still in its secondary stage. Tertiary recovery is partly still experimental and subject to the development of new and better recovery technology. The invention of new technology, however, has always been a driving force in the petroleum industry. It has time and again rendered estimates about the size of reserves invalid and caused upward revisions.[13]

From an Arctic point of view the supply of gas is especially significant since so much of Arctic reserves are believed to be gas. In 2008, world proved reserves of gas stood at 185.02 trillion cubic meters, having increased from148.01 trillion cubic meters in 1998, or by 25.0 percent. US reserves of gas increased between 1998 and 2010 from 4.65 trillion cubic meters to 7.7 trillion cubic meters, or by 65.6 percent whereas Russian reserves increased from 43.52 trillion cubic meters in 1998 to 44.8 trillion cubic meters in 2010, or by 2.9 percent. In 2010, the US and Russia stood for 21.7 and 13.0 percent respectively of total world consumption of gas.

The increase in world gas reserves is partially a result of technological innovation. It will affect Arctic regions more than others and those of Russia in particular since so much of the latter's gas reserves are supposed to be in her Arctic territories. A reference was made above to the increase in US gas reserves of 65.6 percent between 1998 and 2010. In 2008, the US produced 7.5 percent more gas than it did in 2007. Some of this increase was due to technological developments that have made it possible to turn to unconventional sources for added supply by exploiting previously almost inaccessible shale gas deposits. By employing two relatively new technologies, horizontal drilling and hydraulic fracturing, the US proved reserves of shale gas increased from 21,735 billion cubic feet in 2007 to 32,825 billion cubic feet in 2008, or by a staggering 51 percent in just a year. Production increased from 1,184 to 2,022 billion cubic feet, or by 70.8 percent. Shale gas reserves in 2007 accounted for 9.1 percent of gas reserves and in 2008 13.6 percent, and are by all accounts likely to increase considerably.

Primary	**Secondary**	**Tertiary**
Recovery up to 15%	Recovery up to 20% to 40%	Recovery up to 60%
Reservoir's internal pressures pushes oil out	Water or natural gas are pumped into the ground forcing more oil out	Chemicals, heat or microbes thin out the remaining oil

Source: Maugeri (2009)

Table 3.9 Three stages of recovery of oil.

The search for and recovery of shale gas has been going on for some time in the US and less so in Canada but is now spreading to countries in Europe such as France, Germany and Poland and to other parts of the world. China and India are among the countries which are thought to have potentially large quantities of shale gas. Because so little drilling has been done outside North America, estimates of the quantities of shale gas that may be found globally vary a great deal. Yet, "[even] the most conservative estimates are enormous projecting at least a 20 per cent increase in the world's known reserves of natural gas".[14]

3.3.3 World oil and gas production and consumption

To be able to form a view of the significance of the natural resources of the Arctic, they must be seen in the wider context of global resources.

Table 3.11 gives an indication of where the greatest increase in the consumption of oil and gas will take place in the future. The outstanding feature is the growth in the consumption of the EMEs (Emerging Market Economies).[15] Although all major groups, except the former Soviet Union, increased their consumption somewhat, the EMEs did so by 41.3 percent.

Country	**Oil million tonnes**	**% Share of total**	**Country**	**Gas million tonnes o.e**	**% Share of total**
Russia	505.1	12.9	US	556,8	19.3
S. Arabia	467.8	12.0	Russia	530.1	18.4
US	339.1	8.7	Canada	143.8	5.0
Iran	203.2	5.2	Iran	124.7	4.3
China	203.0	5.2	Qatar	105.0	3.6
Canada	162.8	4.2	Norway	95.7	3.3
Mexico	146.3	3.7	China	87.1	3.0
UAE	130.8	3.3	Indonesia	73.8	2.6
Venzuela	126.6	3.2	Algeria	72.4	2.5
Kuwait	122.5	3.1	Netherlands	63.5	2.2
Iraq	120.4	3.1	Malaysia	59.8	2.1
Nigeria	115.2	2.9	Egypt	55.2	1.9
Total world	3913.7	100.0	Total world	2880.9	100.0

Source: *BP Statistical Review of World Energy*, June 2011, pp. 10 and 24

Table 3.10 Top oil and gas producers in 2010.

The two biggest EMEs are China and India, and Table 3.12 tells a great deal about the weight they carry as oil consumers now and are likely to do in the future. The US and Japan decreased theirs, Russia increased hers whilst India increased her consumption by 48.6 percent between 2000 and 2010, and China by no less than 91.2 percent during those ten years.

In 2010 world energy consumption increased by 5.6 percent, the highest rate since 1973. After two years of decline, global trade in oil grew by 2.2 percent of which Asia Pacific net imports made up nearly 90 percent. The imports of China grew by 14.6 percent.[16]

Table 3.13 shows that in 2010 the US alone consumed 21.1 percent of world oil and 21.7 percent of world gas, China 10.6 percent of the oil and 3.4 percent of the gas whilst Russia consumed 3.7 percent of the oil but 15.0 percent of the gas.

Despite media headlines and ominous predictions to the contrary, the world would seem not to be running out of oil and gas any time soon. In a speech to the World Petroleum Council in June 2008, the Chief Executive of British Petroleum (BP) said that the "world has more than 40 years of proven oil reserves, 60 years of natural gas and 130 years of

Groups of states	**2000**	**2005**	**2007**	**2010**	**% Growth 2000–2010**
	Million tonnes				
European Union	699.3	723.1	708.4	662.5	–5.3
OECD	2217.1	2303.6	2276.3	2113.8	–4.7
Former Soviet Union	180.4	185.4	194.3	201.5	11.7
Other EMEs	1354.5	1604.9	1731.0	1914.3	41.3

Source: BP Statistical Review of World Energy, June 2011, p. 11

Table 3.11 Growth in oil consumption by major groups of states.

Country	**2000 Million tonnes**	**2010 Million tonnes**	**Change Million tonnes**	**Change %**
US	884.1	850.0	–34.1	–3.9
Japan	255.0	201.6	–53.4	–20.9
China	224.2	428.6	204.4	91.2
Russia	129,7	147.6	17.9	13.8
India	106.1	155.5	49.4	46.6

Source: *BP Statistical Reviewof World Energy*, June 2011, p. 11

Table 3.12 Oil consumption changes by country over time.

coal. Put another way, we have produced around 1 trillion barrels of conventional oil, we're sitting on another 1 trillion barrels of proven reserves and yet another trillion of non-proven resources. On top of that there are vast quantities of unconventional hydrocarbons[17] including oil sands, heavy oil and unconventional oil and gas. And then there are hydrocarbon basins as yet unexplored".[18]

3.3.4 International shipping trends

As the world economy recovers there will be no shortage of shipping to shift goods and merchandise around the globe. The first years of the new millennium were quite remarkable from a shipping point of view. The primary cause of the development was the strong economic growth of the emerging market economies, especially China, and the subsequent demand for shipping.

The good times created among ship owners an optimism that led to a massive ordering of new ships as reflected in Figure 3.1. When the global recession set in, public demand for

Country and groups of countries	**Oil consumption million tonnes and rank**	**% Share of oil consumption**	**Gas consumption million tonnes and rank**	**Share of gas total %**
United States	850.0 (1)	21.1	621.0 (1)	21.7
China	428.6 (2)	10.6	98.1 (3)	3.4
Japan	201.6 (3)	5.0	85.1 (4)	3.1
India	155.5 (4)	3.9	55.7 (8)	1.9
Russia	147.6 (5)	3.7	372.7 (2)	15.0
S.Arabia	125.5 (6)	3.1	75.5 (6)	2.6
Germany	115.1 (7)	2.9	73.2 (7)	2.8
S.Korea	105.6 (8)	2.6	38.6 (9)	1.4
Canada	102.3 (9)	2.5	84.5 (5)	3.0
World total	4028.1	100.0	2858.1	100.0
European Union	662.5	16.4	443.3	15.5
OECD	2113.8	52.5	1397.6	48.9
Non-OECD	1914.3	47.5	1460.5	51.1
Former Soviet Union	201.5	5.0	537.1	18.8

Source: *BP Statistical Review of World Energy*, June 2011, pp. 11 and 25

Table 3.13 World top consumers of oil and gas in 2010.

and spending on finished goods plummeted, followed by a dramatic fall in the demand for primary commodities such as oil, gas and ores, and processed commodities such as steel. This in turn caused a drop in demand for bulk carriers and tankers. It also led to an over-supply of container ships, the like of which one has to go back to the early 1980s to find. In May 2009 almost 10 percent of the world container fleet lay idle. The Baltic Dry Index, which records bulk ship charter rates, fell almost 97 percent between July and December 2008 before recovering somewhat in early 2009. Figure 3.2 shows the fate suffered by dry bulk freight rates in 2008.

As in the case of container vessels, owners of bulk vessels had by April 2009 laid up 314 vessels or about 5 percent of available tonnage of bulk vessels above 10,000 dwt. According to Det Norske Veritas (DNV), orders for 325 bulk carriers, 47 tankers, and 78 container vessels (altogether 37 million dwt) had been cancelled worldwide by May 2009.[19]

By the end of 2009 the world fleet consisted of almost 55,000 ships. The largest category was general cargo vessels (including other dry cargo) with 33 percent, tankers of all sorts constituting 26 per cent, dry bulk 14 percent and containers nine percent. Measured in dwt transport capacity, the world fleet at end of 2009 is almost 1.3 billion dwt.

The distribution is 39 percent tankers (all types), 34 percent dry bulk, 13 percent containers and only seven percent for general cargo vessels.

Figure 3.3 indicates that a rapid increase in the world fleet is expected. IHS forecasts a yearly increase of 6–7 percent for the period 2009–2013, somewhat dependent on the number of orders that are being cancelled. With the financial crisis coming on top of massive over-contracting, international shipping is facing some difficult years.

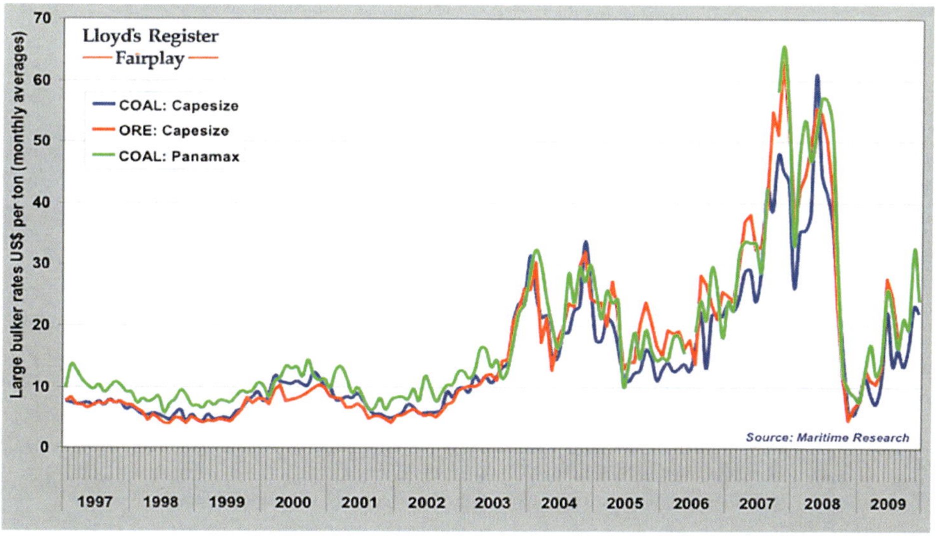

Source: IHS

Figure 3.1 Orders for new vessels.

3.4 Arctic resources

3.4.1 Oil and gas

How much of world oil and gas reserves can be found within the Arctic? And how much remains to be found?

The answer depends on how the Arctic is defined. None of the states that until now have been referred to as Arctic states are wholly "Arctic". With the exception of Russia, which has most of her sea territory in the Arctic, the largest parts of the national territory of the four countries on which we have focused until now are located in the climatically temperate zone of the world.

The problem is that the most diligent search for a generally accepted definition of what constitutes the Arctic will fail. Several definitions may exist within a single country, and this state of affairs is not made clearer when the term "Arctic" is used interchangeably with other terms such as the *North*, *High North*, *High Arctic*, the *Northern* Areas, and the like. "The delimitation of the Arctic ... varies depending upon the perspective from which one approaches it".[20] We shall, however, try to follow as closely as possible the definition of the Arctic adopted in the Introduction to this study.

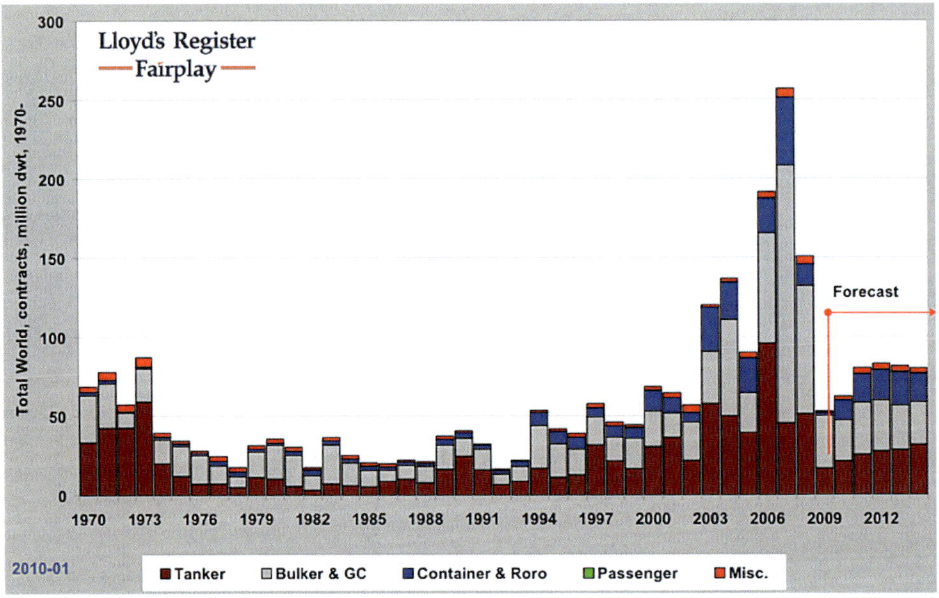

Source: IHS

Figure 3.2 Dry bulk freight rates 1997–2009.

Oil and gas are produced in four of the states that possess land or sea territories defined as being parts of the Arctic. These are Russia, Alaska, the Northwest Territories of Canada, and Norway. Large discoveries of oil and gas in the Arctic began with the Tazovskoye Field in Russia in 1962 and the Prudhoe Bay Field in Alaska in 1967. About 61 large oil and gas fields have been discovered inside the Arctic Circle territories of these four countries. Of these, 15 fields have yet to go into production; 11 are in Canada's Northwest Territories, two in Russia, and two in Alaska. Of the 61 large fields, 42 are located in Russia. Thirty-five of these large Russian fields (33 natural gas and two oil) are located in the West Siberian Basin. Of the remaining eight large Russian fields, five are in the Timan-Pechora Basin, two are in the South Barents Basin, and one is in the Ludlov Saddle. Of the 18 large fields outside Russia, six are in Alaska, 11 are in Canada's Northwest Territories, and one is in Norway.[21]

Tables 3.14 and 3.15 show how Arctic states' oil and gas reserves have developed over the last 20 years. As far as the Arctic states are concerned, the absence of data for 1987 from Soviet times (and Russian data for 1997) and the addition of oil sands to Canadian reserves in later years make it difficult to identify a trend in proved reserves. There is a decline in Norwegian reserves since the peak year of 1997, and a substantial decline in US reserves of 17 percent during the last 20 years or so. As both tables suggest, any depletion of total oil and natural gas reserves of the Arctic states due to production and consumption

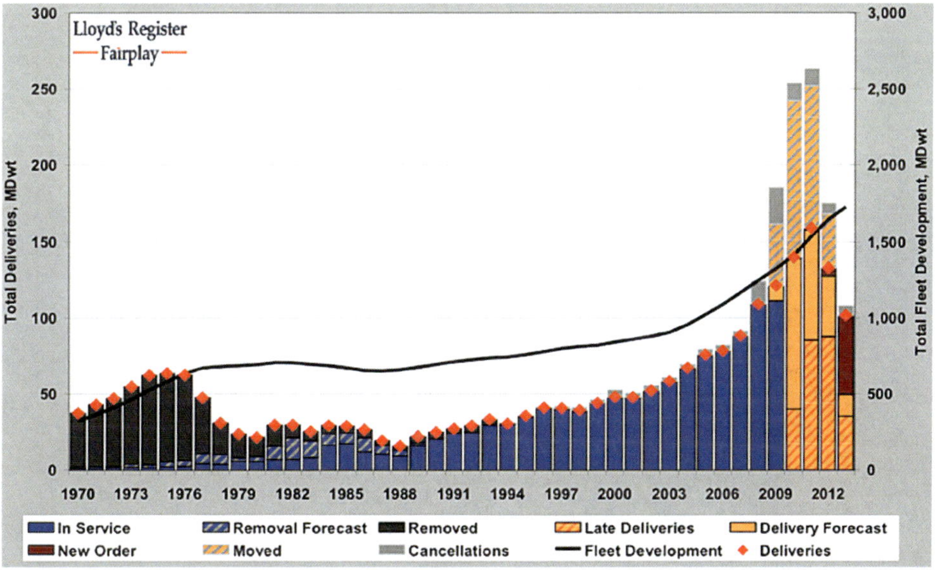

Source: HIS

Figure 3.3 World fleet development 1970–2013.

are replaced by new proved reserves. Table 3.14 also clearly demonstrates the overwhelming position of Russia in regard to natural gas reserves. In 2007 Russian reserves made up about 81 per cent of the total reserves of the Arctic states and 25 per cent of the world total reserves.

Table 3.10 above told us that of the four Arctic states in question, Russia and the United States are the largest and third largest producers of oil. Together with Canada and Norway, they contribute about 28 percent of total world production. The table also tells us that they are the four top producers of gas contributing no less than about 46 percent of total world output, and that the Middle East is a much larger producer in the oil sector than in the natural gas sector.

Table 3.16 shows the share of the Arctic of world proved oil and gas reserves.

About 5.3 percent of the oil and about 21.7 percent of the gas are found in the Arctic. Today, the Arctic produces about 10 percent of the world's oil and about 25 percent of its natural gas.

Compared to other parts of the world, the level of exploration in the Arctic has, by all accounts, been modest but more than sufficient to indicate rich deposits of various resources,

Source: Knut Espen Solberg, DNV

Figure 3.4 Framstredet

including oil and gas. Much has been found and exploited but much more remains to be exploited.

3.4.2 Undiscovered oil and gas resources in the Arctic

A great deal of the attention paid to Arctic resources stems from the promise of undiscovered hydrocarbon resources. In May 2008, the USGS completed a new assessment of undiscovered resources north of the Arctic Circle (Circum-Arctic Resource Appraisal, or CARA for short). Only those geologic areas were included which were considered

"... to have at least a 10 per cent chance of one or more significant oil and gas accumulations. For the purposes of the study, a significant accumulation contains recoverable volumes of at least 50 million barrels of oil and/or oil-equivalent natural gas. The study included only

Country	**End of 1990 Billion barrels**	**End of 2000 Billion barrels**	**End of 2009 Billion barrels**	**End of 2010 Billion barrels**
United States	33.8	30,4	30.9	30.9
Canada	11.2	18.3	32.1	32.1
Russia	n/a	59.0	76.7	77.4
Norway	8.6	11.4	7.1	6.7
Arctic states total	n/a	119.1	146.8	147.1
World total	1003.2	1104.9	1376.6	1383.2

Source: *BP Statistical Reviewof World Energy*, June 2011, p. 6

Table 3.14 Proved reserves of oil 1990–2010.

Country	**End 1990 trillion cubic meters**	**End 2000 trillion cubic meters**	**End 2009 trillion cubic meters**	**End 2010 trillion cubic meters**
United States	4.8	5.0	7.7	7.7
Canada	2.7	1.7	1.7	1.7
Russia	n/a	42.3	44.4	44.8
Norway	1.7	1.3	2.0	2.0
Total Arctic states	n/a	50.3	55.8	56.2
Total world	125.7	154.3	186.6	187.1

Source: *BP Statistical Review....June 2011*, p. 20

Table 3.15 Proved reserves of natural gas of Arctic states and the world over time.

Arctic region of	At end 2007 billion barrels of oil	% Share of total reserves of oil	At end 2007 tcm natural gas	% Share of total reserves of natural gas
United States	4.2	20.0	0.18	3
Canada	2.0	0.1	0.16	10
Russia	59.2	75.0	38.07	99
Norway	0.2	2.0	0.21	7
Total Arctic regions	65.6	45.9	38.41	64

Source: *BP Statistical Revieof World Energy*, June 2008 and *Arctic Oil and Gas* 2007, p. 32

Table 3.16 Arctic region's share of proved reserves of oil and gas.

those resources believed to be recoverable using existing technology, but with the important assumptions for offshore areas that the resources would be recoverable even in the presence of permanent sea ice and oceanic water depth ... So-called nonconventional resources...were explicitly excluded from the study".[22]

The area inside the Arctic Circle makes up about 6 percent of the Earth's surface or about 21 million km^2 of which almost 8 million km^2 is onshore and more than 7 million km^2 is on the continental shelves under less than 500 m of water. "The extensive Arctic continental shelves may constitute the geographically largest unexplored prospective area for petroleum remaining on Earth".[23]

The assessment showed that the Arctic might contain 90 billion barrels of undiscovered oil, 1,669 trillion cubic feet of natural gas, and 44 billion barrels of undiscovered natural gas liquids. The Arctic is thus supposed to account for about 13 percent of undiscovered oil, 30 percent of undiscovered natural gas, and 20 percent of undiscovered natural gas liquids in the world. This represents about 22 percent of all undiscovered, technically recoverable oil and gas resources in the world. Expressed in oil-equivalency terms, undiscovered natural gas is thought to be three times more abundant than oil in the Arctic.

As much as 84 percent of these resources are expected to be offshore and to be relatively close to the shore in areas that are not subject to territorial disputes, except parts of the East Barents Sea (Basin). In the view of the USGS, most of the undiscovered resources are concentrated between the shoreline and the 500 meter contour line and within the 200 nautical miles limit.[24]

About a third of the undiscovered oil is in the province referred to by the USGS as Arctic Alaska which, together with four other provinces (Amerasia Basin, East Greenland Rift Basin, East Barents Basin, and the West Greenland Basin), accounts for about 70 percent of the undiscovered oil resources. This includes substantial parts of the Chukchi Sea and parts of the Beaufort Sea. Most of the undiscovered natural gas, about 32 percent, is thought to be in the northern part of the West Siberian Basin of Russia which, together with the East Barents Basin and Arctic Alaska, accounts for more than 70 percent of the undiscovered natural gas resources.[25]

Oil provinces	Million barrels of oil	Natural gas provinces	Billion cf of gas
Arctic Alaska	29,960.94	West Siberian Basin	651,498.56
Amerasia Basin	9,723.58	East Barents Basin	317,557.97
East Greenland Rift Basin	8,902.13	Arctic Alaska	221,397.60
East Barents Basin	7,406.49		
WestGreenland-East Canada	7,274.40		
Sub total (70.3%)	63,267.54	Sub total (71.3%)	1,190,454.10
Total Arctic (100.0%)	89,983.21	Total Arctic (100.0%)	1,668,657.84

Source: US Dept of Interior, US Geological Survey, USGS Fact Sheet 2008-304

Table 3.17 Location of undiscovered resources in the Arctic.

About 65 percent of the undiscovered oil, 26 percent of the natural gas and 37 percent of the natural gas liquids are supposed to be resting (so far) in North America, whereas 34 percent of the oil, 73 percent of the natural gas and 62 percent of natural gas liquids are to be found in Eurasian parts of the Arctic. Of the total resources (oil equivalent) 63 percent are in Eurasia and 36 percent in North America (Table 3.17).

The area in the Arctic most accessible and presumably less costly from an exploration and recovery point of view may be the Barents Sea. According to the USGS, the Russian and Norwegian Barents Sea shelf may contain about 11 billion barrels of undiscovered oil, 11 trillion cubic meters of undiscovered natural gas and 2 billion barrels of undiscovered natural gas liquids. Most of this – about 68 percent of the oil, 85 percent of the natural gas and 65 percent of the natural gas liquids – is estimated to be in the Russian-controlled East Barents Basin's province. The resources of the Barents Sea shelf are assumed to be recoverable regardless of the presence of sea ice or depth of water.[26]

Another interesting area is offshore West Greenland. In the summer of 2010 the Scottish oil company Cairn Energy will start the first drilling for oil in Greenland's Western Basin.[27] These are sedimentary provinces (see Figure 3.5) which are three times larger than the North Sea Basin and only six oil and natural gas exploration wells have been drilled off West Greenland.[28]

Exploring for oil and gas in the Arctic is, however, financially very demanding. Statistics based on US experiences give a partial but telling picture. In 1960, the average US cost of drilling crude oil and natural gas wells, including dry holes, was $261,100 per well, in 1980 it was $680,400, and in 2006 $1,803,000 per well.[29] In 2007, the cost for oil wells only was $4 million as compared to $2.2 million a year earlier. The average cost per natural gas well was $3.9 million in 2007 as compared to $1.9 million dollars in 2006 (Table 3.18).

Data from Alaska will give an indication of exploration costs in the Arctic.[30] As Table 3.18 shows, in 2005 the cost of drilling an onshore well in Alaska was estimated to be about 640 percent higher than the average cost of drilling wells in the US as a whole.

These were onshore wells. The cost of drilling an exploration well offshore in Alaska's Chukchi Sea has been put at $60 million. Drilling a well in the shallow waters of the eastern Gulf of Mexico costs about $7 million whereas a US onshore well may cost as little as $82,000. This is before the cost of production infrastructure and many other factors that demand financial resources are taken into account, especially as far as offshore exploration and production are concerned.[31]

Drilling costs are influenced by a wide variety of factors that can and do vary greatly from place to place. Lacking detailed enough data makes it a hazardous business to argue that the case of Alaska is representative of the rest of the Arctic. But conditions in the rest of the Arctic have more in common with those in Alaska than with other regions. Remote locations, harsh weather and winter conditions, environmental requirements and many other factors combine to make the costs of finding and developing oil and gas in the Arctic a very expensive proposition indeed.

Yet costs have to be seen in relation to demand. If found in substantial quantities, and the demand is strong enough, Arctic oil and gas resources will be exploited, as they already have been for a long time. Although it has dropped to more "normal" levels, the recent high price of oil, reaching up to $147 a barrel in 2008, was not caused by speculators. Beginning in 2004, rapidly increasing demand from countries like India and China that coincided with increased demand from OECD countries and the US in particular, was the principal culprit. Current and future demand will come from economic development and levels of economic growth never experienced before. Whereas the economic development and industrialization of Europe and North America in the past involved somewhere between 50–200 million people, economic development today involves billions of people. Over the next decade, more than half a billion people will reach the $5,000 a year income level which is a huge leap compared to times not far back. In other words, there is nothing to suggest that the demand for energy will not continue to ceaselessly grow.

In their 2008 reference scenario, the *International Energy Agency* (IEA) predicts that China and India will "account for just over half of the increase in world primary energy demand between 2006 and 2030", and fossil fuels will account for 80 percent of the world's primary energy mix in 2030. Collectively, non-OECD countries will account for 87 percent of the increase in world primary energy. Their share of world demand will increase from 51 percent to 62 percent. "Their energy consumption overtook that of the OECD in 2005." [32]

The reference scenario projections of the IEA for the years 2007–2030 also calls for massive investments in energy infrastructure of some $26 trillion (in year 2007 dollars) of which nearly half goes to oil and gas, mainly for exploration and development. "Just over

Year	US average Per foot	US well depth 10,000 feet	Alaska per foot	Alaska well Depth 10,000 feet
2000	$ 111	$ 1.100.000	$ 283	$ 2.830.000
2005	$ 294	$ 2.940.000	$ 1.880	$ 18.800.000

Source: Composed of data from the Energy Information Administration, US Dept of Energy

Table 3.18 Cost of drilling onshore wells – US average and Alaska.

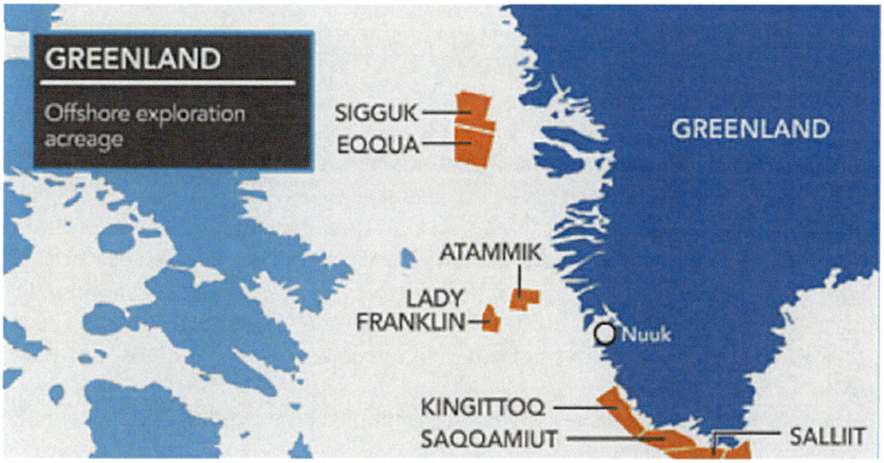

Source: http://www.cairn-energy.plc.uk

Figure 3.5 Cairn Energy exploration offshore West Greenland.

half of projected global energy investment in 2007–2030 goes simply to maintain the current level of supply capacity: much of the world's current infrastructure for supplying oil, gas, coal, and electricity will need to be replaced by 2030." To pay for these investments, the IEA assumes that the crude oil price will average $100 per barrel (in real 2007 dollars) in the period 2008–2015 after which it will rise above $120 in 2030. Combined with the oil-demand projections, the investment projections of the IEA points to persistently high levels of consumer spending on oil in both the OECD and non-OECD countries.[33] Put simply, consumers, whether of the individual, corporate, or public variety, will have to pay more because the costs to the investors and producers will be higher – much higher.

These is no one set of factors that will influence how, where and when the Arctic will become an arena of much more intensive exploration and extraction activities of oil and gas than today. Technology, climate change and environmental regulations will also weigh heavily, not to mention a variety of political, economic, and social factors.

3.4.3 Other mineral resources

Beyond existing and prospective petroleum resources, deposits in the Arctic of other mineral resources are substantial by any standards. Yet many known reserves of minerals are not exploited because they are found in areas to which access is poor or very difficult. To arrive at a reasonably accurate picture of the share of the Arctic as a repository and producer of minerals is difficult because of the uncertainty as to what constitutes the Arctic, as there is a lack of data and information, and different recording and reporting practices.

This is simply to say that the data should be treated with caution in this area as in the other resources areas dealt with in this chapter.

Russia

The largest amounts of minerals are found in Russia. About 20,000 mineral deposits have been explored and over 30 percent has been mined. The country is a leading, or a very significant producer, of many minerals and minerals-derived products. Russia is the fourth largest producer of steel after China, Japan, and the United States, and together with Japan the world's leading exporter of steel. In 2005 the export of metals, precious stones and related products stood at $37 billion or 17.7 percent of total exports, second only to the export of oil and gas which was valued at $141 billion or 67 percent of the total. The share of Arctic Russia in the production and export of minerals of Russia is a question to which we shall return later.

The Russian minerals industry is highly concentrated although less so now than in the past. Although very substantial indeed, the resources are not inexhaustible, as Table 3.19 indicates.

It is difficult to determine with accuracy how much of Russian minerals production is found in the Arctic. There is better information about some areas and regions than others.

Northwest Russia stands out in this context. The most abundant mineral resources are located on the Kola Peninsula where over 700 different minerals have been found. The Murmansk Oblast has more than 200 deposits of 40 types of minerals. There are large reserves of nickel-copper ores, alumina, rare metals, titanium, construction materials and semi-precious stones, to mention just a few. It is the biggest producer of phosphate fertilizer in the world. The total resources of phosphor-bearing ores are estimated to be 16 billion tons, including proven reserves of 5.8 billion tons. The region contains 687 million tons of proved reserves of apatite ores, and ten iron ore deposits with total reserves of 3.2 billion tons including 1.8 billion tons of proved iron ores reserves. Titanium resources are believed to amount to more than 10 billion tons. The resources of building materials are huge, including decorative building stones, sand and sand-gravel formations and some clay deposits.

The Arkhangelsk Oblast has total bauxite reserves of 120 million tons, and the north Onezhsky bauxite mine is the largest in Europe. Total reserves of diamond deposits are estimated at 130 million carat of diamonds of unusually high quality. The Republics of Karelia and Komi also have their share of valuable mineral resources and rare metals such as gold, silver, platinum, diamonds, nickel, chromium, cobalt, titan, wolfram, molybdenum and so on.[34]

The United States

The total value of the non-fuel mineral production of the United States in 2006 was $66.5 billion which was an increase of more than 20 percent compared with 2005. The value of Alaska's non-fuel minerals production in 2006 was about $2.9 billion or 4.4 percent of the US total. This was slightly more than double the value of the production in 2005. In 2007, the value increased to $3.3 billion. This made Alaska the fifth largest producer by value of minerals in the United States. Metallic minerals account for about 95 percent of the total value of of Alaska's non-fuel mineral production. In 2007, nearly all the value of the metals came from the production of zinc ($2.1 billion), lead ($389.5 million), gold ($511.2 million) and silver ($270.4 million).

In 2008 the value of exports totalled $3.9 billion which was 11 per cent down on 2007. The falling values of some minerals were partly to blame. Because of high mineral prices in recent years, the value of Alaska's export of minerals increased strongly until 2008. Alaska's main export markets are Japan, China, Canada, South Korea, Germany and Switzerland.

Canada

Canada is one of the leading mining nations in the world. Mining and mineral processing industries generated 3.4 percent of the national GDP in 2007 and the country is a leading exporter of minerals and mineral products. Practically all mining takes place outside the Arctic areas of Canada.

Ontario's mineral output amounted to 27.9 percent of the total value followed by British Columbia with 16.7 percent. The Northwest Territories contributed 4.8 percent and the Yukon and the Nunavut Territories 0.1 percent each.

Norway

About 80 million tons of various mineral resources valued at $1.5 billion were produced in 2007. Norway is a leading producer of some industrial minerals such as olivine which accounted for 66 percent of world production. The total export value for the mineral industry was about $871 million. Northern Norway and Svalbard account for about half the value of non-petroleum minerals production.

3.4.4 Forests and forestry

In 2005, just under 4 billion hectares of forests covered about 30 percent of the world's total land area. The situation is much the same today. Russia, Brazil, Canada, the US and China together account for more than half the total forest area.[35] Among the Arctic states, Russia alone has 20 percent of the world's forest area, Canada another 10 percent, and the US in excess of 7 percent.

But only 20 percent of the forests of the Arctic states are in their Arctic territories. Arctic Russia has 90 percent of the wood volume of the Arctic, with somewhat over 5 percent found in Alaska. In parts of Siberia and elsewhere in Russia, production of wood is partly or totally prohibited for reasons of habitat conservation. The situation is similar in Alaska, where large areas have been declared nature reserves for the purpose of biodiversity and ecosystem protection but nonetheless the state does have a small forestry industry with exports worth $133 million in 2008. However, the story about Arctic forests and forestry is overwhelmingly a Russian story.

Russia has by far the largest forest resources of any country in the world, with almost 50 percent of its land area covered by forest, amounting to 809 million hectares of forest. The northwest of Russia is one of the country's most important forestry regions. Forests cover about 86 million hectares of the region, or about 10 percent of the Russian total. The forest industry constitutes a dominant component of the economy of northwest Russia.

Commodity	Year of depletion
Bauxite	Beyond 2025
Chromite	2010
Coal	Beyond 2025
Copper	2016
Diamond	2013
Iron ore	Beyond 2025
Natural gas	Beyond 2025
Nickel	2015
Oil	2015
Phosphate	Beyond 2025
Platinum group metals	2018
Tin	2015
Tungsten	2016

Source: U.S.Geological Survey, *Minerals Yearbook 2005*

Table 3.19 Year in which reserves may be depleted.

Table 3.20 shows the mix of products of the Russian forestry industry as well as the composition of exports. Whilst Russia is the top world exporter of round wood (unprocessed logs), the share made up by value-added forest products is low by comparison and also by international standards. This is also reflected in the value of exports of forest products which in 2004 made up only 4 percent of global forest exports despite the large proportion of world forests found in Russia.

The export destinations of Russian forest products are divided almost equally between Europe and Asia. This is partly a function of the geography of Russia which has significant forest resources both in Europe and Asia. About 22 percent of Russian forests are in Europe and 78 percent in Siberia and the Russian Far East. Europe and China account for about 40 percent each of round wood imports.[36] Wood pulps exports from Russia are almost evenly divided between the two markets, Europe and China. By comparison, most of the other value-added products, sawn wood, and wood based panels and paper and paper board, are exported to Europe.

Russian exports of forest products make up a significant proportion of all exports from the country. Forest products account for about 3.5 percent of the value of all merchandise exports. Due to the abundance of forest resources, Russia has a big natural advantage in the production of forest products. This potential production far exceeds what the domestic market can absorb, which means that future growth in the forest sector is most likely to come from exports of forest products. Growth, however, does not depend only on high levels of investment in capital and labor but also in the production of more processed and value-added products.

3.4.5 Fish and fisheries

When it comes to global fisheries, the Arctic areas are moderately important. The fisheries of the states to which they belong (including Greenland and Iceland) figure overall more prominently than their constituent Arctic parts. To better appreciate their role in the fisheries sector, we need to place them in the global fisheries context. But first we shall look at some general trends in world fisheries. Table 3.21 has two columns for each year: the first column describes the world or global state of affairs, the second the world situation minus the contributions of China.

Some features of this table stand out. Per capita food fish supply remained much the same in the period 2005 to 2009 despite an increase of about 10 percent in fish available for human consumption. An increase in world population of some 300 million during the period explains why this state of affairs came to be.

Another noticeable feature is the changing relationship between capture and aquaculture fisheries. In 2005, capture fisheries made up about 67 percent of total fisheries; in 2009,

Statistic	Russian Federation	% Share of world total	Rank in the world
Forest area	808.8 mill ha	20	1
Production			
Industrial roundwood	134.0 mill m^3	8	3
Sawn wood	21.5 mill m^3	5	3
Wood based panels	7.2 mill m^3	3	5
Wood pulp	6.9 mill mt	4	7
Paper and paperboard	6.8 mill mt	2	12
Exports			
Industrial roundwood	$US 2338 mill	25	1st
Sawnwood	$US 1518 mill	5	5
Wood based panels	$US 536 mill	2	15
Wood pulp	$US 703 mill	3	7
Paper and paperboard	$US 1244 mill	1	17
All forest products	**$US 6405 mill**	**4**	**8**

Source: From Kilman and Whiteman (2006), p. 2

Table 3.20 Key Russian forestry sector statistics.

	2005 Million tonnes		2007 Million tonnes		2009 Million tonnes	
Production inland	**World**	**Ex China**	**World**	**Ex China**	**World**	**Ex China**
Capture	9.4	6.7	10.0	6.6	10.1	7.5
Aquaculture	26.8	6.5	30.7	7.6	35.0	10.1
Total inland	**36.2**	**16.7**	**40.6**	**18.7**	**43.1**	**20.8**
Marine						
Capture	82.7	70.3	79.9	67.5	79.9	67.2
Aquaculture	17.5	6.7	19.2	7.5	20.1	8.1
Total Marine	100.1	77.0	99.2	75.0	100.0	75.3
Total capture	92.1	77.5	89.9	75.2	90.0	75.1
Total aquaculture	41.9	16.2	49.9	18.5	55.1	21.0
Total World fisheries	**136.4**	**93.7**	**139.8**	**93.7**	145.1	**96.1**
Utilization						
Human consumption	107.3	70.4	112,7	73.5	117.8	75.5
Non-food uses	29.1	23.2	27.1	20.2	27.3	20.5
Population (billions)	6.5	5.2	6.7	5.4	6.8	5.5
Per cap food fish supply (kg)	16.5	13.5	16.9	13.7	17.2	13.7

Source:*The State of World Fisheries and Aquaculture 2010*, FAO Rome 2010.
This is an adapted version of Tables 1 and 2, pp. 3–4

Table 3.21 World fisheries and aquaculture production and utilization.

about 62 percent, a reduction of 5 percent. Aquaculture increased its share from 30 percent to 38 percent during the same four years.

The third and perhaps most important feature of Table 3.21 is the position of China in the overall picture. China is by far the largest fish and fish products producer in the world, with a production in 2005 of some 42.7 million tonnes which accounted for 31.3 percent of total world production, as against a share in 2009 of about 33.8 percent or 349 million tonnes. One should note, however, that "...capture fisheries and aquaculture production statistics for China may be too high … and that this problem has existed since the early 1990s". As mentioned in the beginning, the case of China is therefore often discussed separately from the rest of the world in the reports from the FAO.[37]

In 2008, two of our Arctic states were among the top ten producers of marine and inland capture fisheries. The United States were placed fourth after China, Peru and Indonesia, with a total production of 4.3 million tonnes. Russia came eighth with a production of 3.4 million tonnes whilst Norway dropped from tenth to eleventh in 2008 when she produced about 2.4 million tonnes.

In terms of Arctic and northern fishing and aquaculture, four large marine northern ecosystems dominate the scene. These are the Northeast Atlantic which comprises the Barents and Norwegian Seas, the Central North Atlantic which includes the waters around Iceland, the Faroe Islands and East Greenland, the waters off North-eastern Canada which takes in Newfoundland and the Labrador area, and the Bering Sea. Of these, North-eastern Canada contributed in 2002 a minor share (some 80.000 tonnes) of the total of about 7.3 million tonnes for all these areas combined, whereas the Bering Sea (2.36 million tonnes), the Central North Atlantic (2.61 million tonnes) and the Northeast Atlantic (2.21 million tonnes) contributed about a third each of the total volume.[38]

In 2004–2005 the northeast Atlantic, including the Barents Sea, provided 40 percent of the total Russian catch and the northwest Pacific 56 percent. Most of the catch consisted of Alaska pollock (44 percent), herring (13 percent), cod (10 percent) and Pacific salmon (9 percent). In 2005, Alaska exported fish and fish products for about $1.7 billion. The most important fish species were pollock, salmon, halibut and cod.

Since then, however, the volume of world capture fisheries has stagnated or even declined after 2002, today's volumes will not differ much from what they were in 2002. Future numbers may be a different matter. Early in 2009, the US North Pacific Fishery Management Council voted unanimously to prohibit industrial fishing in the nearly 200,000 square miles of US waters in the Chukchi and Beaufort Seas. The US Commerce Department is said to be expected to approve the recommendation. This is done for preservation, environmental and management reasons. If so, Alaska pollock, the largest US fishery by volume, will be taken off the statistics.[39] The US initiative was quickly picked up in Canada where calls were made for a similar Canadian management plan.[40]

Under the influence of climate change some fish stocks, such as cod and herring, may be more plentiful in Arctic waters in the future. Moderate warming could improve conditions for these fish stocks as higher temperatures and reduced ice cover might lead to higher productivity. But the effects will not necessarily be uniformly positive. Some other species, such as northern shrimp, may be negatively affected. Species whose principal habitat was in Sub-Arctic waters might also move north, further complicating the situation. What the consequences of climate change will be for marine fish stocks in the Arctic is highly uncertain.[41]

The total catch of capture fish in the Arctic could amount to somewhere around 7 million tonnes, as it did in 2002. Most of this fish is landed in harbors of the countries from which the fishing vessels hail. Norway has a number of them. In Northwest Russia, by far the most of the catch is landed in Murmansk, Belomorsk and Arkhangelsk. In the Russian northwestern Pacific, Vladivostock and Nakhodka receive up to hundreds of thousands of tonnes of fish. There are also a number of secondary ports. In Alaska, Dutch Harbor-Unalska and Kodiak are by far the most important, with the former receiving 412 thousand tonnes and the latter 119 thousand tonnes in the year 2003. Iceland has two landing sites of importance for fish – Neskaupstadur with about 240 thousand tonnes and Vestmannaeyjar with about 201 thousand tonnes. In Canada, most fish catches are landed in harbors in Newfoundland. To these should be added fishing vessels from EU countries that operate in the north Atlantic and Barents Sea waters and carry most of their catch back to harbors in their country of origin.

The Arctic countries of this chapter have populations that from time to time consume fish and fish products. Some countries are net importers and others are net exporters and consume more than they catch and vice versa. In 2003, the US exported 3.07 million tonnes of fish and fish products and imported 4.7 million tonnes. In the same year, on the other hand, Russia exported 1.4 million tonnes and imported 815 thousand tonnes, Norway exported about 2.4 million tonnes and imported 420 thousand tonnes, and Iceland exported 678 thousand tonnes and imported 67 thousand tonnes.

3.5 NWP: Economies, resources, shipping and competing modes of transport

As mentioned earlier, in 2007 the Arctic produced about 10 percent of the world's oil and about 25 percent of its natural gas.[42] About 5.3 and 22 percent of the world's proven oil and gas reserves respectively are in the Arctic. The biggest producers are Russia and the US (Alaska). According to the United States Geological Survey's (USGS) latest assessment of 2008, about 22 percent of world undiscovered oil and gas could be found in the Arctic.

3.5.1 The economy and resources of the North America Arctic

At the end of 2010, Canada had 32.1 billion barrels of proven oil reserves. If, however, her oil sands deposits are included, Canada has proven reserves of 143.1 billion barrels, second only to Saudi Arabia and Venezuela. More than 99 percent of Canada's oil exports go to the US. The oil production comes mainly from the Western Canada Sedimentary Basin (WCSB), the oil sands deposits in Alberta, and offshore fields in the Atlantic.

As in the oil industry, Canada's natural gas production is concentrated in the WCSB, especially in Alberta. And as with the case of oil, what is not domestically consumed of natural gas production is exported to the US. Canadian natural gas made up 90 percent of US import of natural gas in 2008 which represented some 16 percent of US natural gas consumption. Although some new finds of conventional natural gas have been made in WCSB, it is thought that production has peaked, and that in the future production will come from unconventional sources like coal bed methane and shale gas deposits of which several potentially promising areas containing huge amounts exist.

Of the Arctic parts of Canada, the Mackenzie Delta in the Northwest Territories could contain some 5–6 trillion cubic feet, equivalent to about 10–11 percent of Canada's proven natural gas reserves. Gas from this source could begin to find its way south if and when the planned Mackenzie Gas Pipeline is built. This would carry the gas to Alberta where it would be fed into the existing natural gas transportation system. A consortium of natural gas companies has been in existence for some years with the aim of constructing the pipeline. Regulatory delays and rising costs have so far put a stop to the start of the project.

Now, new doubts stem from the increase in natural gas reserves in the US in general, and the development of huge shale gas reserves in particular which serve to undermine the economics of the pipeline project.[43] After several years of hearings the federally appointed Joint Review Panel delivered in December 2009 the results of its examination of the socio-economic and environmental impact of the pipeline project, and a decision about whether to go ahead is expected in the autumn of 2010. Several more hurdles have to be passed, however, before construction can begin, not least the uncertainty as to whether or not Federal Government subsidies will be forthcoming without which the entire project is in serious doubt. The final bill is expected by some to reach Canadian $20 billion.[44]

As part of Canada's Northern Strategy, in July 2009 the government announced a new project called *Geo-Mapping for Energy and Minerals* to learn more about the geology of Canada's North, including the Canadian Arctic Archipelago. According to the document, "[the] large-scale projects already underway barely scratch the surface of the North's immense store of mineral, petroleum, hydro and ocean resources … [The] full extent of the natural resources potential in the Arctic is still unknown." [45]

Among all states with Arctic territories, the US can claim on behalf of Alaska an area that can readily be defined "Arctic" without too many reservations. The Alaskan economy is dominated by the petroleum industry and the Federal government which each provides about a third of all jobs in Alaska. More than 80 percent of the state of Alaska's revenues come from the oil and gas industry and about 20 percent of US oil production is from Alaska. Most oil production comes from the North Slope with a small amount coming from the Cook Inlet which also produces some natural gas. All of it is consumed domestically. Alaska's traditional resource industries are the seafood, mining and timber industries. In 2006, Alaska produced 4.53 percent of US nonfuel mineral production. Zinc, gold, lead, silver, sand and gravel were the principal products. One mine, the Red Dog Mine, accounted for nearly 59 percent of the annual production value of the state's mining industry.[46] Although locally important producers of minerals, both the US and Canadian parts of the Arctic are relatively minor contributors to the global supply of non-petroleum resources.

3.5.2 NWP: Shipping and Arctic resources

Oil and gas

The shipping activities of the Northwest Passage (NWP) are on a very small scale. In the year 2005 there were only 60 shipping movements in the Canadian Arctic, most of which were carrying the commodities of daily life such as food and fuel to the local communities. Some nine voyages by five cruise ships were part of this very modest level of activity. Transit traffic through the NWP is infrequent, to say the least.

There is little likelihood that oil and gas will be shipped through the NWP from ports in Arctic Canada in the next few years. It is expected that the Mackenzie Gas Pipeline will be built within the next ten years or so but this will depend heavily on the price of gas in North America and will in any case obviate the need for shipping.

In the 1970s and early 1980s, the Canadian government invested in Arctic oil and gas exploration and important discoveries were made in the Mackenzie-Beaufort Sea basin and in the Arctic islands. But once the effects of the oil shocks of the 1970s wore off, interest in the Arctic resources fell and the government withdrew its support for exploration. But although current production is very small, Canada's Arctic region is rich in oil and gas development opportunities. As much as 1.7 billion barrels of oil and 880 billion m^3 of gas have been discovered of potential resources of 8.4 billion barrels of oil (discovered plus undiscovered) and 4.3 trillion m^3 of gas. This represents about 25 percent and 33 percent respectively of Canada's remaining resources of conventional crude oil and natural gas. Some exploration activity is now under way. In 2007, exploration permits were issued to some companies to explore for oil and gas in the Beaufort Sea. In 2008, six companies received permits to explore the Beaufort Sea and the Mackenzie delta for oil and gas.

As far as Alaska is concerned, the North Slope holds about 4.2 billion barrels of US proved oil reserves or about 20 percent of US total proved reserves. In 2007, Alaska contributed 20 percent of US oil production but only about 3 percent of natural gas production. The main mode of transport of these resources out of Alaska is not shipping but pipeline from Prudhoe Bay in northern Alaska to Valdez, Alaska from where it is shipped by tanker to the lower 48 states.[47]

The Alaskan Arctic coast is thought to hold at least 27 billion barrels of oil and 1 trillion m^3 of natural gas that have remained undisturbed because of cost pressures and various regulatory and political factors. In August 2008, BP and Conoco-Phillips announced that they had joined forces to build the Alaska Gas Pipeline. The project will consist of a gas treatment plant on Alaska's North Slope and a large-diameter pipeline that will travel more than 700 miles through Alaska and into Canada through the Yukon Territory and British Columbia to Alberta. The pipeline will be the largest private sector construction project ever undertaken in North America.[48] As of early 2012, no pipeline seems likely to be built in the next few years.

Other mineral resources

The closing of some mines in the Northern Territories and Nunavut of Canada meant that no mining production took place that would involve Arctic shipping. Mines in a pre-production stage are under development but will not lead to substantial cargo volumes. As far as the future is concerned, the most promising cargo potential is to be found in the development of an iron ore project at Mary River on Baffin Island. This is expected to be operational by 2020 and to generate shipment levels of about 12 million tons per year. Assuming 200,000 dwt vessels, about 60 loads would be in transit per year through the NWP. Other mining developments could require shipping services but the quantities are likely to be relatively small due to the nature of the minerals that are being exploited. In some cases, no more than 6–12 vessel transits are expected during an extended season. Other means of transport operating over land are likely to be used to carry the output away. Some, like the output of gold, diamonds, and uranium in some places, are more readily flown out by aircraft that are already a part of the supply chain.[49]

Alaska's export of minerals and precious metals amounted to 16 percent of the state's total exports. The minerals export was valued at $511 million. The most important mineral both in volume and value terms is zinc. The Red Dog Mine in Northwestern Alaska is the world's largest zinc mine and accounts for more than 70 percent of the total value of Alaska's mineral production. The mine has its own loading facilities and the shipping is season-

Source: WorldAtlas.Co

Figure 3.6 Map of Canada.

al because of ice conditions. The output is stockpiled at the port facility and is shipped out between July and October to markets in Asia, Europe, and Canada. Coal to South Korea is loaded on board at the Port of Seward.

What is not sent by pipeline of Alaska's exports, as in the case of oil and gas, is shipped south to export markets in other parts of the world through Pacific Ocean maritime corridors and not through the Northwest Passage.

The maritime traffic in the North American parts of the Arctic is likely to continue into the foreseeable future at the same low level of traffic it has in the past. Resources are carried out of the regions mostly by means of transport other than shipping. Starting from an already low level, the Canadian *Arctic Shipping Assessment* of 2007 envisages a very modest increase in shipping activity in the period until 2020.

The melting polar ice has led some to suggest that the Northwest Passage (NWP) will prove to be an economical route compared to the Panama Canal for super-sized freighter traffic from the east coast of Asia with cargoes destined for the east coast of North America. Today, the ships have either to unload on the west coast to allow the cargo to be transhipped to the east by rail or truck, use the Panama Canal, or sail the great distance from East Asia across the Indian and Atlantic Oceans to reach the North American east coast. Apart from escaping the congestion that frequently plagues the Panama Canal, using the NWP route is also about 7000 km shorter, the exact distance being dependent on the ports in question.

Source: WorldAtlas.Com

Figure 3.7 Map of Alaska.

Three factors speak against the NWP, however. The first is that there is no ice in the Panama Canal, the second is the expansion of the capacity of the Panama Canal that will allow ships carrying up to 10,000 containers in so-called post-Panamax vessels as opposed to the current Panamax vessels that can carry a maximum of 4,400 containers, and the third is that melting sea ice notwithstanding, ice conditions and the influence of ocean currents will still remain very difficult and unpredictable with extreme year-to-year and seasonal variability, thus making it next to impossible for shipping companies to guarantee reliable transit times on which the economic viability of the industry relies.

Transportation is not, however, a one-way street with goods and commodities only leaving the Arctic. So far in this chapter, the story about shipping has dealt only with the transport of resources *out* of the Arctic or in transit. Lest we lose sight of the other side of the coin, it must be emphasized that the development of Arctic resources might also require the importation *into* the Arctic of huge production facilities and great amounts of equipment and products at both high and mundane levels of technological sophistication, including food and other necessities of daily life. This is part of destination-Arctic shipping, the future volume of which is very difficult to predict since it depends so heavily on investment decisions in the petroleum sector in particular. To the extent other types of goods are carried into the Canadian Arctic and Alaska, the volumes are relatively limited mainly due to the small population residing in these areas. In 2007, about 35,000 tons of cargo were carried by water transport and loaded and unloaded on the Northwest Territories. In Nunavut, some 159,000 tons were carried by one form of water transport or another. The small populations can to a large extent account for the small volumes. The Northwest Territories has a population of about 42,000, Nunavut about 28,000 and Yukon about 31,000. Alaska has a population of about 690,000. The amount of shipping activity needed to carry the necessities of life for populations as small as these is limited and most of the supplies needed are not freighted through the NWP.

3.6 NEP: Economies, resources and competing modes of transport

3.6.1 Economy and resources of Arctic Russia and Norway

Oil and gas

Russia is the world's leading natural gas producer and in 2010 had the largest natural gas reserves at 23.9 percent of total world reserves. Almost 90 percent of the country's natural gas came from northern West Siberia where three enormous gas fields supplied almost 70 percent of the gas production. In the middle of the decade these reserves were in decline, however. Increasing domestic demand due to economic growth and long-term export commitments to Europe have generated a need for increased gas production which has prompted the search for additional supplies to fields further north and east where the physical environment is harsher and more challenging leading to higher development costs. One of these areas is the Yamal Peninsula and its adjacent offshore areas.[50] However, higher exploration and recovery costs may put the search for new Arctic gas resources at a competitive disadvantage.[51] Add to this another complication that may put the development of Arctic gas resources at a further disadvantage. The great strides made in US shale gas production are also beginning to affect Europe. Potential European shale gas production in particular is seen as a way of making the EU less dependent on Russian gas. In 2006, for example, Russia supplied Germany with 36 percent of her natural gas consumption, Italy 25 percent, France 20 percent, Austria 74 percent, Finland 100 percent, and Switzerland 12 percent.[52] This degree of dependency on Russian gas has caused much concern in Europe.

Because of an ongoing dispute with political overtones between Russia and Ukraine about gas prices, deliveries to Europe have in the past suffered serious disruptions because the pipeline that carries Russian gas to Europe runs through Ukraine.[53] The new emphasis on and hopes attached to shale gas in Europe is in no small measure due to a wish to become less dependent on Russian gas supplies.

Russia, however, will continue to export large quantities of gas from fields already in production well into the foreseeable future simply because the rest of the world will continue to need Russian gas. With 23.9 percent, 5.6 percent, and 18.2 percent of world proved reserves of gas, oil and coal respectively, Russian resources are indispensible. The Russian Arctic is the home to most of these resources, much of it still in Western Siberia, declining production notwithstanding. About 91 percent of her natural gas production and about 80 percent of her natural gas reserves are in the Arctic. It is also estimated that 90 percent of Russia's offshore reserves of hydrocarbons are in the Arctic.[54]

The most important project is the development of the Shtokman gas and condensate field located some 600 km north of Murmansk on the Russian sector of the Barents Sea shelf. The reserves of the field are assessed to be about 3.8 trillion cubic meters of gas and 53.3 million tonnes of gas condensate. To develop the field a special purpose company, Shtokman Development AG, was formed in February 2008 with Gazprom owning 51 percent, Total of France 25 percent and Statoil of Norway 24 percent of the company's shares. In 2008, the construction of two semi-submersible drilling rigs began at the Vyborg Shipbuilding Plant, the first of which was completed in the fourth quarter of 2010 and the second in the first quarter of 2011. Phase 1 of the Shtokman project is planned to produce 23.7 billion cubic meters which will be transported to the mainland by pipeline where part of it will be used to produce 7.5 million tonnes of LNG, and 11 billion cubic meters will be sent by pipeline south. LNG offloading terminals are to be located in the northern part of the Orlovka Bay.[55] On February 6, 2010 the board of Shtokman Development decided

	1995	**2000**	**2008**	**2010**
Transport – total	8,814	7,907	9,451	7,643
of which				
Railway	1,028	1,047	1.304	1,206
Motor (road)	6,786	5,878	6,893	5,235
Pipelines	783	829	1,067	1.062
Marine	71	35	35	37
Inland waterways	145	117	151	102
Air	0.6	0.8	1.0	1.1

Source: Table adapted from Federal State Statistics Service, Russia. www.gks.ru/bgd/regl/, Russia in figures – 2011

Table 3.22 Freight by transport mode (Russia) (million tonnes)

to postpone the so-called Final Investment Decision (FID) on the gas pipeline until March 2011. In December 2011 the decision was again postponed, this time until March 2012. No pipeline gas production is likely before 2016. All postponement decisions have been justified in part by reference to changed market conditions,[56] of which the achievement of the US of near self-sufficiency in gas production is the most important. According to a partner in Shtokman Development AG, the "...project as it is today is not commercially viable".[57]

Among the states with Arctic territories, none is so heavily dependent on oil and gas exports as Russia. The IMF and the World Bank estimate that the oil and gas sector made up about 64 percent of Russia's export earnings in 2007, and attracted 30 percent of all foreign investment in the country. The growth of supply of oil in the decade ahead is likely to come from a few major fields outside the Arctic (the Sakhalin projects), whilst the production in others will help offset the fall in production from mature fields in the Arctic parts of Russia. An important factor in Russian oil and gas production is current government taxation on production and export earnings which, if anything, tend to inhibit attempts to increase output. However, the whole taxation regime is supposedly under review with a view to make it less burdensome on the industry.

As noted earlier, in 2008 the USGS published a revised assessment of Arctic undiscovered oil and gas which was said by the USGS to amount to 22 percent of the world's undiscovered oil and gas resources. Most of the Arctic's undiscovered oil and gas is thought to be found in offshore fields in Russian-controlled waters. It has been estimated that of the recoverable oil and gas on the Russian continental shelf, oil make up 13 percent and gas 87 percent. The Kara Sea is supposed to house 44 percent of these fields, the Barents Sea 25 percent, the Sea of Okhotsk 9 percent, and the Pechora Sea 5 percent.[58]

The extent to which these resources will soon be exploited will depend on a number of factors, including their precise location and accessibility, the general global demand and supply situation, the cost of exploration and recovery in the Arctic as compared to other parts of the world where the remaining 78 percent of the undiscovered oil and gas resources are located and much less financially demanding to develop.

Other mineral resources

Russia is a leading producer of many important minerals. The largest amounts of Arctic non-fuel mineral resources by far are found in the Russian Arctic. In 2006, there were 25 mines operating in Arctic Russia. As described earlier in this chapter, the Kola Peninsula contains a remarkable abundance of various minerals, amongst which are copper, iron, nickel, cobalt, titanium, rare metals, ceramic raw materials and precious stones. In 2007, the mineral raw material sector produced about 30 percent of the Russia's GDP and about 70 percent of the country's budget revenues. The oil and gas sector alone accounted for about 20 percent of the GDP. It also accounted for more than 60 percent of export revenues and 30 percent of all direct foreign investment. Russia's heavy reliance on oil and gas exports in particular and on ferrous and nonferrous metals makes her economy extremely vulnerable to fluctuations in world prices of oil, gas, minerals and metals. The biggest consumer of Russia's exports of energy products in 2007 was the EU which imported 63 percent of Russia's oil exports and 65 percent of her gas exports. Russia was also the leading supplier of oil to the EU and provided 27 percent of the EU's oil imports and 44 percent of total EU gas imports.[59] It should not be ignored that the extraction of minerals in the Russian Arctic has also been attended by very high to extremely high levels of pollution, which is affecting the continued life of mining operations, not to mention human settlements and the natural environment surrounding these operations.[60]

Norilsk Nickel, the world's leading nickel producer, has mining and processing operations on the Kola Peninsula and on the Taymyr Peninsula in Western Siberia. It is also a leading producer of copper. In 2007, independent audits put proven and probable reserves of nickel in the two locations at more than 6 million tonnes. Annual production of nickel products amounts to about 250,000 tonnes. Proven and probable reserves of copper contained in deposits on the Kola and Taymyr Peninsulas exceed 9 million tonnes. Annual production of copper is about 430,000 tonnes. Long-term plans of Norilsk Nickel call for an increase in ore production on the Kola and Taymyr Peninsulas from 21.8 million tonnes in 2006 to 26 million tonnes in 2015. Russia has more than 20,000 mineral deposits of which about 40 percent is in production. In the past 20 or so years, however, the average metal content of ores has decreased by between 30 and 50 percent. The percentage of ores and coal from which it is difficult to extract sufficient metal content to make the mining economical has increased from 15 to 40 percent. A large percentage of the reserves classified as economic using the reserve classification system of Russia would be classified as non-economic if subjected to market economic criteria. In 2007, 34 percent of lead, 49 percent of tin, and from 15 to 30 percent of apatite, coal, copper and titanium would be classified as non-economic resources.[61]

Oil and gas will remain the most important sector of the Russian mineral industry but at a slower rate of growth because of depleting reserves and lack of sufficient investment. Increases in the production of non-ferrous metals are expected most of which will be due to added aluminium production. New copper mines will be opened, thus leading to an increase in copper production. The Russian metallurgical industry, however, is reportedly handicapped by production technology that is inferior to that used in industrially more advanced countries. This has affected the competitiveness of the products since it requires Russian industry to spend comparatively more to produce goods. To remain competitive in world markets, the metallurgical industry would need resource-saving technologies and more high value-added products.[62] The economies of the Arctic rely overwhelmingly on the extraction and some basic processing of mineral resources which are then transported to Sub-Arctic regions to become part of manufacturing industries and processes in many countries. Only about the Russian Arctic can it be said that its mineral resources have given rise to local industrial processing activities of some magnitude. Northwest Russia consists of ten regions of which not all are usually included in the Arctic. The oil and gas industries of the Northwest Russia are nationally not very significant. This may change, however. Russia's most promising proven but untapped gas resources are in the Barents Sea. Considerable oil and gas resources are likely to be found in Western Siberia including the Kara Sea. Arctic offshore exploration and production have scarcely begun as is, indeed, also the case of Arctic North America. Exploration activities are planned or under way, construction of new refineries and pipelines are under consideration, and if plans are to be trusted, Arctic regions in both North America and Russia stand on the threshold of an entirely new stage in their economic and commercial development.

3.6.2 Transportation infrastructure of the Russian Arctic

The role of shipping in the Arctic is determined by a number of factors essential among which are, of course, the availability of suitable cargoes of various kinds and the presence or absence of competing transportation alternatives. Transportation systems, however, are not static or unalterable quantities. What the most efficient and economical means of transport are today may not be so tomorrow, and what is the most efficient in one geographical arena or locality may not be so in another. The climatic conditions under which they have to operate make practically all kinds of transportation in the Arctic a more difficult and costly affair than in Sub-Arctic latitudes. .

Table 3.22 gives a picture of how goods and commodities are transported in Russia. Between 1995 and 2010, the volume of transported merchandise decreased by 1171 million tonnes, or by 13.3 percent in 15 years.

Shipping, on the other hand, carried only about half the volume in 2010 than it carried in 1995. Table 3.22 tells us that the share of the freight volume of railways and pipelines increased whilst the share of road transport decreased between 1995 and 2010, and that in Russia, as elsewhere, road transport dominates.

In 2008, mineral products (oil and gas) accounted for 69.6 percent of the value of Russia's export. Metals and precious stones (and articles thereof) made up another 13.3 percent. In 2010, Russia exported 186 billion cubic meters of natural gas. Most of the gas was recovered from fields in the Arctic, and practically all of it went to Europe by pipeline. Most of the rest of the exported natural gas went by pipeline to the CIS states Ukraine, Belarus, Azerbaijan and Georgia.[63]

Table 3.23 below shows the crude oil exports in 2006–2007 by outlet. Among important new plans are those that will transport oil by pipelines to Murmansk and Varendey. This is illustrated in Appendix 2. If these plans are realized, the export of oil from the two ports will increase considerably.

Pipelines

Russia has a vast pipeline network which is operated by Transneft – a state monopoly. In 2008, the length of Russia's pipeline network was 228,000 km, of which 165,000 km were gas pipelines, 47,000 km oil pipelines, and 16,000 km oil products pipelines. In some scenarios the already-existing available global pipeline capacity, combined with rises in the global LNG liquefaction capacity, is thought to lead to less than an optimal utilization of inter-regional pipeline capacity from around 60 billion cubic meters in 2007 to close to 200 billion cubic meters in the period 2012–2015, or a drop in the utilization rate from 88 percent to less than 75 percent.[64] How much of this underutilization of capacity applies to Russia is not clear.

A number of new natural gas pipelines linking Russia with Europe and Asia are nevertheless planned or under construction. Gazprom plans to double the capacity of the *Yamal-Europe II* pipeline, which carries gas from Russia to Poland and Germany via Belarus, from 1 to 2 trillion cubic feet. The expansion is expected to be completed in 2010 at a cost of US$10 billion.

The *South Stream Project* has Gazprom and ENI as partners and plans to send gas for 900 km under the Black Sea and then across Bulgaria with one alternative going northwest

Transport mode	1995 %	2000 %	2008 %	2010 %
Railway	11.6	13.2	13.8	15.8
Roads	77.0	74.3	72.9	68.5
Pipelines	8.9	10.5	11.3	13.9
Marine	0.8	0.4	0.4	0.5
Internal waterways	1.6	1.5	1.6	1.3

Source: Calculated from data in Table 3.21

Table 3.23 Transport mode share of freights.

and the other through Greece and Albania to link up with the Italian gas network. This project is expected to be completed in 2013.

The so-called *Blue Stream* connects Russia to Turkey and has been in operation since December 2002. The capacity of pipeline is being expanded.

The *Nord Stream Pipeline* (or *North European Gas Pipeline*) extends for 3218 km from Russia to Germany via the Baltic Sea. The project is estimated to cost $11 billion, twice as much as originally planned. The pipeline will carry about 28 billion cubic meters in two parallel pipelines. In January 2010, the construction of the compressor station, which is the starting point for gas supplies via the Nord Stream pipeline, was completed. This is the last facility of the onshore section of the pipeline.[65] The first of the two offshore pipelines in the Baltic Sea was completed in 2011 and in November 2011 the first gas was exported to the EU via the pipeline.[66]

Natural Gas for China is another possible pipeline project that is on the drawing board.[67] When Putin visited China in April 2009 he reportedly signed deals with China worth some $3.5 billion. No substantial progress was made regarding the most important deal, Russian gas supplies to China. Only a preliminary agreement was concluded that opened the crucial question of price and the source of the supplies of gas.[68]

As with so many other countries, Russia found the year 2009 very painful. Her production of natural gas fell by 17 percent which led to a fall in export of gas. Russia's new energy strategy, adopted on 13 November 2009, calls for an increase in gas production from 664 billion cubic meters in 2008 to 885–940 billion cubic meters in 2030. Production in Northwest Russia will increase from 46 billion cubic meters in 2005 to 141 billion cubic meters in 2030, in large measure due to Shtokman. Gas exports are projected to increase from 241 billion cubic meters in 2008 to 349–368 billion cubic meters in 2030, most of it going to Asia.[69]

Railways

The railway system is an important carrier of resources and merchandise. Russia has 86,000 km of railroad tracks. Of the 1.3 billion tonnes delivered by the state railways in 2008, 200 million tonnes or 18 percent were oil.[70] Oil has until recently reached the sea ports of Arkhangelsk, Vitino, and Murmansk by rail from where it has been exported by sea. But

a

b

Source: Knut Espen Solberg, DNV

Figure 3.8 Barents Sea

it is at present not clear what the overall situation is in regard to shipments from the Arctic ports. According to one centrally placed Russian official, crude oil deliveries by rail to the port of Vitino in the White Sea and to Murmansk on the Barents Sea have stopped.[71] The transhipment facilities in the Kola area receive almost all their oil from smaller shuttle tankers sailing along the NSR. The only significant quantities of oil sent out of Russia by rail go to China (see Table 3.23 above). On present indications the role of railways in Arctic oil and gas transport would seem to be diminishing rather than increasing.

Inland waterways

Russia has more than 100,000 km of rivers that are regularly used for navigation and transportation of goods and merchandise. In 2010, 102 million tonnes of cargo was transported on the inland waterways, including 3.7 million tonnes of oil products which went to terminals at various ports along the NSR from where it was sent by shuttle tankers to transhipment facilities in the Kola area.[72] Rivers and canals are mostly used for delivering goods and supplies to the northern regions of Russia. The largest rivers in Siberia and the Far East are the Ob, Irtysh, Yenisey, Lena, and Amur rivers. All of these also serve the oil and gas industries. The river fleet consist of more than 30,000 ships with a total deadweight of more than 12 million tonnes. The average age is more than 25 years. According to the *Transport Strategy–2030* the cargo volume on the inland waterways will reach 203 million tonnes in 2020 and 262 million tonnes in 2030, and some 97 new cargo ships are to be delivered to the river fleet during the period 2010 to 2015.[73]

Shipping

How much oil will be transported in Arctic seas will depend on how much is produced from old fields, on how much new exploration will be done, and on the development of the Varandey terminal which has been designed to have a capacity of 12 million tonnes. The projection is that in the near future (2012–2015) annual volumes of oil to be exported westwards from Murmansk will not exceed 12 million tonnes. "The analysis of real oil transportation schemes in Northwest Russia [leads to the conclusion that the many planned oil transhipment terminals] ... are hardly going to be implemented in the foreseeable future. They [were] abandoned long before the landslide of world oil prices."[74] The most important offshore transhipment facility is the Belokamenka in the Kola Bay; the floating storage and offloading vessel *Belokamenka* is its main unit, which can handle some 12 million tonnes.

The Northern Maritime Corridor (NMC) and the coast of northern Norway plays a potentially important part in the shipping activities of the Arctic: firstly, because of Norwegian offshore oil and gas activities; secondly, because of the suitability of Norway's coastline and relatively sheltered fjords as sites for land and offshore transhipment facilities of Russian oil; and, thirdly, because of what might be found in unopened and unexplored areas of the Norwegian and Russian Barents Sea continental shelves in the future. Several locations for oil transhipment facilities and terminals have been established and tried since 2002. Bøkfjord near Kirkenes and Sarnesfjord near North Cape were the first for which permission was given to carry out transhipment of oil from Russia. The latter continues as a transhipment site operated by Kirkenes Transit. In 2007, eight Ship-to-Ship (STS) operations involving 424,782 tonnes of condensate took place, and in 2008 nine STS operations which yielded 501,319 tons of condensate. The Kvalsund municipality in West Finnmark has had plans since at least since 2006 of establishing a port for loading oil and to act as a service port for the oil and gas industry. The plans involve facilities that can load oil on to modern supertankers (VLCC). Similar plans concern Sørøya in Hammerfest and Hasvik municipalities which envisage the construction of rock cavern storage facilities for oil and oil products and a terminal for transhipping oil from the Goliat field in the Barents Sea and from the Russian Barents region.[75]

In Norwegian waters of the Barents Sea, Statoil is the operator of Europe's first export facility for liquefied natural gas (LNG). The Snøhvit field is also the world's northernmost facility of its kind and is located some 140 kilometers north of Hammerfest. It has no surface installations and the output is transported by pipeline to Melkøya outside Hammerfest where it is turned into LNG. A total of nine wells were to be drilled of which six were drilled during 2004–2005 and put on stream in 2007. Further wells are to be drilled and are due to begin production in 2014–2015. Much of the production is aimed for the US market and will thus be in competition with LNG from Shtokman when and if this field begin to produce.[76] Another significant field is the Goliat oil field in the Barents Sea, some 85 kilometers northwest of Hammerfest. With about 174 million barrels of reserves and a life expectancy of 15 years, the reservoir is scheduled to begin production by 2013 with ENI as the operator. Other significant oil (Skrugard and Havis) and gas discoveries (Norvarg) made on the Norwegian shelf are expected to be producing by 2020.

Whilst the capacity of Russian Arctic oil and gas terminals have increased and should, according to plans, increase substantially by 2015, the actual volumes of oil and oil products shipped through these terminals have so far tended to be rather small, and have in some important cases decreased. In 2008, the terminals in Ob Bay, Varandey, Arkhangelsk, Vitino and Murmansk, with a combined capacity of 35,600 thousand tonnes, shipped 9,050 thousand tonnes of crude oil and oil products. In 2004, the same terminals shipped 11,450 tonnes. In 2015, the capacity of these terminals is supposed to be 42,500 thousand tonnes. The total capacity in 2015 of all planned oil terminals in the Russian Arctic (and not just the five included above) is supposed to amount to 117,500 thousand tonnes.[77] It is an open question whether in 2015 there will be sufficient supply of oil to utilize fully or even substantially the projected capacity of these terminals.

In 2007, Russia exported around 4.4 million bbl/d (587,450 tonnes/d) of crude oil and more than 2 million bbl/d (267,022 tonnes/d) of oil products. Most of the Russian oil exports are routed via Transneft-controlled pipelines. Some 7 percent, or 300,000 bbl/d (40,000 tonnes/d) of oil was shipped by sea or rail. Oil is delivered by rail to Murmansk where it is transhipped. In 2007, roughly 270,000 bbl/d (36,048 tonnes/d) of crude oil and products were shipped from the Murmansk area. The capacity of the terminal at Varandey has been expanded and will allow for increased shipments of oil from the northern parts of Timan-Pechora. Most of the oil transported by sea from Russia goes from the port of Primorsk in the Baltic Sea and the port of Novorossiysk in the Black Sea.[78]

The amount of oil transported by sea from terminals in the Russian Arctic, including the Barents Sea region, has increased substantially during the last decade. In 2002, some 4 million tonnes of oil was transported from terminals in the Russian Arctic along the coast of Norway to markets further south and west. In 2004, this rose to almost 12 million tonnes and then dropped in 2005 to 9.5 million tonnes only to increase to more than 11.5 million tonnes in 2008 and to 18.6 million tonnes in 2010. The completion and opening of the new Varandey offshore terminal in 2008 was an important event. This terminal can by itself double the annual volumes of oil shipped for export through the Barents Sea. In a 5–10 year perspective it is thought that exports of oil by ship from Russian Arctic terminals could reach 100 million tonnes a year.[79]

From the point of view of maritime transport Russia is divided into three operational arenas or regions. There are the ports and harbors of the Northwest region (including the Baltic Sea) which in 2008 handled 215 million tonnes of cargo, or 47 percent of the turnover; the Southern Basin (the Black Sea) ports which handled 150 million tonnes, or 35 percent, and the Far East region ports which handled 80 million tonnes, or 18 percent. Current plans stipulate that the capacity of Russian ports will increase by 454 million tonnes by 2015, thus doubling today's capacity. The Russian-controlled merchant marine consists of 1,486 ships (89 percent under Russian flag) with a total deadweight of 17 million tonnes (36 percent under Russian flag). The federal transportation program calls for an increase by 2015 to 22.6 million deadweight tonnes.[80] Russia's biggest shipping company, Sovcomflot, is a leader in Arctic shipping, has the largest Arctic shuttle tanker fleet, and operates the largest ice-class LNG carrier fleet. It owns some 147 vessels amounting to 10.38 million dwt and has another 15 vessels on order which will add another 1.39 dwt. The average age of the fleet is seven years, and consists of 118 tankers, eight specialized vessels, eight LNG and LPG carriers, and seven dry cargo vessels. The company has started carrying cargo east along the Northern Sea Route (NSR) to Japan.[81]

The Russian ice-breaker fleet consisted in 2008 of 28 ice-breakers working in the Arctic, including seven nuclear ones. The newest, the *50 Let Pobedy,* was completed in 2007, is 159 meters long and 30 meters wide, is 3,500 tonnes deadweight, and is designed to break through 2.8 meter-thick ice. The icebreakers are, of course, essential to the operation of the Northern Sea Route. Some projections envisage that the volume transported along the NSR will increase from 1.5 million tonnes in 2002 to 50 million tonnes in 2020.[82]

3.7 NEP: Shipping

Oil and gas

Seen from the point of view of the two principal Arctic sea routes, the North East Passage (NEP) and the Northwest Passage (NWP), the survey above indicates in no uncertain terms that much the greater volume of the petroleum resources in the Arctic, discovered and undiscovered, is located in Arctic Russia and thus along and in more or less close vicinity of the Northern Sea Route (NEP). So far, only Russian onshore resources are being tapped. When the Russian petroleum industry, as with Norway's, in earnest moves offshore, the NSR and its cousin, the Northern Maritime Corridor (NMC), may take on the lively appearance of the main street of a market town through which everything of any importance has to be moved.

Source: GraphicMaps.com

Figure 3.9 Map of Russia.

Western Siberia is the largest Russian oil and gas region with two-thirds of the country's discovered oil reserves. More than two-thirds of discovered gas reserves are located in the northernmost area of Western Siberia.[83] The Russian continental shelf is said by the Ministry of Natural Resources and Ecology to contain *extractable* reserves of hydrocarbons to the tune of 10.8 billion tons in oil equivalent terms, and to contain 98.7 billion in oil-equivalent terms of *recoverable* resources.

The Barents and Kara seas are supposed to contain 4.5 billion tons and 3.7 billion tons, or 41 percent and 34 percent respectively, of the extractable reserves and 30 percent and 41 percent respectively of the recoverable resources. The Russian Arctic shelf is about 4.5 million km^2 and some 75 percent of it is thought to contain extractable hydrocarbon resources. The Barents, Kara, Laptev, East Siberian, Chuckchi, and Bering Seas together are thought to account for about 85 percent of these resources. The other two shelf areas with significant resources are the Sea of Okhotsk and the Caspian Sea.

Western Siberia is the principal center of the Russian oil industry. More than 53 percent of the oil reserves are in the region, and since the mid 1980s it has produced two-thirds of the Russian oil. Before 2002, the volume of oil transported by sea from the Barents region along the coast of Norway was insignificant. In 2002, however, there was a very substantial increase when some 4 million tons were shipped westwards. This volume was doubled the next year to 8 million tons. The trend continued and between 2005 and 2010 annual amounts ranging from 9.5 to 18.6 million tons were shipped. According to some observers, by 2015 more than 100 million tons of oil and gas could be shipped along the northern coast of Russia and Norway.

Situated along the NSR are a number of ports and terminals for loading oil and gas from Tiksi on the Laptev Sea in the east via Dudinka, Dikson, and Ob Bay on the Kara Sea, Varandey, Koguev, and Indiga on the Pechora Sea, Arkhangelsk, Onega Bay and Vitino on the White Sea, to Teriberka, Murmansk and Pechenga on the Barents Sea. All of these and others are receiving crude oil, oil products and gas condensate by ship, pipeline, railway and river routes, and sending loads for export directly or via offshore transhipment terminals in Norwegian and Russian ice-free Barents areas. See Chapter 5 for a detailed description of ports and terminals along the NSR.

Minerals

As in the case of petroleum resources, the Russian Arctic territories are the major repository of non-fuel minerals among Arctic states. The transport volume of oil and gas along the NSR increased quite dramatically between 2002 and 2008, from about 4 million tons to about 11.5 million tons, albeit from a less than impressive initial level. Minerals as a ship's cargo are much less prominent and cannot record anything comparable. One of the main sources of cargoes is the Norilsk Nickel company that generates some 1.2–1.3 million tons of minerals, mainly nickel and copper, and sends it away through the port of Dudinka. The port of Murmansk is the largest ice-free port in the northern part of Russia and much of the cargo that enters or departs the NSR and the NMC goes through it. The volume of cargo leaving Arctic Russia via the NSR is expected to increase by very substantial amounts by 2015. In 2006, the port of Murmansk handled some 27 million tons of coal, timber and mineral ore. The same year, the Port of Arkhangelsk handled about 10.5 million tons of oil, oil products, coal and timber. The share of minerals in these volumes is uncertain. In 2010, it is expected that the freight going through northwest seaports may nearly double compared to 2006. Both ports have been and are undergoing a modernisation process that will substantially increase capacity.

Norway's coal mines at Svalbard have estimated reserves of between 35 million and 45 million tons. The production in 2007 amounted to 322,000 tons. Svalbard's coal and Sweden's iron ore, most of which is shipped through Narvik, are NMC rather than NSR cargoes

Arctic Russia has a much higher population than other parts of the Arctic. Fairly substantial volumes of goods and merchandise have to be transported every year into the region to sustain human life and settlements. But there are clear indications that the Russian Arctic or parts thereof is losing population. Since the year 2000, the Russian population of the Barents region has decreased by 462,000 or almost 11 percent. The Murmansk Oblast recorded the biggest decline of 10.4 percent, the Komi Republic by 9.3 percent, the Arkhangelsk Oblast by 9.2 percent and Karelia by 6.5 percent. The decline has been steady and continuous since the start of this century and if it continues will influence the freight picture as well as the economy of Arctic Russia.

Forests and forestry

Russia has the largest forest reserves in the world with about 20 percent of global forest resources. But Russia provides only about 4 percent of global forest products. Her annual new forest growth significantly exceeds her current harvesting capacity. Her softwood (mostly pine, spruce and birch) grows at northern latitudes where slow growth gives high wood density and fewer knots. This in turn means high timber value.

Yet Russian forests are an underdeveloped resource. There are several reasons for this. Harvesting costs are low but transport costs are high. Roads and railways are not located well in relation to the forest resources. The forest industry is highly fragmented. Over 20,000 forestry operations exist but there are very few saw mills, and these mostly have equipment over 50 years old. About 95 percent of the timber is harvested manually. The skilled workforce needed for operating modern timber harvesters is not there. The industry is organized to supply the basic round wood product which, as we saw above, has dominated Russian exports.

In late 2007, a New Forest Code was introduced which was designed to alter this state of affairs significantly. Previously, investment in the Russian forestry sector had not been an attractive proposition, especially for foreign investors. The new code provides for strategic investors to qualify for 49-year licenses to harvest. Strict reforestation requirements have been introduced. Punitive tariffs of up to 80 percent put on the value of uncut logs for export became effective in 2009. VAT and other tax benefits are being introduced to encourage the importation of modern foreign machinery and equipment. The main purpose of all this is to make the export of round wood logs economically unattractive and to encourage the industry and investors to focus on value-added timber products.

To emphasize these matters here is simply to point out that the lead time between the initial decision to invest under the New Forest Code in a modernizing Russian forestry industry and the appearance of tangible and profitable results of the investment is probably considerably shorter than in the petroleum and mining industries. Everything else being equal (which, of course, they never are), investments made today in the forest industry in Arctic areas of Russia are likely to show positive economic returns much sooner than investments made today in Russian Arctic offshore oil and gas. This may affect Arctic shipping prospects about which more will be said in later phases of this book.

In terms of Arctic shipping, forest and forestry products are mainly shipped from the Murmansk and the Arkhangelsk ports although it is difficult to determine in what quantity. Aggregate figures are available but product-specific figures are in short supply. A great deal of forest products is also shipped from the port of St Petersburg in the Baltic.

Alaska's forest industry is small and wood exports make up only 4 percent of the state's exports. Of Alaska's export to Japan worth $1.1 billion in 2008 only 3 percent were forest products.

3.8 Arctic shipping routes, Arctic shipping, and Arctic resources

The world financial crisis that took hold in the second half of 2008 and 2009 hit Russia very hard and led to a much reduced demand for Russian oil and gas. Global economic growth is expected to resume during in 2010 and with it the supposedly steadily increasing demand for natural gas. The low carbon content of natural gas compared to oil and coal is expected to lead to a preference for this energy source and a comparatively greater increase in demand. Russia is planning in the years ahead to increase her production of natural gas as she develops the Bovanenkovo gas field on the Yamal Peninsula and the Shtokman field in Barents Sea in particular. Proven world gas reserves totalled 180 trillion cubic meters – equal to about 60 years' consumption at current rates. More than half of these reserves are located in three countries –Russia, Iran and Quatar. Estimates of remaining recoverable reserves are as high as 850 trillion cubic meters. In the next few years, however, the International Energy Agency expects a glut of gas supply caused by the recession of 2008 and 2009, which depressed demand, and the recent rapid development of unconventional gas resources in the United States and Canada. [84] The last decade has witnessed a high rate of increase in natural gas consumption. But there is no lack of reserves. World wide, the reserves-to-production (R/P) ratio is 63 years. The highest ratios by region are 48 years for Central and South America, 78 years for Russia, 79 years for Africa, and over 100 years for the Middle East.

Arctic shipping cannot stay unaffected by these broader international shipping and economic development trends. The global financial and economic crises of the last four to five years have added new uncertainties to a part of the world that already has more than its fair share of uncertainties. To say something sensible about the future of shipping in Arctic waters against this background is a difficult undertaking.

The North East Passage

Of all potential Arctic sea routes, the NEP is much the most promising Arctic shipping route but is also the one most vulnerable to changing global and national economic conditions. The expected economic recovery among countries which have traditionally accounted for the bulk of global demand for consumer and finished goods is likely to be uneven. The EU (27) economies grew by 2 percent in 2010 and the US by 3 percent quarter. The recovery that seemed under way in Europe in 2010 faltered in 2011 under the weight of the debt crisis. A slow and uneven pace of recovery is likely among the advanced economies in which case the demand for the resources of the Arctic, including oil and gas, from traditional customers may take time to recover to earlier levels. This gap between supply and demand is not likely to be offset in the near future by an increase in demand from countries like China, India, Brazil and the ASEAN ones in particular. What seems to be a widely-accepted view is that the economic downturn of 2008–2009 will not be easily overcome, and that global economic growth in the time ahead will take place at more moderate levels than in the past 10–20 years. Maritime trade, like other important sectors, will take time to reach activity levels of the past. If after 2010 economic growth returns to long-term historical average growth levels, the overall effect of the economic downturn on transport activity is assessed to delay growth in this sector by about five years.[85]

Given this state of affairs, the question is how the economic recovery will affect the Arctic and the activities on which the Arctic depend, and what might be the spill-over effects on Arctic shipping. Oil and gas are central factors in this context. Whereas the pace of economic growth dropped somewhat in countries like China and India, in 2009 they continued to grow at a very brisk pace, that is, 8.9 percent and 5 percent respectively.

As shown in Table 3.3 above, most of the advanced economies, however, were in a recession, as indeed the entire global economy is supposed to have been for the first time since the 1930s. Before a recovery gets under way in these countries, the estimates suggest a period of anaemic or little growth during which the demand for oil and gas is likely to be subdued. Global oil demand fell in 2009 for the second year in a row, the first time it has performed like this since 1983. The economies that are expected to lead the world in economic growth belong to the non-OECD club. But the demand coming from the OECD countries will remain a very important driving force. The North Atlantic area is still and will remain an essential component of the world economy, and the extent to which Russia will be able to invest, develop and economically extract its oil and gas resources in particular depend very much on how and when these countries return to a path of stronger economic growth. The sales of gas from fields like Shtokman, targeted at the US market in particular, will in future have to compete with increasing production of gas from unconventional sources in the US and Canada, shale gas in particular. In other words, the outlook over the next four to five years, despite increasing demand, is decidedly uncertain. Indeed, this was precisely the reason given for the postponement of the Shtokman investment decisions.[86]

There are other uncertainties that may affect Arctic shipping. The estimate that 22 percent of the world's undiscovered oil and gas resources are to be found in the Arctic only begs the question of where the other 78 percent might be. As we learned above, more than one half of the world's proven gas reserves are in three countries – Russia, Iran, and Quatar. An educated guess is that much of the 78 percent undiscovered resources is to be found in the Caspian Sea area, the Middle East and in South America. But there are other candidates. In a 2006 US Department of Interior study of undiscovered technically-recoverable oil and gas resources on the US outer continental shelf, the Gulf of Mexico was assessed to be the most promising and was thought to contain some 44.92 billion barrels of oil and some 232.54 trillion cubic feet of undiscovered gas, or 54 percent of the total potential. It was followed by Alaska with 26.61 billion barrels of oil and 132.06 trillion cubic feet of gas, or 31 percent of the potential (presumably included in the Arctic 22 percent of the USGS), with smaller quantities likely to be found on the Atlantic and Pacific continental shelf areas.[87]

But it is also possible that a good deal might be found in Sub-Arctic parts of Russia. Should this be so, a case may be made that exploiting Sub-Arctic gas rather than offshore Arctic gas is less costly and more profitable. Much the same argument can be made if the exploitation of undiscovered offshore Arctic gas finds itself in competition with the exploitation of undiscovered gas in areas with a much more clement climate and easier access.

The latest financial crisis is not the only factor influencing demand for oil and gas in particular. Climate change and global warming are likely to become increasingly important items on the international agenda. Earlier in this chapter the point was made that today's advanced economies achieved their present state of economic development employing more or less the same mix of energy resources, the main component of which was a strong reliance on hydrocarbon resources. If today's developing economies are to achieve similar levels of economic development they will, sooner rather than later, have to rely on a different mix of energy sources involving much less reliance on hydrocarbon resources. Countries like China and India will not be able to escape the rising expectations of their enormous populations for a better quality of life. The stage of economic development this implies will have to employ a mix of energies rather different from today, and climatically and environmentally much less destructive. What the most important ingredients of this mix will be is a very complicated question. Renewable energy sources will certainly be important in that mix. It will in any case take time before a climatically better mixture is developed. One change, the beginning of which is already discernible, is a shift away from oil towards gas because of lesser CO_2 emissions. Gas emits 50 percent less CO^2 than coal and 30 percent less than oil. The speed and magnitude of the change from the one to the other is an unknown quantity but speaks in favor of increased emphasis on the search for and exploitation of new gas resources, including those that may be found in the Arctic.

How and how much of increased volumes of gas will be transported in the future from the Arctic is nevertheless uncertain. Although China today imports relatively little gas as a share of her overall consumption, the general understanding is that she will import large quantities of Russian gas in the future. From where in Russia this gas will come is, however, an open question. The mode of transportation is in any case likely to be pipelines that are on the drawing board. Today most of the oil imported from Russia is transported by rail.

Currently maritime transport in the Euro-Arctic region is primarily a matter of shipping oil along the NSR from old and new operating fields in Western Siberia to transhipment terminals, mostly in the Barents Sea area of Russia, but also in northern Norway. The capacity of the tankers involved in carrying oil from the production fields in the east to the terminals in the west is relatively modest. The transport from the established terminals in the Barents Sea region to foreign destinations is carried out by very large tankers (VLCC). If and when the planned offshore activities start in earnest, other types of ships will be required to

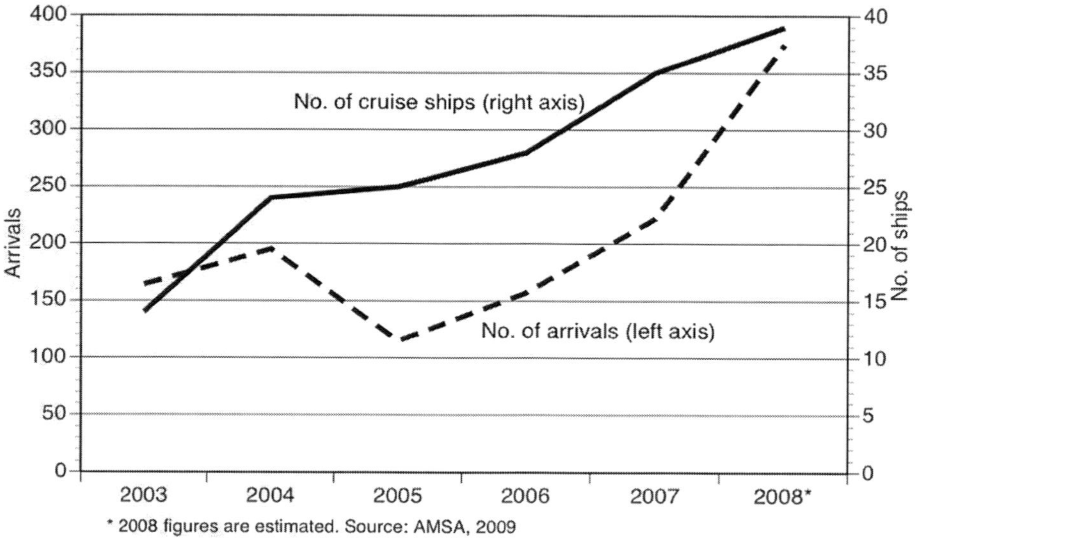

Figure 3.10 Cruise ship arrivals in Greenland ports and harbors, 2003–2008.

supply and service these activities. Although far from being a novelty, the transit sailings of the Beluga ships from the Pacific to the Atlantic via the NSR took place under circumstances and at a time that attracted widespread attention. The apparent success of this transit and the new transit journeys that took place in 2010 and 2011 are likely to have a demonstration effect and perhaps lead to other shipping companies and shippers doiing the same. Again, important drivers in this context are the advanced economies and fastest growing economies of Asia – Japan, Korea and China – both in their capacities as importers and exporters.

The Transpolar Passage route

Through the analysis of its Arctic Container Project (ARCON), Det Norske Veritas (DNV) has essentially said much of what need be said at this time about the TPP from a shipping opportunities point of view. "At present the ice conditions are too heavy for the benefits to exceed the expected additional costs, especially during winter time, and further reductions in ice extent and thickness is required to release the full potential ... There is a potential for profit from Arctic trade routes, but not in the near future".[88] It would not be unreasonable to conclude that it could take at least another 20 to 30 years before the conditions along the TPP prove attractive enough for shipping companies and shippers to use the TPP.[88] There are no resources in the vicinity of the route and the future promise of the TPP lies in its potential as a future transit route between the Atlantic and Pacific oceans.

3.9 Arctic cruise activity

One type of shipping that benefits from climate change and less ice-infested waters and is likely increasingly to do so in the future is the cruise industry. It is perhaps the type of shipping that has been the least affected by the financial and economic turmoil of the last few years. Cruise ships are relatively self-contained entities. They carry their "cargo" (people) into the Arctic and bring the same "cargo" out again without having to load and reload and rely much on shore services and logistics support. Once at sea, they have a degree of operational freedom and flexibility that other types of shipping do not enjoy.[89]

The modern cruise industry has only a 50-year history, starting with the Norwegian Caribbean Line in the early 1960s out of Miami. Since the beginning, the cruise industry has been growing with impressive growth rates in the 8–10 percent range each year, but it is still a fairly small part of the total leisure market, with a penetration rate of only 3 percent of the North American market and only 1 percent of the European leisure market[90].

Over 100 years before the formation of the modern cruise industry, cruise-like trips were made to the high north. The industrial revolution created wealth, and with wealth came a market for luxurious passenger trips with British tourists touring the Norwegian fjords all the way up to North Cape. The first cruise route was set up by the British P&O when they introduced the S/Y *Ceylon* in 1882 for pleasure cruising. They characterized the ship as the first "ocean going steam yacht of large tonnage, specially fitted for and adapted to pleasure cruising".[91] The destination of this ship was the west coastal fjords of Norway and up to the North Cape. This route was later copied by Norwegian companies and is today known as Hurtigruten, operating a daily service with 11 ships from Bergen to Kirkenes.

Cruise activities in Arctic waters are predominantly made in ice-free waters in the summer season, and the main traffic areas are northern Norway, Svalbard, Iceland and the west

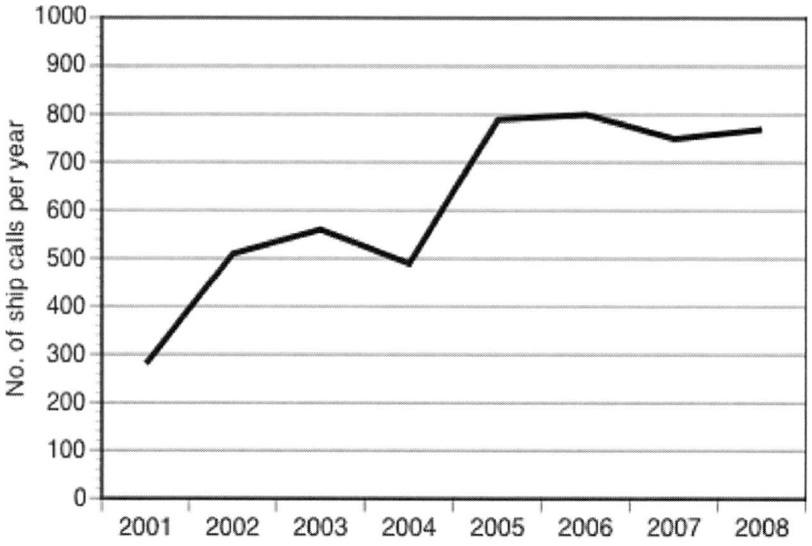

Source: St.meld (White Paper) 22, 2008-09, section 10.4.1

Figure 3.11 Number of ship calls in Longyearbyen 2001–2008.

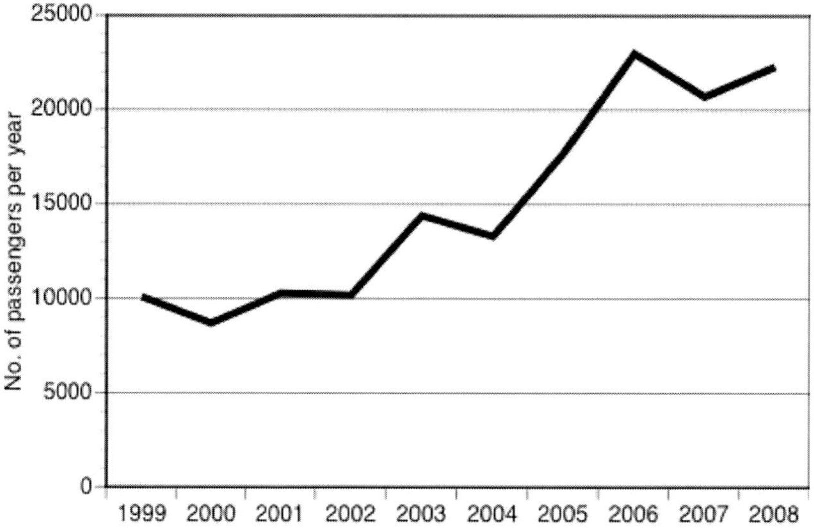

Source: St.meld (White Paper) 22, 2008-09, section 10.4.1

Figure 3.12 Number of cruise passengers arriving at Longyearbyen, 1999–2008.

coast of Greenland. The only regular cruise-like activity in ice-covered waters are trips from Murmansk to Franz Josef's land and the North Pole made by Russian nuclear icebreakers, an activity that has been going on since 1990.[92] Occasionally expedition-like cruises are made both in the NWP, where 27 trips have been made in the period 1984–2004,[93] and in the NSR (see Figure 3.12 below).

Arctic cruises is a fast growing business. The market is definitely a specialist market, although the big cruise operators from time to time make sailings to the edge of the ice in Greenland and Svalbard.

3.9.1 Cruise activity around Greenland and Svalbard

Complete data for all Arctic cruises are hard to obtain, but some indications of the development in later years can be found in data from Greenland and Svalbard.

Figure 3.9 shows the development of cruise ship activities in the ports of Greenland. The number of arrivals of cruise ships in Greenland ports increased by an average of 48.9 percent per year from 2005 to 2008.

Another indication of an increased interest for sailings in the high north is data from Longyearbyen. Figure 3.11 shows the number of ship calls, while Figure 3.12 shows the number of cruise passengers arriving in the port. The average growth rate 2001–2008 of passengers arriving is 14% per year.

No good source exists for the exact number of passengers that go on Arctic cruises. AMSA refers to an independent survey indicating that in 2004 1.2 million passengers went on an Arctic cruise and that this number more than doubled in 2007.[95] This implies that more than 15 percent of all cruise passengers had an Arctic destination, since the world total number of cruisers is estimated at around 15 million. Such a number seems much too high, even if ferry passengers are included. Hurtigruten has 11 ships on daily trips to the north, but in total they carry a little over 100.000 passengers per year, which would be less than 0.5 percent of all Arctic cruises if this survey is to be trusted.

Ice class	**Number**	**%**
Icebreaking capability	8	3.6
Ice class 1A and 1A Super	12	5.4
Ice class B and C	41	18.5
Ice class II	44	19.8
Ice strengthened	117	52.7
Total	222	100.0

Source: www.sea-web.com

Table 3.24 Distribution of passenger vessels with ice class.

Ship name	Operator	Passengers	Area of operations
Birka Paradise	Birka Line	1800	Baltic Sea
Bremen	Hapag-Lloyd Cruises	164	Antarctica, Arctic
Clipper Adventurer	Clipper Cruise Line	122	Antarctica, Greenland, Canadian Arctic
Cristal	Core Marine Ltd	1409	Mediterranean
Hanseatic	Hapag-Lloyd Cruises	184	Antarctica, Arctic
Kristina Regina	Kristina Cruises Oy	245	Red Sea, Black Sea, Norwegian fjords, Iceland
Louis Majesty	Core Marine Ltd	1850	Mediterranean
Orion	Travel Dynamics	130	Antarctica
Polar Star	Polar Star Expeditions	90	Antarctica, East Greenland
Prince Albert Ii	Silversea	158	Antarctica, Spitsbergen, Iceland, Greenland, Newfoundland
Professor Khromov	Far Eastern Research	48	Oceanographic trips, Pacific and Indian Ocean
Turama	Sete Yacht Mgmt	70	Mediterranean cruises

Source: www.sea-web.com

Table 3.25 Passenger vessels with ice class 1 A and 1 A super

3.9.2 Ships used in Arctic expeditions

The ships being used for Arctic cruises come in five categories:

- icebreakers (used to access the extreme north);
- research vessels (often refurbished to obtain close to cruise quality);
- other ice class vessels (a few purpose-built cruise vessels with high ice class);
- ice-strengthened vessels;
- vessels with no ice capability.

According to HIS database, there are 3,755 passenger vessels in the world fleet, of which 222 (5.9 per cent) have ice class.[95] The distribution of these ships according to their ice class category is given in Table 3.25. Of the 3,755 passenger vessels, 554 are cruise vessels, of which 73 (13.2 percent) have ice class, a ratio about three times as high as for non-cruise passenger vessels.

Table 3.25 lists the 12 cruise vessels that have been built according to ice class 1A and 1A super.

Three of these ships are large ships, but they are not used for Arctic operations. One of the ships, *Orion*, operates in Antarctica, *Professor Khromov* is only used as a research vessel, *Turama* is only operating (currently) in the Mediterranean while the remaining six all have Arctic destinations.

When it comes to cruises in ice-covered waters, more ships are operating in Antarctic than in Arctic waters. Most specialist operators will typically offer both. When it is summer in the northern hemisphere, Arctic cruises are offered and in the winter Antarctic is the preferred area.

3.9.3 Main specialist operators

The US company Quark Expeditions is a leading operator of cruises in ice-covered waters. Their fleet consists of mostly of former Russian icebreakers or research ships:

50 Years of Victory is the world's largest icebreaker with two nuclear reactors with a combined power of 75.000 HP. With 64 cabins in five categories the ship can accommodate 128 passengers. It is mainly used to go to the North Pole.

Akademik Ioffe is an adventure ship equipped with both zodiacs and kayaks. It operates mostly in the Canadian Arctic and Greenland, has a shallow draft and can accommodate 109 passengers.

Akademik Shokalskiy is a former research vessel that also carries kayaks, can accommodate 47 passengers and is mostly used around Spitsbergen, Iceland and Greenland.

Akademik Sergey Vavilov is similar to Akademik Ioffe and can accommodate 104 passenger.

Kapitan Khlebnikov has navigated the Northwest Passage more than any other ship. It can accommodate 112 passengers.

Quark Expeditions organizes cruise trips through the NSR, starting in Alaska, then sailing to Anadyr on the east coast of Russia, where passengers can also embark. Then the ships go to the Bering Strait, Wrangel Island, New Siberian Islands, Severnaya Zemlya, Franz Josef Land and disembarkation in Murmansk. In total, a trip of 27 days costs US$ 27,000–47,000. Quark Expeditions also offers a 66-day complete Arctic circumnavigation costing US$ 65,000–113,000. The route is given in Figure 3.13.

Another major operator for cruises in ice is the Dutch company Oceanwide Expeditions BV. They have also chartered-in Russian ships, but are currently refurbishing a former Dutch research ship to become their main attraction. The current fleet operated by Oceanwide Expeditions consists of:

Professor Molchanov, a modern, formerly oceanographic research vessel, built in Finland in 1983. It can accommodate 53 passengers and carries Zodiacs.

Professor Mulkanovskly is a sister ship to Molchanov.

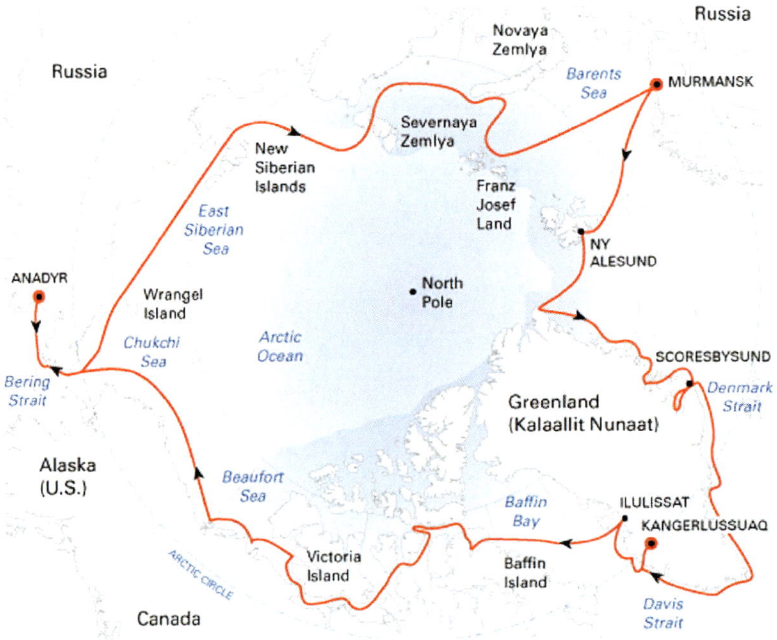

Figure 3.13 Quark Expeditions' 66-day cruise around the Arctic.

The *Noorderlicht* was built as a three-masted schooner in Flensburg in 1910, purchased by Oceanwide in 1991 and re-rigged as a two-masted schooner. She can accommodate 20 passengers.

Antarctic Dream is a Dutch-built passenger vessel, formerly owned by the Chilean navy. It can accommodate 84 passengers.

Plancius was an oceanographic research vessel built for the Royal Dutch Navy in 1976 and operated by the navy until 2004. The vessel is currently in the final stages of being completely rebuilt to accommodate 110 passengers. The ship will carry zodiacs and will operate in both in Antarctic and Arctic waters. The charters for the two Russian ships will expire in 2010.

In Arctic waters Oceanwide Expedition focuses mainly on Spitsbergen and Greenland. Lindblad Expeditions, headquartered in New York, is a subsidiary of Lindblad Travels and organizes exploration cruises worldwide. Lindblad has a unique strategic alliance with National Geographic, and operates five National Geographic vessels. In Arctic waters, Lindblad Expeditions has three main offers:

1. Norwegian fjords up to Svalbard (16 days)
2. Exploring Svalbard (11 days)

3. Svalbard – North Cape – Novaja Zemlja – Franz Josef's Land – Svalbard (26 days)

The trips are made with the ship *National Geographic Explorer*, a Norwegian-built passenger ship with ice class 1C and a passenger capacity of 154 in 81 cabins. The trip into Russian Arctic is a new offer and *NG Explorer* will then be the first western ship to make cruises in Russian waters. Lindblad also offers Alaska cruises, but only in the southern parts of Alaska in ice-free waters.

The New Zealand-based company Heritage Expeditions offers cruises in the Russian Far East, mostly in the Bering Sea, but also in the Chukchi Sea: Nome, Alaska – Bering Strait – Wrangel Island – Cape Vankarem – Provydenia and return to Nome (15 days). In Cape Vankarem this cruise offers close contact with the indigenous people – the Chukchi. The ship used for this trip is the Finnish-built expedition ship *Spirit of Enderby*, fully ice classed with Russian flag and accommodation for 48 passengers. The ship carries zodiacs and hovercrafts for exploring shallow waters.

A New Zealand marketing company called Wild Earth Travel organizes Arctic trips to Svalbard, the North Pole and in the NWP. For the cruise around Spitsbergen, the US vessel *M/S Quest* is used (52 passengers, ice class 1B), for the North Pole it is *50 Years of Victory* as mentioned above and for the NWP cruises the ship *Clipper Adventurer* (122 passengers, ice class 1A; see also Table 3.25 above) is used. The company is marketing two NWP trips: Into the NWP (Kangerlussauq – Karrat Fiord – Devon Island – Bellot Strait – Queen Maud Gulf – Coronation Gulf to Kugluktuk) and out of the NWP (Kugluktuk – Parry Channel – Grise Fiord – Kap York to Ilulissat, Greenland).

The Canadian-based Cruise North Expeditions employs the Russian-built, ice strengthened vessel *Lyubov Orlova* (112 passengers) for various expeditions in the Baffin Sea, but also into the NWP with two itineraries: One from Resolute Bay to the Coronation Gulf and return by going around the islands of King William (where Gjoa Haven is) and Somerset back to Resolute Bay. The other goes from Resolute Bay southwards to Kuujjuaq south of Hudson Strait, or the other way around from Kuujjuaq north and west to Resolute Bay.

Hapag-Lloyd Cruises operates two vessels with ice class 1A* – the *Bremen* and the *Hanseatic* as mentioned in Table 3.25. They focus mostly on Antarctica, but also have trips to the Arctic. They have both conducted trips through the NWP and the Canadian authorities have expressed great concern that these trips were made without proper notification to the authorities. Although the ships have high ice class, the NWP is complicated and dangerous, so the trips were characterized as highly hazardous.

The Norwegian coastal liner operator *Hurtigruten* has 13 ice classed ships, 11 of which are in daily operation between Bergen and Kirkenes. The two remaining ships are in season used for Arctic cruises: cruises to Spitsbergen are made with *Nordstjernen*, built in 1956 with 400 passenger capacity; Greenland cruises are made with *Fram*, a modern ship built in 2007 with a 400 passenger capacity – one of the larger ships in Arctic operations.

Silversea Cruises, an international company of Italian origin, operates the vessel *Prince Albert II*, which has ice class 1A* with DnV. The ship was built in Finland in 1989 and can accommodate 158 passengers. The ship makes cruises in the Antarctica, Tromsø-Longyearbyen as well as trips from Iceland, via Greenland to Newfoundland.

Polar Star Expeditions, which is a Canadian company, operates the vessel *Polar Star*, built in 1969 with ice class 1A* for 105 passengers, operated by Karlsen Shipping. The ship is mostly used in Antarctica, but sailings are also made from Svalbard to the east coast of Greenland and then to Iceland.

The Murmansk Shipping Company has a couple of passenger vessels offering trips in the Arctic. The ship *Polaris* goes to Iceland, Greenland and Spitsbergen, but also to Franz Josef Land and to the Solovetsky Islands in the Onega Bay. *Polaris* has capacity for 76 passengers.

3.9.4 The current status of cruises and expeditions in Arctic passages

Cruises – or expeditions – are currently made within all three passages of the Arctic, but the main cruise activity is in the corridors outside the passages, particularly around Svalbard and Greenland.

NEP and cruises

Cruises are currently offered for the entire NSR as part of a complete 66-day Arctic circumnavigation cruise, but cruises are also offered into the western (Novaja Zemlja, Franz Josef's Land) and eastern (Chukchi Sea) extremities of the NSR.

NWP and cruises

Regular cruises are offered for trips into Queen Maud Gulf and Coronation Gulf .Another offer is from or to Resolution Bay, covering parts of the NWP, but there have also been instances of complete NWP passages with cruise vessels through the NWP.

TPP and cruises

The icebreaker *50 Years of Victory* is making fairly regular trips from Murmansk to the North Pole and back.

3.9.5 Arctic cruises in the future

The Arctic will remain a remote and wild area of breathtaking beauty with a unique flora and fauna. If climate change leads to less ice, more of the Arctic will be accessible to cruise ships. The cruise industry is a concept industry which is always on the lookout for new themes and destinations. It is to be expected that if the ice is retreating some cruise operators will follow the edge of the ice. In a longer-term perspective, Arctic cruises would seem to be a growth industry with a great potential.

In the shorter perspective, specialist operators will continue to offer cruises in all three passages. With powerful icebreakers one can reach the North Pole or circumnavigate the entire Arctic Ocean. With expedition ships and ice classed passenger/cruise vessels, trips are offered into the NWP and in the NSR trips go to Novaja Zemlja and Franz Josef's Land in the west and to the Wrangle Island in the Chukchi Sea in the east. The vessels are fairly small with 50 to 400 passengers – very small compared to the largest cruise vessels of more than 5,000 passengers. The larger vessels concentrate on the blue waters around Svalbard and Greenland, which is the target for most Arctic cruises, while the smaller vessels go on expeditions in ice-covered waters. Arctic cruises are a small niche product, indeed, compared to the mass-markets of the Caribbean or the Mediterranean.

The typical cruise ship for Arctic operations carries zodiacs and hovercrafts to enable exploration of shallow waters. Inevitably such cruise activities will one way or the other interfere with the cultures of the indigenous peoples of the Arctic, disturbing hunting and fishing grounds and even be a danger to people in small crafts like canoes or kayaks. In this respect the cruise operators have a social responsibility to perform.

The cruise vessels used in Arctic operations are very much self-contained units that are not dependent on land infrastructure. This is fine when everything goes smoothly, but is a major problem the day an accident occurs. The coastal states face a big challenge when it comes to providing adequate search and rescue service if cruise activities expand in the Arctic.

3.10 Some concluding remarks

A few concluding remarks are in order.

- The Arctic is rich in resources. Some, like oil and gas, have great strategic value from an economic development point of view and are and will continue to be in great demand. Although substitution is an option, no realistic alternative energy source is in a position to replace the functions that oil and gas are performing in the global economy today.
- Because of their intrinsically high income-generating value, oil and gas are also saddled with a higher conflict potential than most other resources. The higher the value of a resource, the greater the risks one is willing to run to obtain it. Because of the unresolved jurisdictional issues in the Arctic, a great deal has been made out of the possibility that much oil and gas may be found offshore in ocean areas of uncertain legal status. Hence, the prospect of the Arctic developing into an arena of high tension and even military confrontation aimed at gaining control over the resources. Yet a barely-noticed part of the USGS' new assessment of undiscovered oil and gas in the Arctic suggests something different. Almost all the offshore oil and gas resources to be discovered in the Arctic are likely to be located in undisputed territories within the 200 miles economic zone of the relevant states. This is state of affairs that is likely to inspire collaboration rather than confrontation.
- A good deal of the mining sector in the Arctic is handicapped by tremendous difficulties in gaining access to the resources. This is particularly the case in Siberia. The good luck of Russia in this regard is that she has mineral resources in quite large quantities in areas where access is easier and the need for huge amounts of investment capital is less. The minerals industry has also been rather successful in recovering from the disasters of the immediate post-Soviet era.
- The centers of the highest rates of economic growth and development are no longer to be found in Europe and North America. The Emerging Market Economies, of which many are struggling to satisfy the needs of huge and growing populations, are likely to gobble up vast quantities of resources. This means *inter alia* that the international trade

in commodities will grow considerably and cargoes will be carried over longer distances, especially if they originate in the Arctic. Yet, some Arctic commodities are likely not to move far. Practically all of Canada's oil and gas just treks over the border to the US, and we know how and how far most of Norway's gas travels and is likely to continue to travel in the future.

- The Arctic has fish resources. What climate change is likely to do to the size and identity of Arctic fish stocks is uncertain. Yet, one little discussed issue should be aired already here. Table 3.21 showed the extent to which pelagic and capture fishing is relatively declining with aquaculture producing by far the greatest share of any increase in fish production globally. This is a trend that has been under way not only for the last four or five years but for the best part of 30 years.
- Russia has forest resources beyond anything seen elsewhere. As our brief survey showed, much of the Russian forest industry produces and exports by and large roundwood which, as far as Europe is concerned, goes to countries like Finland and Sweden where the logs become value-added products sold on the international market (not unlike the Norwegian export of salmon to Denmark). Forests and forest products in Russia have the potential of being great income earners, especially if investment in value-added products is emphasized.
- The cruise industry is likely to continue to be fairly unaffected by the many trials and tribulations that other sectors of the shipping industry are going through. With larger ice free areas cruises into the Arctic are likely to become more popular both among cruise operators and the cruising public.

Notes

1 Timmons and Heather (2004)
2 Wikipedia, http://en.wikipedia.org/wiki/Oil_reserves
3 *U.S. Geological Survey Minerals Yearbook – 2005*, 7.11
4 *BP StatisticalReview of World Energy* (2008), p .6, notes
5 This survey of terms and definitions is taken from Wikipedia, *op. cit.*
6 Bambulyak and Frantzen (2009), p. 8
7 International Energy Agency (2008)
8 *BP Statistical Review of World Energy*, June 2009, p. 1
9 Mouawad (2010)
10 *BP Statistical Review of World Energy*, 2009, p. 1
11 *BP Statistical Review of World Energy*, 2011, pp. 6 and 20. For a different assessment which sets world oil reserves at 850–900 billion barrels, see University of Oxford (26 March 2010). World oil reserves at 'tipping point'. *Science Daily*. Retrived 31 March 2010 from http://www.sciencedaily.com/releases/2010/03/100324225511.htm
12 See Maugeri (2009), pp. 36–43
13 Maugeri (2009), pp. 40–41
14 Krauss (2009)
15 South and Central America, Africa, Middle East, non-OECD Asia, non-OECD Europe
16 *BP StatisticalReview of World Energy* (2011), pp. 2–3
17 Uncoventional oil resources in the world exceed considerably the amount of conventional

oil reserves but are much more difficult and expensive to develop. *Wikipedia. Oil Reserves*,http://en.wikipedia.org/wiki/

18 18.See http://www.bp.com/genericarcticle, Speech date 30 June 2008, World Petroleum Council, Madrid

19 Joint Transport Research Centre of the OECD and the International Transport Forum, *Transport Outlook 2009, Globakisation, Crisis and Transport*, Discussion Paper No 2009-12, May 2009, preliminary version.

20 For an informative discussion of the issue within a Norwegian context, see Sawhill (2008), p. 2.

21 Philip Budzik, (2009)

22 CARA (2008)

23 CARA (2008)

24 AAPG (2008)

25 AAPG (2008)

26 U.S.Geological Survey (USGS), Fact sheet 2009-3037, June 2009

27 Lundgren (2010)

28 Budzik (2009), p. 8.

29 American Petroleum Institute, *2006 Joint Association Survey on Drilling Costs* (May 2008), quoted by Energy Information Administration, US Department of Energy, http://eia.doe.gov/emeu/aer/txt

30 *Analysis of Crude Oil* (May 2008)

31 *Petroleum Intelligence Weekly*, 5 January 2009. See also Christoph Seidler, "New Estimates for Drilling Costs. The Exorbitant Dream of Arctic Oil", *Spiegel Online* , 26.01.2011

32 World Energy Outlook (2008), p. 4

33 World Energy Outlook (2008), p .6

34 The information in this section has been gathered from http://arcticentre.ulapland.fi/barentsinfo/.

35 State of the World's Forests (2007), p. 64.

36 Valkky, Nousiainen, and Karjalainen (2008), pp. 44–45

37 FAO, *The State of the Worlds's Fisheries and Aquaculture, 2010*, FAO Rome 2010

38 See Glomsrød and Aslaksen (2006), p. 33

39 Joling, Dan (2009)

40 Canada.com. *Arctic Fishing ban report places pressure on Canada*, report by Randy Boswell, 14 February 14 2009. The same report refers to Foreign Minister Gahr Støre as having picked up the US proposal to the effect that Norway also would now consider an Arctic management plan in view of climate changes, apparently oblivious of the fact that Norway has had one for a considerable time already

41 ICIA (2004), pp. 62–64

42 See Section 3.3.1 above

43 US Energy Information Administration, Country Analysis Briefs 2009, *Canada. Oil-Natural Gas*. See also Shawn McCarthy (30 August 2009)

44 CBC News (30 December 2009), www.cbc.ca/canada/north/story/2009

45 Kosich, Dorothy (28 July 2009) www.mineweb.com/mineweb/view/mineweb/en/

46 US Department of the Interior, USGS, *2006 Minerals Yearbook, Alaska*, July 2009, p. 43

47 Beauregard-Tellier (2008)

48 www.gulfoilandgas.com, 4/8/2008

49 The section that follows on Canada relies on material in *Canadian Arctic Shipping Assessment* (2007), Mariport Group Ltd. for Transport Canada

50 Gazprom, Yamal Megaproject, www.gazprom.com/production/projects/mega-yamal

51 See above for data about the costs of developing resources in the Arctic
52 US Energy Information Administration, Country Analysis Brief, *Russia Natural Gas*, May 2008
53 BBC News, 01/07/2009
54 Glasby and Voytekhovsky (2009) http://www.geochemsoc.org/publications/geochemical-news/gn140jul09/
55 Gazprom Release, "On meeting dedicated to Shtokman fild development", 28.10. 2010
56 56 Gazprom, "Shtokman Development AG Board of Directors approve was forward on Shtokman", 08.02.2010, www.gazprom.com/press/news/2010/february
57 www.barentsobserver.com, "Statoil impatient on Shtokman", 07.11.2011
58 US Energy Information Administration, Country Analysis Briefs 2009, *Russia Oil*
59 US Department of the Interior, USGS, 2006 Minerals Yearbook, July 2009, *Commonwealth of Independent States, Russia,* p. 7.15
60 See Glasby and Voytekhovsky (2009)
61 USGS, Commonwealth of Independent States, Russia, *op. cit*, pp. 7.16, 7.18, 7.23.
62 USGS, Commonwealth of Independent States, p. 7.24
63 See BP Statistical Review, June 2010, p. 28
64 BP Statistical Review, June 2008, p. 13
65 *Gazprom*, Release 15.01.2010, "Gazprom launches Portovaya compressor station construction", www.gazprom.com/press/news
66 www.gazprom.com, *Nord Stream*
67 US Department of Energy, Energy Information Administration, *Russia – Natural Gas,* May 2008
68 China Baobel International Investment Group, *Progress Stalled on Russian Gas Deal with China*, 12.04.2009. http://en.worldmr.net/shownews
69 www.barentsobserver.com, 08.12.2009
70 Bambulyak and Frantzen, (2009), p. 35
71 See Chapter 4.3 by Michael Grigoriev in Bambulyak and Frantzen (2009), p. 72
72 For a detailed account of ports along the NSR, see Chapter 5 of this study
73 Bambulyak and Frantzen (2009), pp. 28–29
74 Bambulyak and Frantzen (2009)
75 Bambulyak and Frantzen (2009), pp. 63–64
76 Statoil, *Facts about Snøhvit*, 23.11.2009, www.statoil.com/en/OurOperations/
77 Calculated from Tables 4.1 and 4.2 in Bambulyak and Frantzen (2009), pp. 37–38
78 US Department of Energy, Energy Information Administration, *Russia – Oil Exports*, May 2008
79 Bambulyak, A. and Frantzen, B. (2009), pp. 37–38 and 42
80 Bambulyak and Frantzen (2009), p. 30
81 *Sovcomflot*. 2010. www.scf-group.com/, and www.barentsobserver.com 06.01.2010
82 Bambulyak and Frantzen (2009), p. 31. See also Chapter 6 of this study for more details
83 Bambulyak and Frantzen (2009), p. 8
84 International Energy Agency, *World Energy Outlook 2009. Executive Summary,* pp. 11–13
85 *NYT*, February 13, 2010, pp. 20–22
86 See p. 11 above
87 US Department of the Interior, Minerals Management Service, MMS Fact Sheet RED-2006-01b, February 2006, *Assessment of Undiscovered Technically Recoverable Oil and Gas Resources of the Nation's Outer Continental Shelf, 2006*
88 Morten Mejlænder-Larsen and Øyvind Espeland (2009), p. 8. See also Chapters 1 and 5 of this book

89 Search and rescue operations in Arctic waters are very difficult, especially in the case of cruise ships in distress because of the many people involved. See Chapter 6 of this book for an extended treatment of the topic
90 Wijnolst and Wergeland (2008)
91 Martinussen (1992)
92 AMSA (2009)
93 AMSA (2009)
94 AMSA (2009), p. 79. No source is given for the mentioned independent survey.
95 www.sea-web.com

4 Environmental challenges and other operating conditions*

The state of Arctic climate change, the projections for ice coverage change and the nature of current marine vessel traffic in the Arctic Region indicate that there will be significant potential impacts on the marine environment from the anticipated growth of marine traffic. The purpose of this chapter is threefold: First, to identify the most important aspects of natural conditions significant for shipping along the three Arctic shipping corridors and connecting corridors in southern waters. Second, to describe shipping as an independent source of pollution. Third, to consider the Arctic transportation routes and connecting southern corridors in light of climate change and various physical characteristics.

4.1 Natural conditions of the Arctic significant for shipping

Climate models have been predicting warming for some time and have indicated that effects of change would be the most dramatic in the Arctic. Signs validating these forecasts are being seen. Arctic average temperatures have risen considerably in the last few decades. Northern polar temperatures are increasing twice the rate of the global mean.[1] Ice concentrations have shown a steady decline. Glaciers are retreating, shore erosion is becoming more pronounced in some areas and rising permafrost temperatures have already had a significant effect on essential infrastructure. An acceleration of these climate trends is predicted during this century, which will have a significant social and economic effect on the Arctic.[2]

Some of the changes associated with Arctic warming and a decreased sea ice presence may have an impact on Arctic marine traffic. For instance, climate change will pose a

* This chapter has been written by Karl Magnus Eger, Ocean Futures.

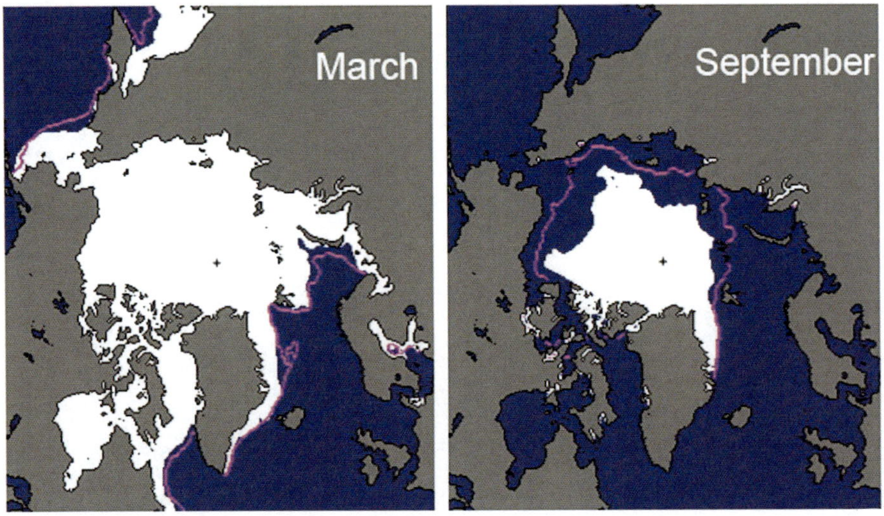

Source: National Snow and Ice Data Center Sea Ice Index

Figure 4.1 Sea ice extent in (left) March and (right) September 2011, illustrating the respective monthly winter maximum and summer minimum extents. The magenta line indicates the median maximum and minimum ice extents in the given month for the period 1979–2000.

serious threat to the permafrost[3] of the Northern Hemisphere. Melting the once permanently frozen solid could have a remarkable effect, altering ecosystems and damaging infrastructure across Canada, Alaska, and Russia. Simulations performed by the Arctic Climate Impact Assessment (ACIA), indicates that over 50 percent of permafrost could thaw by 2050 and as much as 90 percent by 2100.[4] The impact on the transportation of goods to the Canadian Arctic by winter roads will not likely be significantly impacted by 2020. However, by 2050 the winter road season will be shorter and both the traffic volume and weight that these roads will be able to sustain will decrease. The condition of land transportation routes in Russia is also affected by thawing of the permafrost. In the early 1990s, 10–16 percent of the sub grade in the permafrost areas of the Baikal Amur railroad line was deformed because of permafrost thawing, and this increased to 46 percent in 1998. The majority of runways in Norilsk, Yakutsk, Magadan, and other cities may be closed for shorter or longer periods due to lack of maintenance.[5] The complexity and cost of building permanent overland roads or rail beds will increase as will the construction costs associated with establishing new Arctic harbor infrastructure. The result of less ice and costlier ground transportation options will likely result in increased marine transportation to support the extraction of gas, oil and minerals from the Arctic region.

Surveys performed by the ACIA programme estimate an annual decline in the Arctic ice of 5–10 percent and reduction in thickness at 10–15 percent over the last decades (for example, see Figure 4.1). Measurements performed with submarine sonar in the central

Arctic Ocean showed a reduction in ice thickness of 40 percent in this area. These trends indicate longer periods of less and thinner ice cover in the Arctic Ocean and possibilities of increased shipping. Climate models show an acceleration of this trend, with periods of extensive melting higher during spring and autumn. Significant retreat of sea ice during the summer will possibly open up new sailing channels and extend the sailing season. Sailing season is often defined as the number of days in the year when it is possible to sail, i.e. days of less than 50 percent ice concentration.[6]

Although the ACIA projections indicates an increasing length of the navigation season for the NSR,[7] it is more uncertaint when it comes to predicting future ice conditions on the NWP based solely on the output of climate models.[8] As Arctic warming continues, it is likely that much of first-year ice in the Canadian Archipelago will break up much earlier than is currently the case. However, this will likely allow old hazardous ice from the permanent ice pack to enter the NWP in greater quantities, posing a threat to shipping at least to the end of the century (see Chapters 1 and 5).

4.2 Pollution in the Arctic and shipping as an independent source of pollution

The extent of rates of pollutants in the Arctic has been studied for years. A study performed by the Arctic Monitoring and Assessment Programme (AMAP) in 2002, concluded that the Arctic is still one of the cleanest places on Earth as few toxins have been produced, used or stored.[9] Due to its relatively small human population and the general absence of large-scale industrialization, the Arctic region does not represent a major source of pollution. In most places, current levels of pollution are low –[PI] considerably lower than in most urban and industrialized areas in the mid-latitudes.

Locally, pollution from mining, industrial smelters, military activities, and oil and gas development has caused harm or posed potential threats to plant and animal life. However, pollution of international significance is generated within the Arctic regions in only a few areas. The most important of these at present are the highly industrialized Kola Peninsula and White Sea regions of north western Russia, and the large metallurgical and wood processing complexes of north-central Siberia. Each of these areas contributes to circumpolar pollution; the prevailing winds carry airborne pollutants over the central Arctic Basin, and the rivers deliver their contaminants to the Arctic Ocean.

Long-range transport of pollutants from sources outside the Arctic, in the atmosphere, rivers, or ocean currents, is also of concern.[10] One can find traces of pollutants throughout the Arctic environment and the ecosystem (i.e. in the atmosphere, in soils and sediments, in snow and ice, in seawater and freshwater, birds, and animals, including humans). Pollution from lower latitudes is carried into the Arctic by atmospheric circulation and ocean currents. Global atmospheric circulation patterns are such that eastward-moving air masses in northern mid-latitudes become polluted near the surface and then could be carried at moderate or higher elevations to the Arctic regions, where they descend and could deposit their impurities. The main pathways by which airborne pollutants reach the Arctic are over northern Europe and Asia and then across the Arctic Ocean to northern Canada and Alaska, although excursions of polluted air from the industrialized Midwest of North America into northern Canada and the Arctic Ocean are not uncommon. Pollutants reach the Arctic through one major point of entry, the northeast Atlantic. Chemically stable or slow-reacting pollutants from industrialized eastern North America are carried by winds or rivers into the Atlantic Ocean and then northward by the Gulf Stream and North Atlantic Drift into the Arctic Ocean. Augmented by drainage and winds from Europe and by north-flowing Siberian rivers, they are carried under the Arctic ice where they remain protected from sunlight and vigorous oxygenation which otherwise would hasten their chemical break-down. By these means, much of the far-traveled and persistent waste products of the industrialized world appear to be ultimately deposited in Arctic regions. A portion of these, to which may be added materials deposited directly from the atmosphere, becomes incorporated in the upper layers of Arctic Ocean waters and returned to the Northwest Atlantic, where they sink to lower ocean depths and are carried slowly southward as the Atlantic Deep Water Current, eventually spreading at depth throughout the World Ocean.[11]

4.2.1 Ship pollution

Ships can pollute the oceans in many ways, for instance, by spills from oil tankers and chemical tankers, and ejection of sulphur dioxide, nitrogen dioxide and carbon dioxide gases into the atmosphere from exhaust fumes. Discharge of cargo residues from bulk

Vessel Category	Fuel Use (t/y)	CO_2 (t/y)	BC (t/y)	NOx (t/y)	PM (t/y)	SOx (t/y)	CO (t/y)
Bulk carrier	224,799	723,885	191	2,135	632	11,965	3701
Container ship	60,026	193,291	51	5,701	169	3,195	988
Fishing vessel	80,741	259,997	69	7,668	227	4,297	1329
General cargo	120,711	388,708	103	11,464	339	6,425	1987
Government vessel/icebreaker	85,694	275,946	73	8,139	241	4,561	1411
Passenger vessel	12,606	40,592	11	1,197	35	671	208
Special purpose vessel	42,689	137,465	36	4,054	120	2,272	703
Tanker	44,514	14,334	38	4,228	125	2,369	733
Tug and barge	6,925	22,301	6	658	19	369	114
Unknown	37 597	121 068	32	3571	106	2001	619
Total	**716 302**	**2 306 591**	**609**	**68 030**	**2014**	**38 125**	**11 793**

Source: AMSA (2008), p. 163

Table 4.1 Estimated emissions in the Arctic 2004 by ship type.

carriers can pollute ports, waterways and oceans. Ships create noise pollution that disturbs natural wildlife, and water from ballast tanks can spread harmful algae and other invasive species. Studies assessing the potential impacts of international shipping on climate and air pollution demonstrate that ships contribute significantly to global climate change and health impacts through emission of greenhouse gases (e.g. carbon dioxide (CO_2), methane (CH_4) or chlorofluorocarbons (CFC)), aerosols, nitrogen oxides (NOx), sulphur oxides (SOx), carbon monoxide (CO) and particulate matter (PM).[12]

The AMSA has developed an activity-based estimate of Arctic marine shipping emissions using empirical data for shipping reported by all Arctic Council member states. Emissions were calculated for all vessel/trips for which data were available for the base year 2004 (Table 4.1). The 3744 voyages analyzed represent about 7.6 million km of distance traveled (or 4 million nautical miles), with about 4 million km (2 million nautical miles) from non-fishing vessels. Collectively, this represents more than 15,000 voyage days for 2004, with about 3,700 fishing days. Results are potentially an underestimation of current emissions, given probable underreporting bias and anecdotal reports of recent growth in international shipping and trade through the Arctic.[13]

CO_2 is produced from marine vessels as a by-product of the oxidation of carbon in diesel fuel. International shipping emitted approximately 800 million metric tons (MMT) of CO_2 in 2000, contributing about 2.7 percent of global CO_2 emissions that year. Arctic contributions for 2004 amount to about 1 percent of the total ship CO_2 emissions, not an amount that would cause significant effects in the global context.[14]

Black carbon (BC) is a component of particulate matter (PM) and is produced by marine vessels through the incomplete oxidation of diesel fuel. International shipping emits between 71,000 and 160,000 metric tons of BC annually, representing about 15 percent of total PM from ships and about 2 percent of global BC from all sources. In the Arctic region in 2004, approximately 609 tons of black carbon was released. Although this is a relatively low amount as compared to global emissions, as shipping traffic increases in the Arctic, the region-specific effects of BC described above means that even small amounts could potentially have a disproportionate impact on ice melt and warming of the region. More research is needed to determine how mitigation benefits may be associated with BC reductions from ships in sensitive regions. However, the AMSA study found that the heaviest CO_2 and[15] BC emissions were found in the Bering Sea region, around Iceland, along the Norwegian coast and in the Barents Sea. There are also moderate emissions along the western coast of Greenland. However, effective reduction of ship emissions can be achieved through the application of feasible and best available technologies, through air emission techniques and, most importantly, through effective implementation of relevant IMO regulations.[16]

4.3 Arctic ecosystems and the impact by shipping activities

An *ecosystem* is a natural unit consisting of all plants, animals and micro-organisms in an area functioning together with all of the physical factors of the environment. In these systems organisms are interdependent, which means they share the same habitat. Central to

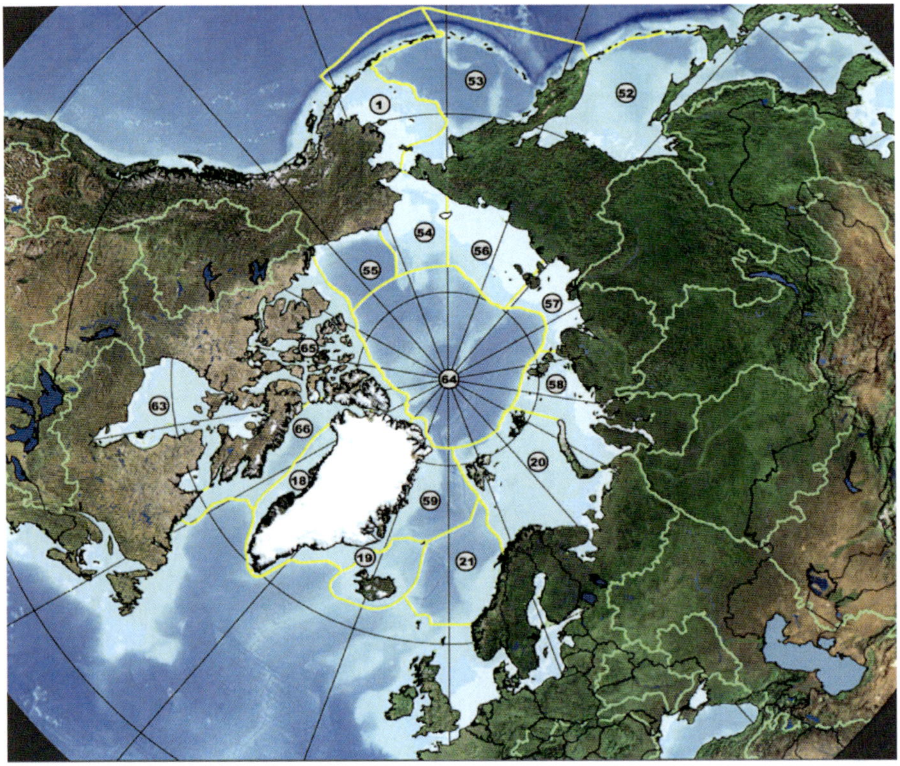

Source: www.lme.noaa.gov/LMEWeb/Publications/tm208.pdf

Figure 4.2 Map of the 17 Arctic LMEs and linked watersheds.

the concept is the idea that living organisms interact with every other element in their local environment. Ecosystems usually form a number of food webs that show the interdependence of the organisms within the ecosystems.[17] The system of Large Marine Ecosystems (LMEs) has been developed to identify areas of the oceans for conservation purposes: "They are national and regional focus areas of a global effort to reduce the degradation of linked watersheds, marine resources, and coastal environments from pollution, habitat-loss and over-fishing",[18] for instance, feeding areas of special importance to seabirds, migration routes for mammals and important fishing areas for indigenous people. LMEs are regions of the world's oceans, relatively large in the order of 200,000 square km or greater, characterized by distinct bathymetry, hydrographic and productivity. The world's 64 LMEs produce 95 percent of the annual marine fishery biomass yields. Most of the global ocean pollution, overexploitation, and coastal habitat alteration occur within their waters.[19] The 17 Arctic LMEs (see Figure 4.2), are various and dynamic systems under stress from global warming and the melting of sea ice.

Arctic marine species are few, but each species has high numbers. Advances in the melting of the Arctic ice have implications for zooplankton, fisheries, fish stocks, marine mammals, and marine birds, which appear to be shifting northward. The number of species generally decreases with increasing latitude. The Arctic flora and fauna, together with the harsh physical environment of Arctic habitats, have resulted in lower species diversity in the Arctic compared to other regions. Conservation of Arctic Flora and Fauna (CAFF) has found that around 3 percent of the global flora and about 2 percent of the global fauna occur in the Arctic.[20]

The Arctic marine environment is in particular exposed to potential impacts from marine activity, such as shipping. Increased shipping in the Arctic could pose a potential threat to the Arctic ecosystem and on the population of species. Many species circulate throughout the Arctic to feed, mate, give birth, take care of their young and moult. As they follow their patterns of living, they're also exposed for various forms of disturbances and implications from shipping activity, whether by the release of substances through emissions to air or discharges to water, accidental releases of oil or hazardous cargo, disturbances of wildlife collisions or the introduction of invasive alien species.[21] As an example, some 280 species migrate and around 450 species are recognized breeding.[22] A number of Arctic marine mammal and bird species are circumpolar, and are represented by several populations and even sub-species (e.g. the bowhead whale, walrus, bearded seal, ringed seal, different gulls, etc.).[23]

Many species that undertake seasonal migrations appear to use environmental conditions as cues.[24] As an example, the migration passages used by mammals and birds correspond largely with the main shipping routes into and out of the Arctic. Presently, there is limited overlap throughout the spring migration as all shipping activity typically will occur later in the year than the animal migration. In the fall, however, there is a larger possibility for interaction between ships and migration species, as both are leaving the Arctic in advance of the formation of the pack ice. The spring migration passages are especially sensitive and exposed for marine mammals and birds. Short feeding seasons mean that if the feeding periods of some animals are disrupted, that animal may not get enough food to survive the winter. These unique characteristics make Arctic species vulnerable to disturbances that shipping activities might lead to.

4.3.1 Marine ecosystems on the Northeast Passage

The following LMEs are defined on the NEP: Barents Sea, Kara Sea, Laptev Sea, East Siberian Sea and Chukchi Sea (see Figure 4.2). According to the NOAA, these LME boundaries refer to a system of conservation purposes only, which means that organisms across the LMEs must be understood as they are functioning together. There are exact delimitations between: (1) the *western part* including the Barents Sea (as part of the NEP) and the Kara Sea (as part of NEP/NSR), and (2) the marginal seas of the *eastern part* of the NEP/NSR including the Laptev, East Siberian and Chukchi seas. Until the mid 1990s there were no data available on the state of ecosystems in winter and spring for the seas running along the NEP/NSR. Based on results of regular monitoring carried out by the Murmansk Biological Institute from Russian nuclear-powered icebreakers navigating the Arctic during the whole ice season, a series of new conclusions on the structure and function of Arctic ecosystems at different tropic levels (plankton communities to marine birds and mammals) has been obtained.[25]

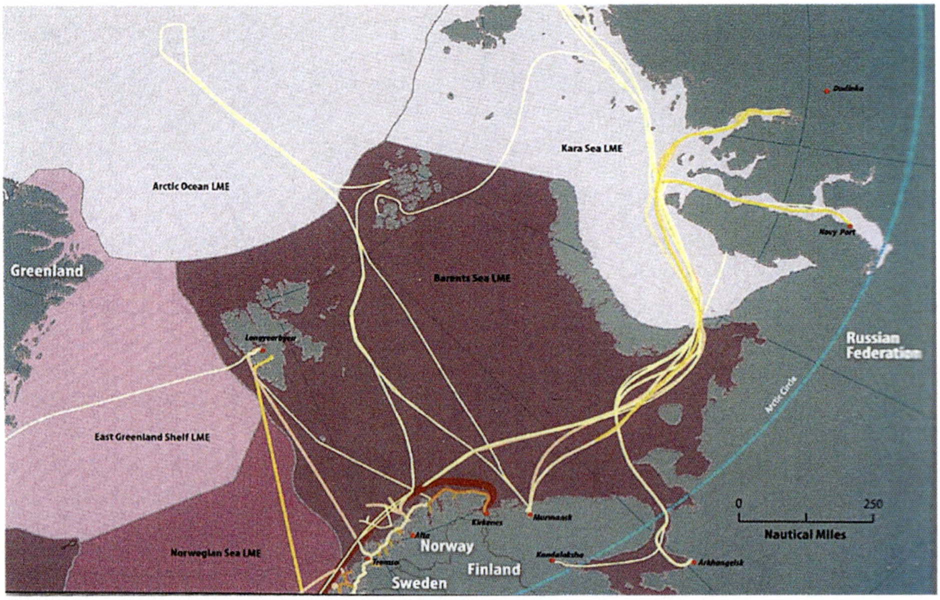

Source: AMSA (2009)

Figure 4.3 NEP/NSR regional traffic and LME map.

Ecosystems on the western part of the Northeast Passage

There is a fairly large volume of shipping in the Barents Sea while the shipping activity is considerable less in the Kara Sea.[26] The Barents Sea LME is considered as a high productivity ecosystem, with biological activity determined mainly by seasonal changes in the temperature and light regimes, advection and ice cover. About 3,000 species have been recorded in the area and it is one of the world's most important fishing areas.[27] Currently, the prevailing maritime activities on the NEP are fisheries in the Barents Sea and cargo shipping. The fish fauna includes about 150 fish species, but less than 30 species are of commercial value. The major species fished are capelin, Atlantic cod and herring, with herring and capelin being major prey of cod. Overfishing has been substantial, but has been declining since 2006.[28]

The main shipping route into the area is along the coast of Norway (see Chapter 1). One main shipping lane goes in inshore waters, and much of the traffic to and from ports in northern Norway follows this route. Traffic to and from Russia follows an offshore route in the open sea to ports in Murmansk, the White Sea and other areas. Transport of oil from Russia is from ports in the White Sea, Murmansk, Pechora Sea (Kolguev, Varandey), and Ob and Yenisei estuaries in the Kara Sea. There is also year-round shipping of nickel ore from Dudinka, a port in the Yenisei estuary (see Chapters 1 and 5). In the western Barents Sea there is a shipping route to Svalbard with seasonal traffic of cargo ships supplying the communities, bulk carriers transporting coal, and cruise ships. There is also a substantial number of fishing vessels that operate year round in the ice-free part of the southern and central Barents Sea, while there is little fishing activity in the Kara Sea (see Figure 4.3).[29]

Vulnerable areas in the Barents and Kara seas have been identified in relation to oil and gas activities, based on where there are aggregations of animals that could potentially be impacted by oil spills or disturbances from activities. However, and apart from oil spill, characteristic disturbance factors (noise, potential collisions with animals, etc.), shipping is not seen as a large threat to the overall ecosystem along the NEP.[30] However, the Barents Sea holds more than 7 million pairs of breeding seabirds, with major colonies on Svalbard, western Novaya Zemlya and along the coast of northern Norway. The southern and eastern Barents Sea is a wintering area for many seabirds and sea ducks that breed further east in the Russian Arctic. The Pechora Sea area and the southern Kara Sea lie adjacent to tundra and wetlands that are important breeding grounds for geese, ducks and shorebirds. Many of these use coastal habitats for staging during spring migration and after breeding when they prepare for the fall migration out of the Arctic. Nevertheless, there is currently no documented evidence of damages to sea species as a result of shipping activities.

Background contamination in the Barents Sea results from transboundary transport of contamination from land-based sources and intensive navigation. Contamination levels in ice free areas of the Barents Sea are significantly lower than those of western and southern European seas. Exceptions are some coastal areas, especially Kola Inlet which houses the largest port and industrial complex in the Arctic (see Chapter 5). High ecological vulnerability is also characteristic of the southeast Barents Sea which receives contaminants from onshore oil and gas production sites located in the Pechora River Basin. Oil and gas production and extraction of chemical, mineral and building raw material have been substantially increasing over the last years (see Chapter 3). Increased offshore oil production presents a potential ecological threat to this part of the Barents Sea. Other maritime activities either do not seriously affect the ecosystem (shipping, other military activities) or have just started developing and are expected to be intensified in the future (oil and gas production on the shelf). Nevertheless, and even if there are no documented negative impacts from shipping activities,[31] shipping is a major activity in the Barents Sea area. Traffic involving fishing vessels and passenger ships could potentially have adverse impacts on the environment through operational discharges to water and air, releases of pollutants from anti-fouling systems, noise and introduction of alien species via ballast water or attached to hulls and local discharges from zinc anodes in ballast tanks. Accidental oil spills have occurred and have been associated with high local mortality of seabirds. However, these incidents have not had material impacts on the population level for the affected species.[32]

Ecosystems on the eastern part of the Northern Sea Route

The number of known species of benthic invertebrates[33] decreases from west to east along the NEP, of which the NSR is the main part. The abundance, diversity, biomass and species composition of benthic invertebrates can be used as indicators of changing environmental conditions. There are more than 2.5 times as many species known from the Barents Sea compared to the Chukchi Sea. This is partly a result of harsher Arctic environmental conditions eastward along the NSR, but also because the benthic fauna of the central and eastern part are some of the least studied animal communities in the world.[34] The fish resources of the NSR play an important role for local communities, but on a global scale fish resources from these Arctic areas are insignificant. The fish fauna are so sparse and difficult to access that no commercial fishing takes place in the open parts of the seas, except from the western Kara Sea and occasionally in the western Chukchi Sea. No offshore fishery takes place in the Laptev Sea. The commercial fisheries of the NSR are restricted to the lower parts of the large rivers and estuaries. However, effects of increased shipping and navigation along the NSR might be both adverse and positive for fisheries. Operational and accidental discharges (such as oil spills) belongs to the first category. Increased sailing can also physically disrupt the fisheries. On the other hand, the NSR could serve as a means for transportation of fish products to markets outside the area, and also ensure supply of fishing gear, equipment, etc., which can facilitate exploitation of fish resources that currently are considered less attractive.[35]

The NSR area comprises many important marine mammals, marine bird species and seabird colonies that use different parts of this area for activities such as nesting, mating, breeding and seasonal migration. During INSROP most of the mammalian species occurring along the NSR were evaluated, focusing on species that could potentially be affected by shipping, and where changes in the populations could occur as a result of such impacts.[36] Polar bear and walrus were the most vulnerable species.[37] Noise, smell and visual impressions from ship traffic, in addition to potential oil spills, could be the largest threat to these species. However, there is no specific case or significant trends from the INSROP study showing permanent damage to these species as a direct result of shipping activities. Although there could be four principal transit routes of the NSR – coastal, marine, high-latitudinal and near-pole routes (see Chapter 1) – most traffic follows the coastal route and the traffic decreases, going from the coastal to the near-pole routes. This indicates that the further from the coast one gets, the smaller amount of marine traffic occurs, with the likely reduced impact on the environment from the marine traffic.

Smaller colonies of seabirds use the mainland coast of NSR for breeding and nesting, but large colonies are also recognized on the Arctic islands, especially on the northeast of Novaya Zemlya, specific parts the New Siberian Islands and Wrangle Island. The most vulnerable period is the time during the breeding period from early April until September, which is also the period when the marine traffic along the NSR is most frequent. Possible oil spills are considered as the main threat to marine bird species.[38] However, there are also positive factors related to shipping and the ecosystems along the NSR. The creation of leads, crevices and turning of ice floes as a result of shipping activity could increase the food availability and have a positive effect on certain seabird populations. Still, the input of this factor is unknown.[39]

4.3.2 Marine ecosystems on the Northwest Passage

The Canadian Arctic is home to a diverse range of wildlife that thrives in a variety of ecosystems. These populations are now under stress in varying degrees due to the changes occurring in the Arctic environment as a result of global climate change. Four of the 17 Arctic LMEs occur in Canadian waters (see Figure 4.2).[40] This includes Hudson Bay where the port of Churchill is located, a prime trading port for destination Arctic shipping and endpoint for the shipping route starting in Murmansk, the so-called "Arctic bridge" (see Chapter 1). The other three LMEs are Baffin Bay/Davis Strait, Arctic Archipelago and Beaufort Sea going through the NWP. A number of geographically restrictive areas are located here used as shipping lanes for various destination reasons. This includes the Bering Strait, the Hudson Strait, the Lancaster Sound and the Aleutian Island chain. This illustrates the use of the Northern Pacific as a connection between the Arctic Corridors (see Chapter 1). Furthermore, these areas are also used by species such as bowhead, beluga and grey whales, fur seals, marine birds and other organisms for seasonal migration. It is where shipping routes travel through these geographically restrictive areas, or chokepoints, that marine mammals are most vulnerable and where there is the greatest potential for conflict between ships and migrating marine mammals. The migration routes used by bowheads and belugas from their wintering areas in the southern extent of the ice and into the high Arctic are also broadly the routes used for destination shipping in the Arctic. In the Bering Strait, for example, the western stock of bowhead whales seasonally migrates through. In this area, they are physically tightened to a relatively small passage and exposed to interactions with vessels transiting this area during the spring and fall. Bowhead and whale migration could potentially also be interrupted by icebreakers; on the other hand, whales could also follow the open leads created by icebreakers.

The Beaufort Sea LME is undergoing rapid changes as annual and multiyear ice coverage shrinks relative to the past 50 years.[41] Increased vessel traffic in these areas is likely to result in greater incidents of pollutant discharges and could increase the risk of disturbance effects such as ship noise and ship strikes on foraging bowheads or other marine mammals. Any vessel incidents in this region would also have the potential to adversely affect major populations of nesting shorebirds, waterfowl and other birds that utilize breeding, nesting and foraging habitats along the coastal Beaufort Sea and along the coast of western Alaska. Neither the Beaufort Sea coast along the northern slope of Alaska nor the Russian side has any infrastructure (see Chapter 5). In case of any incidents in this area, there are no responses, rescues and debris for cleanup.

The North Pacific Great Circle Route between western North America and eastern Asia is a high volume shipping lane that passes through the Unimak Pass in the Aleutian chain (see Chapter 1), passing in close proximity to important marine mammal resting place, rookeries and nesting sites for marine mammals and seabirds. In addition, close to commercial fishing grounds and one of the biggest protected essential fish habitats in the world.[42] The AMSA study calculates that approximately 2,800 vessels passed along this route in 2004. This region seasonally supports populations of shorebirds, nesting sea birds, herring and other marine resources as well as millions of salmon during their migrations to streams of origin. This route passes through the U.S. Alaska Coastal Maritime Wildlife Refuge, which provides nesting and foraging habitat seasonally for millions of seabirds and year-round habitat for thousands of marine mammals. The populations of numerous marine species in this region are declining and in danger of potential extermination. Commercial fisheries in the region, where the Great Circle route passes, provide a large proportion of the annual landings by the US fishing industry. Salmon, halibut, herring, crab, ground fish and many other fisheries are processed annually. In 2004, Alaska fish landings were 2.43 million metric tons, valued at US$ 1.17 billion.[43] This segment of the Northern Pacific Corridor, mentioned above, also passes across two Large Marine Ecosystems (LMEs): the East Bering Sea LME and the West Bering Sea LME.

There are three LMEs of Alaska within the US exclusive economic zone: The Gulf of Alaska , the East Bering Sea and the Beaufort Sea LME. This also illustrates the connection between the NPC and NWP. The East Bering Sea LME includes the Aleutian Islands Sub Area and is the most productive area in the US in terms of fisheries, followed by the Gulf of Alaska LME. Together, these two systems have produced 55 percent of US fisheries landings in recent years. Increased vessel traffic of the eastern route that passes through the eastern Bering Sea will potentially increase interactions with this region's rich fishery resources, fishing communities and numerous fishing vessels and support vessels. Spills due to accidental or illegal discharge from vessels could drift ashore to western Alaska areas where seasonal herring and salmon fisheries occur.

Regulatory restrictions have been around since the 1970s,[44] with the intent to protect local communities and the environment of the Canadian and US Arctic. However, and due to the changing climate, there is now the potential for a significant increase in resource development activity, which would lead to increased destination Arctic shipping. Whether this increase in activity will have severe or minor impacts on the local environment is yet to be determined. However, any increase in activity brings with it a corresponding increase in the risk of injuries to the environment from regular ship operations and accidental or urgent situations. Due to the current relatively low levels of shipping activity occurring on the NWP, any increase in activity in this region will be significant.

Canada is prepared to work in partnership with the US on an LME demonstration project in the Beaufort Sea. The Beaufort Sea LME is a high-latitude, mainly ice-covered LME bordered by northern Alaska and Canada, with an extreme environment driven by major seasonal and annual changes. New developments for the production of oil and gas are considered. It is clear from existing assessments that the resident populations on both the Canadian and the US coasts of the Beaufort Sea LME are undergoing major socioeconomic changes caused by the significant increase in the rate of ice melt in the ecosystem.[46]

4.3.3 Marine ecosystems on the Transpolar Passage

The central Arctic Ocean LME is centred on the North Pole and is bordered by the landmasses of Eurasia, North America and Greenland (see Figure 4.2). The LME has a permanent ice cover that extends seasonally between 60° N and 75° N latitude and is subject to a rapid climate change with the ice cover shrinking in thickness and extent. The National Aeronautics and Space Administration (NASA) reported on 13 September 2006 that, in 2005–2006, the winter ice maximum was about 6 percent smaller than the average amount over the past 26 years. The sea ice extent in September 2007 was about 20–25% below the long-term mean.[47] Low temperatures, ice cover and extreme seasonal variations in light conditions are some of the physical characteristics that slow down biological processes, limit the productivity of Arctic ecosystems and make them more vulnerable to contaminants. The Arctic Ocean primary production strongly depends on the ocean's sea ice cover (SIC). Over the last decade, the Arctic SIC extent and thickness decreased significantly. The SIC area in 2007 and 2008 was 20–25 percent smaller than ever before.[48]

There is a limited number of Arctic species of commercial importance in the Central Arctic Ocean LME. Fish fauna is not well studied, partly because of the lack of commercial fishery, but there a some few observations.[49] For instance, Arctic charr occur throughout the

NWP, and have been sighted further north than any other fish species. In the summer, many stocks of Arctic charr migrate to the sea where they have a larger resource base to exploit and thus are able to grow faster. While at sea, they feed on crustaceans and small fish. Under extreme winter conditions, they hardly feed at all. Sea mammals abound and are still exploited. However, the Arctic LME does include waters seasonally ice free and regularly commercially fished, both in the Northwest Atlantic including Baffin Bay and Davis Strait (connecting the NWP in the east), and in the Northeast Atlantic, north of Iceland and towards Svalbard (connecting the Fram Corridor and the TPP).

In terms of shipping, no commercial cargo ship has yet crossed the central Arctic Ocean.[50] The Arctic Container Project (ARCON) concluded that the Arctic sea ice cover has always been and will continue to be a significant physical barrier to developing a global trade route, even if climate models show a continuous decrease in September ice extent.[51] Currently, the ice conditions are too heavy for the benefits to exceed the expected additional costs. Furthermore, the ice conditions will continue to be heavy during winter and spring seasons, even in 2050, and the route is not expected to be completely ice free in summer.[52] In that respect, there has been little discussion about the environmental impacts caused by shipping on the TPP.

4.4 Effects of oil spill in Arctic waters

Oil in the Arctic marine ecosystem originates mainly from two sources: drilling activity and oil spills during transportation. Drilling activity causes long-term exposure and thus chronic effects on Arctic marine ecosystems, such as changes in species composition, dominance and biomass, while oil spill effects are acute and can cause severe damage locally.[53] However, the oil products, whether they originate from shipping or drilling activities, destroy all aspects of the environmental integrity of the marine ecosystems including fisheries, marine mammals, corals, ocean and shore birds, and the coastal wildlife and thus lead to changes in, for example, behavior (feeding, activity and motility, avoidance reactions, etc.), growth, and reproduction.[54]

Oil spills in ice are more complicated to address than oil spills in open waters. Should this occur in the wrong place at the wrong time, for example in the marginal ice zone, in polynyas, or during the high production period, the impacts could be serious. Shallow waters are the most sensitive to such pollution, and these areas are important to organisms at all levels of the Arctic food chain. Apart from the normally long distances from existing infrastructure, the oil is less accessible in ice-covered waters. The slow rate of biological degradation of oil at near-zero temperatures has led biologists to suggest that oil spills in the Arctic Ocean might remain there for periods of 50 years or more. The dynamics of the ice pack combined with the long life of the oil could allow an oil spill to have a major effect on the albedo in certain regions of the Arctic. It is estimated that the transit time of an oil spill on the fringes of the Beaufort Sea around the circumference of the Beaufort gyre would be about 7–10 years. Several mechanisms act both to diffuse the oil and to put the oil on the surface of the ice. Therefore, as the source continued, the area affected by the spill would grow. By the time the original spill site returned to its original approximate geographic coordinates, a considerable area of the Beaufort Sea could have its albedo changed. The significance of the resultant albedo change for the Arctic heat balance is moot, but most scientists agree that the Arctic environment will require special precautions to minimize the risks of accidental oil spills.[55]

According to the ARCOP study,[56] it seems that long-term effects of oil spills on an open water ecosystem may be less important than generally assumed, and restricted to a relatively small area. However, there is still uncertainty about the effect of oil in ice-covered environments. Oil associated with sea ice may have a notable effect on the ice biota, particularly as oil tends to become trapped between ice floes, and within brine channels in ice, and thus lengthens the contamination time. Even after several years of an oil accident, the recovery process of a littoral or benthic ecosystem may still be incomplete and, in the worst case, may not ever reach the original state.

4.4.1 Oil spill response in the Arctic

Timely and effective response to oil spills requires *containment*, *recovery* and *restoration*. Oil spills in polar seas are especially damaging because the natural conditions severely constrain an effective response. As a part of normal operations, ships produce a range of substances that must eventually be eliminated from them through discharge into the ocean, incineration or by transfer to port-based reception facilities. The MARPOL convention has effectively reduced pollution in the marine environment by regulating the release of regular discharges. The amount of legally discharged oil under MARPOL indicates that oil should not pose a significant threat to the local ecosystem as long as the laws are strictly followed. Currently, there are limited port reception facilities to dispose of oily sludge along the main shipping routes in Arctic waters, i.e., the NEP/NSR and the NWP. As a consequence of this limited port side infrastructure, as well as the cost of disposing of waste using port reception facilities, there are incentives for illegal dumping of wastes produced on board ships.[57]

The risk of accidental release of oil and other contaminants grows with any increase in shipping activity that involves the use of oil or other chemicals. The fact that most recovery equipment and personnel are normally remotely located hinders a timely incident response, and freezing conditions at the incident site further inhibit recovery operations. However, according to the Arctic Council Assessment of Oil and Gas, oil pollution in the Arctic is currently at a low level. This conclusion can hardly have included statistics on oil accidents among cruise ships in recent years. For the time period 1993–2007 cruise ships on polar voyages committed 21 fuel spill incidents. Many of those incidents were the result of serious accidents including sinking, groundings and collisions.[58]

There are currently inadequate techniques for recovering spilled oil in ice-covered environments. The solutions that today are available for oil spill recovery in the Arctic include mechanical methods, bio-remediation, in-situ burning and dispersants.[59] The severity of contamination is not only dependent on the type of organisms exposed to oil, but also on the type and volumes of spilled oil, the speed of oil, weathering processes, oil combating measures taken, and the location of the spill. Whether large oil spills require more time than smaller spills, the use of oil dispersants has usually more severe consequences on organisms than mechanical and/or biological oil combating. On the basis of the extreme conditions of the environment in the Arctic, responding to oil spills is a key challenge along the various sailing routes, especially where ice is present. The AMSA study suggests that strong prevention must be of primary concern, while response measures, being both unreliable and untested, should be secondary. One of the main suggestions proposed was:

> "Prevention of oil spills needs to be the highest priority in the Arctic for environmental protection. Therefore, Arctic states, in collaboration with the industry, must support enhanced funding, research, and technology transfer to prevent release of oil into Arctic waters."[60]

4.4.2 Oil spill response systems on the Northeast Passage

Transportation of oil and gas along parts of the NEP is expected to increase in the years ahead, particularly in the Barents Sea. There will always be a risk connected **to** oil and gas activities and shipping in the Barents Sea. However, shipping contributes considerably more to the overall risk of acute oil pollution than the oil and gas industry. Nevertheless, in spite of an expected increase in the volume of shipping, the implementation of measures such as traffic separation schemes and vessel traffic service centers will reduce the risk of oil spills associated with shipping.[61]

When it comes to responsibility for oil spill contingency operations on the NEP, Russia has agreements with Norway for the Barents Sea and with the United States for the Bering Sea.[62] The Russian oil spill contingency response consists of two subsystems: state and private. The federal system has divided its activities between the Federal State Unitary Enterprises (FSUE) of Murmansk and Sakhalin, the Russian Transport Ministry and their branches in Arkhangelsk and Petropavlovsk-Kamchatski. Operative zones and responsibility management are shared between Murmansk's and Sakhalin's "Basin emergency and rescue managements", and include seaways from the boundary of Norway to the La Pérouse Strait, which includes the NSR, and part of the NPC and the NMC (see Chapter 1). However, the federal plan also requires support by special marine divisions and equipment from other regions: in the western part from the Baltic and Black Seas, and in the eastern part from Nakhodka and Sakhalin. In addition, the plan envisages a request for rendering assistance in accordance with the Intergovernmental Agreement: in the western part of the Arcti: from Norway and Finland, and in the eastern part from the US.

Private systems include response from oil companies and operators of oil terminals on the NEP (i.e. Varandey, Murmansk, Kandalaksha and Arkhangelsk). The establishment of basin emergency and rescue managements and the available equipment used for the elimination of oil spills was mainly carried out in the 1980s and was based on risks existing at that time, when oil in the Arctic was transported episodically and in small volumes.[63] Currently, response systems of a majority of oil companies and operators of oil terminals in the Arctic are developed, but according to domestic legislation the owners and operators are only responsible for oil spill response within the zone of operation. Despite improvements to the equipment, the operative response systems are not sufficient to minimize the risk in such a huge region. In addition, the federal Emergency and Rescue Managements need additional facilities in order to increase the emergency response should any accident should occur. At present, such emergency measures lack the necessary state funding.

The AMSA study's suggestions are clear[64]: Elimination of oil spills at the federal level should be carried out by the joint efforts of several basin emergency and rescue managements, as well as by forces and means of oil companies. Assistance from polar countries may be requested and obtained first of all from Norway and Finland in the west and from the USA and Canada in the east. For the delivery of needed equipment, "Federal plan of the prevention and liquidation of oil spills" enforced in 2003 envisages the use of transport aviation and airports capable of accepting planes in areas close to places of probable accidents. Equipment to the places of spills would be delivered from large airports (Murmansk, Arkhangelsk, Norilsk, Tiksi, Yakutsk and Pevek) by helicopters and sea ships. To ensure the readiness for the oil spill elimination on seaways of the NSR it is necessary to:[65]

- Establish bases for the equipment used to eliminate oil spills in Amderma, Tiksi and Pevek. These bases should include multi-purpose ships capable of being immediately ready to respond to emergency and rescue operations on the NSR.
- To conclude agreements with private companies operating on the NSR about rendering assistance in the event of oil spillage, these agreements clearly specifying terms and financial issues of such assistance.
- To develop federal law on the prevention of the pollution of sea providing for the creation of the trust fund to ensure readiness to the elimination of oil spills.

4.4.3 Oil spill response systems on the Northwest Passage

Currently, a satisfactory implementation of oil spill response on the NWP is absent. This also reflects the lack of adequate infrastructure in the region (see Chapter 5). However, the northern part of Canada is estimated to hold about 30 percent of the remaining potential of oil and gas in the Arctic. Two-thirds of this total is offshore (see Chapter 3).[66] Plans for the development of oil and gas in the Mackenzie delta is one example. An increase in future oil production will potentially lead to increased drilling activity, as well as to an increased volume of shipping of the resources. Both activities will likely increase the risk of oil pollution, but also the need for emergency response.

Legislation such as the Canadian Environmental Assessment Act has introduced standards that impact performance requirements for all aspects of oil and gas exploration and development, including risk for oil spills. Strategies for enhancing Canada's capacity for oil spill response are currently under development. However, in 1977 Canada and the US agreed to establish a joint marine contingency plan for the Beaufort Sea. The plan sets out regional contacts and procedures in case of oil spills and other noxious substances and has periodically been revised, most recently in 2003. However, if any ship accident, which includes oil spill, should occur at this stage, there is no emergency unit with the necessary facilities to perform quick and timely response. One of Canada's foremost experts on the northern areas, Franklyn Griffiths, addresses oil spill response cooperation between the US and Canada as one of the main challenges of the Arctic policy. Currently, the support system of potential oil spill incidents is missing. He argues that the issue is covered by the political discussion about which parts of the Arctic to be controlled by whom. Recently Griffith suggested:

> "Canada should forget about trying to convince the U.S. to concede Canadian control of the passage and instead pursue a policy of conflict-avoidance and partnership-building with the U.S. on an array of Arctic matters ... such as an adequate oil-spill response system."[67]

However, history shows multiple oil spill accidents in the western part of the Arctic. One example, the Exxon Valdez oil spill in Prince William Sound in 1989, is the largest marine spill of the Arctic thus far. It exemplifies consequences to the ecosystem after an oil spill. The vulnerability of sea otters was documented clearly during the Exxon Valdez, where an estimated 2000–3000 animals perished.[69] Twenty years after discharge, there are still some seabird species in the area that are not back to their natural state. Regardless, many lessons were learned from this accident, including the important one that some cleaning methods used to remove the oil led to greater environmental damage than the oil itself.[68] Although the likelihood of oil spills in the Canadian and US Arctic are expected to increase during the next years, no accidents in recent years have shown significant damage to the ecosystems. One example from July 2003 was a 20-tonne container filled with a hazardous chemical used in cement washed off the western coast of Bering Island. Approximately 15 tonnes of the chemical leaked, creating a 400 square meter oily spill. A later survey counted 46 dead birds and one dead seal.[70] This can hardly be claim to be a significant degradation of the ecosystem.

4.5 Variability of natural conditions and shipping possibilities in the Arctic

Climate warming of the magnitude forecast by climate models will result in a sea level rise of 45–50 cm during the 21st century. This rise will come predominately from the thermal expansion of ocean waters as they become warmer. The melting of sea ice will not contribute to level rise as it is floating in the sea and is already displacing its own mass. The melting of sea ice will, however, decrease the overall salinity of the Arctic Ocean. Sea level rise combined with higher storm force winds will pose increased threats associated with storm surges and coastal shore erosion. Nearly all predictions or climate simulations indicate that the sea level will continue to rise beyond this century as a result of the long time lag associated with greenhouse gas. The implications for the Arctic are complex and not yet fully understood by the science community. A review of recent assessments, observations and studies indicates there remains much to understand about the present and future trends in Arctic sea ice. The operating conditions for Arctic ships will remain challenging.

It is likely that Arctic sea ice will become more mobile, especially in the spring, summer and autumn, as the cover continues to retreat from Arctic coastlines. Arctic coastal seas might experience increased ridging of seasonal sea ice, potentially creating more difficult operating conditions of marine navigation. The observed records of sea ice extent on the NSR and the NWP present high inter-annual variability. Projected reductions in sea-ice extent are likely to improve access along these transport corridors. Moreover, projected longer periods of open water are likely to foster greater access to all coastal seas around the Arctic Ocean. While voyages across the central Arctic Ocean will possibly become feasible this century, longer navigation seasons along the Arctic coasts are more likely.[740] The next sections will present the main challenges related to natural conditions combined with shipping on the Arctic transport corridors.

4.5.1 Natural conditions and navigation through the Northeast Passage

The ice conditions along the NEP are dynamic, leading to large annual, seasonal and regional variations and the NEP has traditionally played a great role both by volume of cargo shipped and by its strategic significance (see Chapters 1 and 2). Various parts of the NSR have been in regular use for coastal Russian cargo movements for many years between July and October. However, during recent summers, the trans-Arctic sailing season extended up to 6 months. The 2011 season, for example, started as early as in June and ended in November (see Chapter 5). In winter, the sailing routes are covered with ice, and even in the summer some parts are ice covered, especially in the Laptev Sea and the East Siberian seas. Difficult ice conditions often exclude the use of the shortest route between two points, and lead to the need of ice breaker assistance (see Chapter 5).

Natural conditions and shipping along the Barents and Kara Seas

An important feature of ice conditions in the Barents Sea is that, unlike the Kara Sea and other Siberian Shelf seas, it is never completely ice covered. In the winter, the ice area in the Barents Sea usually comprises 55–60 percent of the total sea area. The constant presence of non-freezing areas in the Barents Sea means that a zero heat budget zone is located in this region in the winter. Based on its geographical location and natural conditions, the Kara Sea is subdivided into two large areas, the north-eastern and south-western, in which ice conditions are poorly related. In the winter, the entire Kara Sea is covered with close drifting ice with fast ice occupying large areas of the coastal areas. Its south-eastern part is covered by ice of local origin with a predominant thickness up to 1.5 meters for 8–10 months a year, whereas in summer it is usually completely ice free. Freeze-up of the south-western Kara Sea usually occurs in October, in the absence of residual ice. The north-eastern part is not completely free of drifting ice during the summer. In the north-western part, young ice forms among residual ice in September. After that and until the summer, the Kara Sea is covered with close drifting ice.[75]

Two ice massifs – Novozemelsky and Severozemelsky – are located in the western area of active shipping, which is fewer compared to the eastern part of the NSR with its five ice massifs. The size of the ice massifs comprises several hundred square kilometers at maximum development. Usually, by late June the Novozemelsky ice massifs occupies about 80 per cent of the south-western Kara Sea (see Figure 4.6). By mid August, the areas of the Novozemelsky massif and close ice in all regions are usually less than 20 percent and in September less than 10 percent.[76] The Severozemelsky ice massifs are identified in the north-eastern Kara Sea along the Taymyr coast and Severnaya Zemlya. In May, fast ice and close drifting ice are located over the entire area. On average, in September it is less than 10 percent and 30 percent in the western and eastern Severozemelsky regions.[77]

The *Kara Gate Strait* is the main shipping lane between the Barents and Kara Seas, connecting the western part of NEP with the NSR (see Chapter 1). The 18 nautical miles strait has a minimum depth of 21 meters and shipping uses an established traffic separation scheme. Maximum fog and low visibility occur during July and August. In winter, the Kara Gate Strait is covered with drift ice. The duration of the sailing season without icebreaker escort is 3–3.5 months.[78] The *Yugorskiy Shar Strait* is located along the south coast of Vaygach Island and is the southernmost entrance from the Barents to the Kara Seas, also connecting the western part of NEP with the NSR (see Chapter 1). It is 21 nautical miles in length, 5.5 nautical miles wide at its narrowest and has depths ranging from 13 to 15 meters in the west and 16 to 30 meters in the east. Hazardous sand shoals are located in the middle and eastern regions. Ice is present in the strait from November to July. The duration of navigation without icebreaker assistance is between 3 and 3.5 months.

Natural conditions and shipping along the Laptev Sea

The Laptev Sea has the largest expanse of fast ice in the world from January to June (see Figure 4.5). The fast ice thickness typically reaches 200 centimeters due to mean mid-winter air temperature of minus 30°C and can grow up to 250 centimeters during severe winters. The concentration of multiyear ice in the Laptev Sea is limited due to wind directions and ocean currents. The total area of summer melt is particularly extensive due to the reduced concentration of multiyear ice. In the western part, the ice drift is southwards and large masses of ice are deposited along the coast of Severnaya Zemlya and the Taymyr Peninsula. Along with the eastward ice deposition from the Kara Sea, the *Vilkitskii Strait* and the Taymyr coast present a serious challenge to navigation at all times of the year.[79] The Vilkitskii Strait is the main connection (60 nautical miles) between the Kara and Laptev Seas and a key strait along the NSR. With its depth of 100–200 meters, ships of any size and draft can pass through. However, the eastern and western approaches to the Vilkitskii Strait can be clogged with ice fields, depending on wind direction; icebreaker escort can be complicated. Fast ice can be present in the strait until August. The *Shokalskiy Strait* is located in Severnaya Zemlya north of the Vilkitskii Strait, and is a second shipping route between the Kara and Laptev Seas. The 80 nautical miles strait (10 nautical miles wide at its narrowest point) has a minimum depth of 37 meters and depth will not be a limiting factor for any ships. Still, ice conditions in the Vilkitskii Strait are generally more favorable.[80]

Natural conditions and shipping along East Siberian Sea

The East Siberian Sea is the shallowest of the Eurasian seas. The broad continental shelf allows fast ice, averaging from 170–200 centimeters thick, to extend as far as 500 kilometers outward from the coast (see Figure 4.5). In winter, the prevailing wind direction is from the south, producing weak ice conditions and potential navigation lanes at the outer edge of the fast ice like in the Kara and Laptev Seas. The East Siberian Sea has the highest fraction of old ice and the Ayon massif has more than 60 percent of old ice and the average thickness may be 250 centimeters in the winter months. The eastern East Siberian Sea is less ice-free than the other regions. In the summer, the winds shift to northerly and the ocean current favors the influx ice from the north resulting in the permanence of the Ayon massif. Ice melting in the East Siberian Sea begins only under the influence of the solar radiation heat and advection of atmospheric heat and about 50 percent of the East Siberian Sea is ice free before the onset of ice formation.[81] Winter freeze-up begins in the north in September and is usually complete by mid-October. The *Dmitri Laptev Strait* is the southernmost passage between the New Siberian Islands and the mainland, linking the Laptev and East Siberian Seas. It is 63 nautical miles long, 30 nautical miles wide and oriented east–west. The strait has a depth of 12–15 meters; however, the eastern approach has depths of 10 meters or less, restricting traffic to ships with less than a 6.7-meter draft. The *Sannikov Strait* is a second

Source: Barents Observer

Figure 4.4 The route north of the New Siberian Islands (yellow line).

passage through the New Siberian Islands linking the Laptev and East Siberian Seas. It is 160 nautical miles long and a minimum depth of 13 meters. From a navigation point of view, the low surrounding New Siberian Islands make visual and radar observations difficult to obtain, particularly during long periods of reduced visibility.[82] During the last two summer seasons (2010 and 2011), tankers owned by the Russian oil and gas company Novatek have been sailing through the NSR – to the north of the New Siberian Islands. Using this route makes it possible to navigate tankers with draft of 12–15 meters.[83]

Natural conditions and shipping along the Chukchi Sea

On the eastern part of the NSR, the southern Chukchi Sea becomes ice free earlier than the other regions. The seasonal variations in the ice conditions are large, resulting in the loss of about 80 percent of its maximum winter extent in the summer season. Important factors influencing the variability are the bathymetry, wind, currents, air temperature and the presence of Wrangel Island. Due to advection of comparatively warm Bering water, up to 10 percent of ice in the south-western sea area melts as early as late May, on average. The Chukchi Sea is almost ice covered from early December to mid-May. However, ocean currents and wind tend to transport old ice from the Arctic to the Long Strait under great pressure, which sometimes presents the greatest obstacle on the route.[84] *Long Strait* separates Wrangel Island from the mainland linking the East Siberian and Chukchi Seas. Two general routes through this wide (75 nautical miles) passage are described: a 120 nautical miles southern route with 20 meters minimum depth, and a 160 nautical miles northern route with 33 meters minimum depth. Draft will not be a problem as long as ships keep

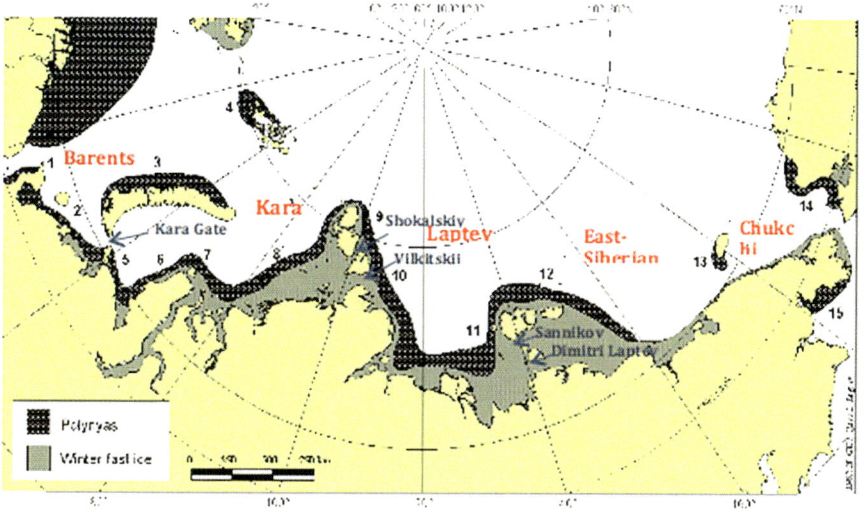

Source: Ragner (2000)

Figure 4.5 Regions of winter fast ice and polynyas on the NSR.

well offshore from the mainland. During winter, fast ice extends. Hazardous ice conditions for navigation can occur when old ice and ice massifs are displaced into the strait region.[85]

Although there is some projected winter retreat of the sea ice edge, particularly in the Bering and Barents Seas where ship access is likely to improve throughout the 21st century, most of the Arctic Ocean is projected to remain ice-covered in the winter. Sailing during the winter season (November–May) is generally much more difficult than in the summer season, due to the thicker ice cover. An important, special feature of winter navigation is the fast ice – stable, immovable ice which is "clinging" to the coastline. Depending on the location of islands and sea depths, the fast ice may extend up to 500 km from the Russian mainland. Fast ice is very difficult to pass through, and normally it is preferable to avoid it by using northerly routes. If offshore winds prevail, during winter one will often find open leads at the edge of the polynyas, which are very suitable for navigation. Typical locations of fast ice and polynyas are illustrated in Figure 4.4. In that respect, the Kara Sea offers the easiest conditions for navigation. Here, the extension of the fast ice is normally small, but the existence of polynyas is also relatively normal. The only areas where polynyas are not often found are the eastern East Siberian Sea and the Chukchi Sea. This is again mainly due to currents pushing ice from the central Arctic Basin towards the coast, thus creating extremely difficult ice conditions.

The fact that both Severnaya Zemlya and the New Siberian Islands normally become enveloped by the fast ice will often force ships to choose a route north of these archipelagos, routes which may expose the ships to extremely harsh ice conditions, in addition to being a detour. The summer season has traditionally been defined as June–October, but

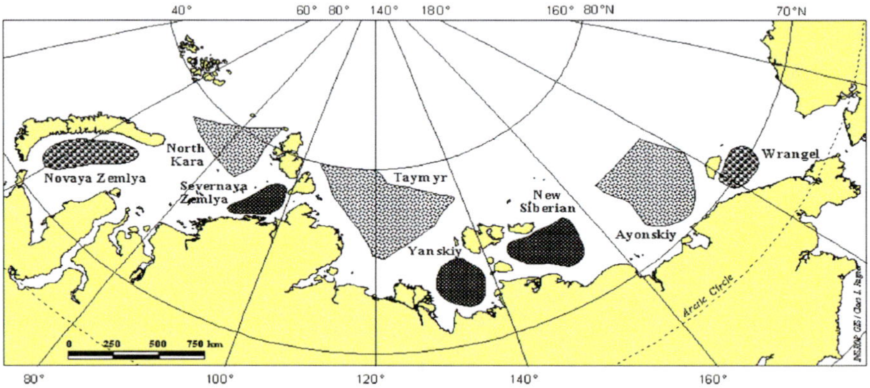

Source: Ragner (2000).

Figure. 4.6 General locations of summer ice massifs on the NSR. Noted are three types of summer ice massifs: Multiyear ice from the Arctic Ocean (North Kara, Taymyr and Ayonskiy); local drift ice (Novaya Zemlya and Wrangel); and fast ice remnants (Severnaya Zemlya, Yanskiy and New Siberian).

technological improvements have gradually allowed an extended summer season, with navigation until December in years of favorable ice conditions. Today, it is only shipping on the Dudinka-Murmansk line that operates on a year-round basis. All other transport is carried out only during summer.

4.5.2 Natural conditions and navigation through the Northwest Passage

Sea-ice conditions on the NWP are very complex. Observations of minimum sea-ice extent in the eastern and western regions of the NWP illustrate the extraordinary inter-annual variability of the ice conditions. Although the trends in sea-ice extent are negative in both regions over the period, the year-to-year variability is extreme and sometimes differs between the two regions. For example, one of the largest observed minimum extents in the western region (1969–2003) occurred in 1991, while in the eastern region the minimum sea ice extent that year was relatively low. While these observations indicate a recent overall decrease in the extent of sea ice in the NWP, the inter-annual and spatial variability is not conducive to planning a reliable marine transportation system.[86] Although climate models for the Arctic, like the ACIA models, indicate a general retreat of sea ice throughout the 21st century, the designated projection models provide no information on ice thickness (a critical factor for navigation). It is important to note that the horizontal resolution of these models (generally 200 kilometers) is not fine enough to take into account the complexity of the Canadian Archipelago. Hence there is more uncertainty when it comes to predicting future ice conditions in the NWP based solely on the output of climate models.

There are potentially seven routes through the Canadian Archipelago, of which three are considered as being practical for routine marine traffic. This includes: (1) M'Clure Strait,

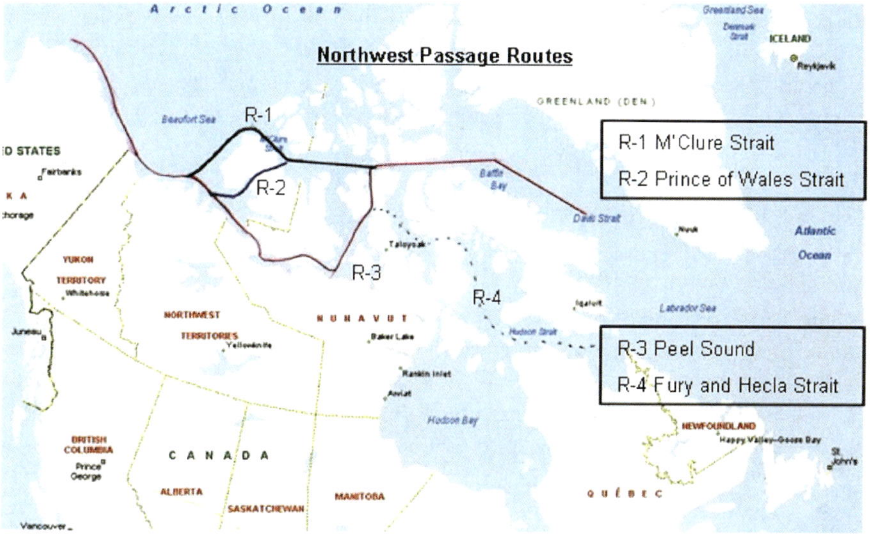

Source: CASA (2007)

Figure. 4.7 Route alternatives on the NWP.

(2) Prince of Wales Strait and (3) Peel Sound. The fourth, Fury and Hecla Strait, is less practical (see Figure 4.7). Although much has been written and said about the impact of climate change and that an ice-free Arctic will likely occur, it is important to note that such ice-free conditions are not likely to occur prior to 2070. Even then, the Arctic will continue to be ice-covered during the winter and large seasonal, annual and year-to-year variations will continue to occur. Ice conditions will continue to impede shipping through a major portion of the year. On the NWP, large quantities of drifting ice will continue despite Arctic warming, and shipping through the Northwest Passage will remain risky.

Natural conditions and shipping through the M'Clure Strait

The first NWP route passes through (from east to west): Davis Strait, Lancaster Sound, Barrow Strait, Viscount Melville Sound, M'Clure Strait, Beaufort Sea, Chukchi Sea, Bering Strait and Bering Sea. This is the shortest and deepest route with no draft restrictions, but is the most difficult due to the severe ice in the M'Clure Strait (120 kilometers wide at the east end, 275 kilometers to the Beaufort Sea and depth at over 400 meters).

Old, hard and dangerous ice is present most of the time, which can seriously hamper progress and is potentially damaging even to ice-strengthened ships. Projections carried out by the Canadian Arctic Shipping Assessment (CASA) anticipates that by 2020, the frequency of an open summer passage will increase.[87] Even in those years when the M'Clure Strait is passable there is a risk of encountering old/multiyear ice. It is likely that this passage will remain ice packed most years. Since the M'Clure Strait is considered to be the constraining point for this route, the historical minimum ice period data that are provided indicate that significant multiyear ice is the norm. The trend for the period 1968–2006 period indicates a gradual decrease in ice coverage. However, little change is expected before 2050.[88] This route, based on the Zone/Date System, is closed to vessels below Arctic Class 3 which can operate there between 20 August and 30 September (see Appendix, Chapter 5). In a warm year, the season could be extended to an estimated 10–15 days. During a cold year it would likely remain closed.[89]

Natural conditions and shipping through the Prince of Wales Strait

The second NWP route passes through (from east to west): Davis Strait, Lancaster Sound, Barrow Strait, Viscount Melville Sound, Prince of Wales Strait, Amundsen Gulf, Beaufort Sea, Chukchi Sea, Bering Strait and Bering Sea. This is an easier deep draft alternate route which avoids severe ice in the M'Clure Strait. However, the passage through the Prince of Wales strait (minimum width of less than 10 kilometers about half way through the strait, 230 kilometers long) has an average depth of 32 meters which tends to be the limiting factor. Normally, this strait is open during September, but there continues to be a threat of encountering some old ice. Projection based on certain changing climate indicators anticipates that by 2020 this route could be open for 8–10 weeks during some summers. Large inter-annual variability will continue and icebreaker escort services will be required. There could be an increase in the presence of old ice, as the decrease in land-fast ice in the western part of the Archipelago would allow more old ice from the Arctic Ocean to pass into channels between the islands. The historical minimum ice data that are provided indicate that significant multiyear ice is the norm. There is a stronger decreasing trend than that of the M'Clure Strait for the same period (1968–2006). The *SS Manhattan* succeeded a westerly transit of this route in 1969. The potential shipping season through this passage will continue to increase through 2050 but the threat of encountering old ice during most years will be significant.-[90]

Natural conditions and shipping through the Peel Sound

The third NWP route passes through (from east to west): Davis Strait, Lancaster Sound, Barrow Strait, Peel Sound, Franklin Strait, Victoria Strait, Coronation Gulf, Amundsen Gulf, Beaufort Sea, Chukchi Sea, Bering Strait and Bering Sea. This is the longest transit and the most frequently used. Compared to sailing through (1) James Ross Strait – Rae Strait – Simpson Strait, and through (2) Prince Regent Inlet – Bellot Strait, the sailing route through Peel Sound is considered the best option.[91] However, navigation of this route is more challenging and the route is limited to ships having a draft of less than 10 meters. Currently this route is passable from mid August to mid September. It is expected that the active shipping for this passage will continue to increase. However, old ice will continue to be a significant navigation hazard and there will continue to be large year-to-year variability. Data from the period of 1968–2006, for conditions at or near the ice minimum, indicate the presence of old ice throughout the shipping season for the Peel Sound. The trend indicates a slight increase in the amount of ice present.[92] Historical data for the ice choke point of this route, which includes the Larsen Sea and the Victoria Strait, show a higher concentration of old ice, which constrains access to the entrance to the Western Arctic Waterway of the Northwest Passage. Once again, there has been a decreasing trend over the last 36 years. Even though this decreasing trend is expected to continue, significant amounts of old ice will persist through 2050 and caution will be required.[93]

Natural conditions and shipping through the Fury and Hecla Strait

The fourth route connecting the Arctic Bridge and the NWP (see Chapter 1), passes through Hudson Strait, Foxe Basin, Fury and Hecla Strait (160 kilometers long), Bellot Strait, Franklin Strait, Victoria Strait, Coronation Gulf, Amundsen Gulf, Beaufort Sea, Chukchi Sea, Bering Strait, and Bering Sea. This route also has a draft limit of 10 meters. It has had light ice years but it will likely remain difficult through 2020. Foxe Basin and the Gulf of Boothia can have significant amounts of ice through mid August. Severe ice conditions at and in the vicinity of the western approaches to Fury and Hecla can be expected. Bellot Strait will generally be passable during the later part of August and through mid September but strong tidal currents will continue to be a factor for both the movement of ice and ships. By 2050, the frequency of ice free conditions throughout this route will increase. However, it is not likely that this route will be of interest to those shipping activities requiring reliable transit times as the probability of requiring an icebreaker escort is much higher than other NWP alternatives. The decreasing trend for this area is more pronounced than the other three routes, but it is clearly evident that significant amounts of old ice are likely to be present most years. Consequently, Route 4 is going to be a challenge most years for a Type B vessel. In addition to frequent congestion in Fury and Hecla Strait, there are potential ice problems in both Bellot Strait and Franklin Strait. Based on the Zone/Date System the season would begin 25 August and close 30 September. During a warm year the closing could be extended by almost 30 days (see Appendix, Chapter 5). During a cold summer season it would likely remain impassable for most vessels.[94] This route is not generally considered as a viable commercial passage for moderate-to-deep draft ships.

Natural conditions and the shipping route to Churchill

The port of Churchill is not a part of the NWP. However, it is an important destination of cargo from international markets, for instance, as the west endpoint of the Arctic Bridge. As well as an important destination point when considering intra-Arctic, Destination-Arctic and Transit route alternatives (see Chapter 5). The shipping route to Churchill is by the Labrador Sea, through the Hudson Strait and then directly across the Bay to Churchill or, alternatively, just south of Southampton Island south-westward toward the western shoreline and then south into Churchill. Usually the route through Hudson Strait opens around the beginning of July, but it can be delayed by almost a month during colder years. Churchill and the western shore of the Hudson Bay are generally open by the first week of July but can be delayed up to four weeks in colder years, or if the prevailing westerly winds are absent. Over the last decade this area has had generally lighter ice conditions than in the previous two decades with navigation conditions that would warrant an extension to the closing of the shipping season by to to three weeks. Climate change will likely increase the probability of a two to three week extension to the closing date, but it is unlikely that there will a significant impact on the opening date for this route through 2020. By 2050, the frequency of an earlier opening date will increase. Large inter-annual and year to year variations will continue throughout the first half of the 21st century.[95]

4.5.3 Natural conditions and navigation through the Transpolar Passage

When considering the natural characteristics and shipping on the TPP, the central Arctic Ocean has no draft limitations, narrow straits or complex archipelago to put constraints on navigation. Nonetheless, the sea ice conditions are the key factor of limitation in terms of any potential commercial shipping routes. No commercial cargo ship has yet crossed the central Arctic Ocean. There are huge uncertainties as well as variations between different climate models trying to predict the development of ice conditions. In terms of any regular shipping on these routes, however, DNV's Arctic Container Project (ARCON) simulations indicate that the ice will be too heavy and the calculated costs too high for any regular transport. Models indicate that the ice conditions will continue to be heavy during winter and spring seasons, even in 2050, and the route is not expected to be completely ice free in summer.

4.6 Some concluding remarks

Chapter 7 of this book will present the comparison of the environmental challenges of the three transportation corridors. Nevertheless, it is appropriate to end this chapter with some considerations.

Firstly, no significant negative impacts on the Arctic ecosystems caused by shipping activities have been documented. Moreover, effective reduction of ship emissions can be achieved through the application of feasible and best available technologies and, most importantly, through effective implementation of relevant IMO regulations.

Secondly, there is generally not a sufficient operative oil spill contingency response in the Arctic. The general lack of preparedness, adequate equipment and facilities for oil spill elimination makes the ecosystems highly vulnerable should any accidental oil spill occur.

Thirdly, a set of detailed natural characteristics makes each corridor unique, as well as more or less favorable in terms of shipping operations. As demonstrated, both the NEP and the NWP face large water depth variations and a complex archipelago which put constraints on the technical construction and capacity of the vessels operating on these corridors. Furthermore, the sea ice conditions vary significantly both between and within each of the three passages, presenting different challenges to navigation. Despite significant variation within each transport corridor, the NEP seems to offer more favorable navigation conditions than the NWP. Furthermore, it is not very likely that the TPP is going to be used as a regular transport corridor. The ice models indicate that the ice conditions will be too heavy for any commercial shipping.

Notes

1 WWF (2008), p. 5
2 IPCC (2007)
3 In geology, permafrost or permafrost soil is soil at or below the freezing point of water (0°C for two or more years). http://en.wikipedia.org/wiki/Permafrost
4 ACIA (2005)

5 ACIA (2005), p. 935
6 ACIA (2004)
7 ACIA (2004)
8 CASA (2007)
9 AMAP (2002)
10 ACIA (2005), p. 564
11 http://www.carc.org/pubs/v18no3/1.htm
12 AMSA (2009), p. 141.
13 AMSA (2008).
14 AMSA (2009), p. 142
15 AMSA (2009), p.144
16 AMSA (2009), p.153
17 http://en.wikipedia.org/wiki/Ecosystem
18 Defined by the US National Oceanic and Atmospheric Administration (NOAA) and the Arctic Council
19 http://en.wikipedia.org/wiki/Large_marine_ecosystem
20 Norsk Polarinstitutt (2009)
21 Østreng (1999a), p.163
22 ACIA (2005), p. 1016
23 ACIA (2005), p. 624
24 ACIA (2005) p. 496.
25 Norsk Polarinstitutt (2009), p. 21
26 AMSA (2009), p.144
27 Norsk Polarinstitutt (2009), pp. 44 and 48
28 Brigham (2007)
29 AMSA (2009), p. 144
30 AMSA (2009), INSROP (1998) WP 99, and Norsk Polarinstitutt (2009)
31 AMSA (2009), p. 146
32 Norsk Polarinstitutt (2009), p. 45
33 Benthic invertebrates are organisms that live on the bottom of a water body (or in the sediment) and have no backbone, and comprise several types of feeding groups, e.g. deposit-feeders, filter-feeders, grazers and predators; www.ozcoasts.org.au/indicators/benthic_inverts.jsp
34 INSROP (1998a), p. 17
35 INSROP (1998a), p. 15
36 INSROP (1998a), p. 45
37 INSROP refers to the term "*Valued Ecosystem Components*" defined as: "*Any part of the environment that is considered important by the proponent, public, scientists and government involved in the assessment process. Importance may be determined on the basis of cultural values or scientific concern*" http://a100.gov.bc.ca/appsdata/epic/documents/p239/d21939/1151444602531_f314149f768a4109abd3ab91484841ea.pdf
38 Ibid
39 Ibid, p. 29
40 AMSA (2009), p. 149
41 Norsk Polarinstitutt (2009), p. 107
42 AMSA (2009), p.136
43 AMSA (2009), p.143
44 In 1970, the Canadian government enacted the *Arctic Waters Pollution Prevention Act (AWPPA)*, which asserts Canadian regulatory control over pollution within a 100-mile zone

45 Library of the Canadian Parliament (26 January 2006), http://www.parl.gc.ca/information/library/PRBpubs/prb0561-e.htm
46 Norsk Polarinstitutt (2009), p. 106
47 NASA (2006)
48 http://www.lme.noaa.gov/index.php?option=com_content&view=article&id=110:lme64&-catid=41:briefs&Itemid=72
49 http://www.lme.noaa.gov/index.php?option=com_content&view=article&id=110:lme64&-catid=41:briefs&Itemid=72
50 AMSA (2009), p. 109
51 Mejlander-Larsen et al. (2009)
52 Mejlander-Larsen et al. (2009)
53 ARCOP (2006), p.4
54 INSROP, WP 76 (1997)
55 http://www.nap.edu/openbook.php?record_id=12024&page=119
56 ARCOP (2005), p.25
57 AMSA (2009), p. 141
58 AMSA (2009)
59 AMSA (2009), p. 138
60 AMSA (2008), p. 4
61 Norsk Polarinstitutt (2009), p. 48
62 INSROP WP 76 (1997b), p. 10
63 INSROP WP 76 (1997b)
64 AMSA (2008), p. 199
65 AMSA (2008), p. 200
66 Norsk Polarinstitutt (2009), p. 83
67 Canwest News Service (August, 2008)
68 http://www2.canada.com/topics/news/features/arcticambitions/story.html?id=de0a569c-1ec6-478c-9fd0-a45622e2fe9a
69 Waldichuk (1990
70 Aleut International (2008),
71 www.aleut-international.org/files/Download/2008_1st_2nd_Quarter%20Newsletter.pdf -
72 Aleut International (2008),
73 www.aleut-international.org/files/Download/2008_1st_2nd_Quarter%20Newsletter.pdf -
74 ACIA (2005), p. 934
75 Johannessen et al. (2007), p. 253–254
76 Johannessen et. al. (2007), p. 258
77 Johannessen et. al. (2007), p. 262
78 Brigham (1999), p. 49
79 INSROP WP 99 (1998), p.12
80 Brigham (1999), p. 50
81 Johannessen et al. (2007), p. 281
82 AMSA (2009), p. 23
83 http://barentsobserver.com/more-traffic-along-the-northern-sea-route.4947500-16149.html
84 INSROP, WP 99 (1998), p. 12
85 INSROP, WP 2 (1994)
86 ACIA (2005), p. 924
87 CASA (2007), p. 27
88 http://ice-glaces.ec.gc.ca/IceGraph/IceGraph-GraphdesGlaces.jsf
89 http://ice-glaces.ec.gc.ca/IceGraph/IceGraph-GraphdesGlaces.jsf

90 AMSA (2009), p. 21
91 AMSA (2009), p. 21
92 http://ice-glaces.ec.gc.ca/IceGraph/IceGraph-GraphdesGlaces.jsf
93 http://ice-glaces.ec.gc.ca/IceGraph/IceGraph-GraphdesGlaces.jsf
94 http://ice-glaces.ec.gc.ca/IceGraph/IceGraph-GraphdesGlaces.jsf
95 http://ice-glaces.ec.gc.ca/IceGraph/IceGraph-GraphdesGlaces.jsf

5 Shipping and Arctic infrastructure*

The purpose of this chapter is to examine the state of available infrastructure significant for shipping on the NEP , the NWP and the TPP. The focus will be on six aspects:

1. Technical standards and requirements for ships operating along the Arctic transportation corridors.

2. Training and operational requirements for crew and ice navigators operating in Arctic waters.

3. Availability of relevant information, communication and support in order to provide safe navigation.

4. Russian and Canadian Arctic ports and port services of importance for Arctic shipping

5. The costs and bureaucratic procedures of Arctic shipping

6. Availability of adequate and appropriate risk coverage for vessels navigating in Arctic waters

5.1 Technical requirements for ships operating in the Arctic

Among a large number of international codes and conventions,[1] the *Convention on Safety of Life at Sea*, 1974 (SOLAS) is regarded as the most important of all international conventions concerning the safety of merchant ships capturing all matters relating to international

* This chapter has been written by Karl Magnus Eger, Ocean Futures. Inputs on technical standards have been provided by Morten Mejlænder-Larsen, DnV. Inputs on costs and bureaucratic procedures have been provided by Tor Wergeland, Ocean Futures (affiliated).

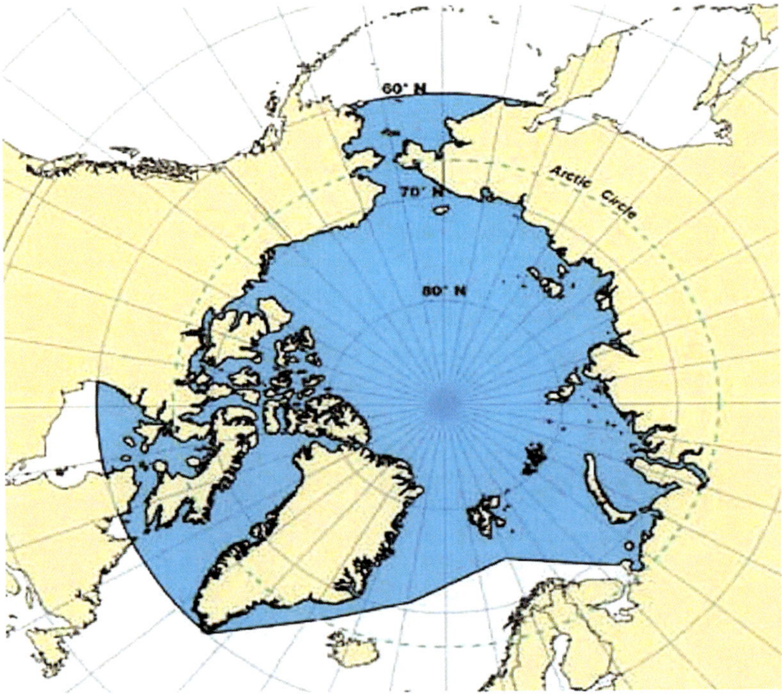

Figure 5.1 Geographical scope of the IMO Guidelines for ships operating in Arctic ice-covered waters.

standards for safety and security at sea in all marine regions.[2] The primary purpose of the convention is to specify minimum standards for the construction, equipping and operation of ships, consistent with their safety. Flag States are responsible for making sure that ships under their flag comply with the requirements, and certificates are approved in the Convention as proof that this has been done.[3] Control provisions allow parties to inspect vessels of other parties if there are clear grounds for believing that the ship and its equipment do not comply with the requirements. Before the convention becomes binding upon states which have ratified it, individual contracting states must formally accept it into their national maritime regulatory regime. Currently, all the Arctic states have ratified the SOLAS Convention.

The IMO Guidelines: Applicable to technical standards in Arctic waters

The SOLAS Convention applies to ships engaged on international voyages, but does not specify any additional construction regulations and equipment required for operations in Arctic waters. However, an important initiative for such regulations took place within the International Maritime Organization (IMO) in the early 1990s, driven by the disaster of Exxon Valdez off the coast of Alaska in 1989.[4] The IMO recognized the need for recommendatory provisions related to ships operating in Arctic ice-covered waters in addition to the requirements contained in existing IMO instruments. In 2002, the *Guidelines for Ships Operating in Arctic Ice-Covered Waters* (the IMO Guidelines) were published. Member governments were invited to bring the IMO Guidelines to the attention of ship owners, ship designers, ship builders, ship repairers, equipment manufacturers, installers and all other parties concerned with the operation of ships in Arctic ice-covered waters. The use of such non-binding instruments within international law of the sea is increasing (see Chapter 6).[5] Moreover, adopting regulations as voluntary rules avoids the need for lengthy procedures and the risk of getting only a limited number of ratifications. Nonetheless, with regard to the IMO Guidelines, it proved to be just as difficult and time consuming to gain universal consensus for a treaty.[6]

Part A of the IMO Guidelines provides safety recommendations for ships operating in Arctic conditions, additional to existing regulations, in particular the SOLAS Convention,[7] as well as other approved industry guidelines.[8] While the SOLAS Convention is formally integrated as part of national regulations, the IMO Guidelines are provided as non-binding regulations. Nevertheless, the IMO Guidelines provide an opportunity to assess and strengthen guidance of construction and equipment for ships operating in Arctic waters.

At present, the sphere of validity of the IMO Guidelines covers the Arctic waters, north of 60°N only. Yet, the Guidelines define *Arctic ice-covered waters* as an area "(...) in which sea ice concentrations of 1/10 coverage or greater are present and which pose a structural risk to ships" (see Figure 5.1 and introductory chapter).[9]

Furthermore, in the context of ship operations, the IMO Guidelines relate to ships operating in Arctic ice-covered waters.[10] The IMO Guidelines, however, provide recommendations for ships in those parts of Arctic waters where ice cover makes navigation complicated. "Ship" is defined in Paragraph 3.23 as "any vessel required to comply with the 1974 SOLAS Convention".[11] This excludes fishing vessels, pleasure yachts, wooden ships of primitive build, cargo ships of less than 500 gross tonnage and naval vessels, whereas passenger ships and cargo ships of 500 gross tonnage or more are subject to the regulations.[12] In the next section we give a presentation on the recent progress of the IMO Guidelines.

The IMO Polar Code

During a meeting in London in 2008, several states agreed to cooperate in order to make the IMO Guidelines a mandatory tool. Currently, this is an ongoing process under review by the Maritime Safety Committee (MSC) of IMO and in light of rapidly developing technology and changes in the characteristics of many traditional ship types and equipment used in marine industrial operations. The initiative was approved as a new agenda item by MSC 86 in May 2009 and the work is coordinated by the Sub-Committee on Ship Design and Equipment (DE). The working group meets during the regular DE sessions and corresponds in between. A target completion date was originally set for end 2012, but it is assumed that some more time will be needed.

The purpose of the code is to:

> "... provide for safe ship operation and the protection of the Polar environment by addressing risks specific for the Polar regions and not explicitly considered by other instruments of the Organization."

The agreed principles of the code made at the DE 53 meeting are to make an add-on code with additional requirements to existing IMO legislation, and the code shall be risk based with functional requirements supported by deterministic requirements where necessary. The code shall ensure the same level of safety for persons, the environment and the ships in Polar waters as in other waters. A correspondence group was formed to progress the work.

The status after DE54 is an agreement of the basic structure of the code and a preliminary agreement around the geographical boundaries of application. A Hazard Identification of the additional Polar Hazards was also carried out.

At the DE 55 meeting in March 2011 it was decided to divide the operation into three different categories:

- Ship type A operating in Polar ice-covered water with more than 10 percent ice cover having a minimum Polar Class strengthening.
- Ship type B operating in Polar open water with less than 10 percent ice cover but where the ice may pose a structural risk. This will require that an assessment is carried out with regard to ice strengthening or other mitigating measures.
- Ship type C operating in Polar open water including ice-free waters. This includes ships that may operate in waters with 0–10 percent ice cover where the ice does not pose a structural risk. No ice strengthening will be required.

The content of the code was also agreed upon and addresses the following:

- Certification
- Design
- Equipment and systems
- Operation
- Environmental protection
- Manning and training To some extent)

There are still on-going discussions and different opinions about the code. Some of the challenges in its development are related to new discussions about geographical limitations, various opinions about the additional risks when operating in Polar waters, and how to mitigate the additional risks. There is also a discussion about the level of need for additional requirements, and additional environmental protection measures. It is not finally agreed with regard to application of the code with regard to ship types, it is only for cargo and passenger vessels and not fishing vessels. The three different ship categories, A, B and C, are also still under discussion and also whether a sailing permit system in addition to certification should be introduced.

Other identified important challenges not to be covered in the code but which have to be considered in other forums or codes is the need for the improvement of existing and development of new charts in both paper and electronic format, more complete systems for ice information, improvement of systems for search and rescue and, finally, improvement of spill preparedness.

Mandatory requirements for additional safe manning, additional qualification and special training are also identified as important issues that should be included in codes already addressing these topics. According to the Norwegian Coastal Administration and in progress of the code, the aim is that the Polar Code is mandatory in 2016. However, it remains to seen whether this milestone will be achieved.

Baltic and Polar ice classes

Ships are generally divided into a number of categories according to type and function. To ensure the safety of all types of ships, each vessel is subject to various rules and standards regarding the appropriate use of the vessel, the environment and other issues, as established by the classification societies and other organizations. Ships intended for operation in ice-covered waters will normally be built with an additional ice-class notation. The different ice classes include requirements for the strengthening of hull, rudder and the propulsion system. Specification of an ice-class notation is voluntary and the actual ice class will depend on type and area of operation. In addition to an ice class, the vessel can be assigned an icebreaker notation. An icebreaker is a vessel for which icebreaking is the main purpose and hence ramming is part of the normal operation.

Today the ice classes are normally divided in two different regimes depending on area of operation and ice conditions. The Finnish Swedish Ice Class Rules (FSICR) have been developed for the Baltic Sea and Bay of Bothnia and are today generally applied for vessels operating in areas with first-year ice up to 1.2 meters. In other areas where first-year ice is dominant, like St. Lawrence and southern part of Sakhalin, the FSICR notations are often applied. For areas with heavier ice conditions, the Polar Class notations will normally be applied. Table 5.1 shows both the Baltic and Polar Class notations. Ships having a

Baltic		**Polar class**	
		PC 1	Year-round operation in all Polar waters
		PC 2	Year-round operation in all Polar waters
		PC 3	Year-round operation in second-year ice which may include multiyear ice inclusions.
		PC 4	Year-round operation in thick first-year ice which may include old ice inclusions
		PC 5	Year-round operation in medium first-year ice which may include old ice inclusions
1.0 m first year ice	ICE-1A*	PC 6	Summer/autumn operation in medium first-year ice which may include old ice inclusions
0.8 m first year ice	ICE-1A	PC 7	Summer/autumn operation in thin first-year ice which may include old ice inclusions
0.6 m first year ice	ICE-1B	PC 1 to PC 6 may be assigned additional notation ICEBREAKER	
0.4 m first year ice	ICE-1B		

Source: IMO

Table 5.1 Baltic and Polar Class notations.

PC7 and PC6 are normally considered as one way equivalent with ICE-1A and ICE-1A* respectively.

The IMO Guidelines have defined "Polar Class Ship" as "(...) a ship for which a Polar Class has been assigned".[13] The IMO Guidelines suggest a classification of Polar Class ships into seven categories according to intended ship operations and the level of ice in the area.

The system is developed to designate differing levels of capability for vessels navigating in Arctic waters. Seven Polar Classes are listed, based on sea ice conditions. Polar Class 7 is the least capable, limited to vessels operating in summer/autumn in thin, first-year ice (with old ice inclusions), whereas ships of Polar Class 1 are to be capable of operating year-round in all Arctic ice-covered waters.

In a complementary and parallel effort with the IMO Guidelines, the International Association of Classification Societies (IACS) has adopted a set of *Unified Requirements* for member societies. The IACS Polar rules are incorporated into the IMO Guidelines.[14] Currently, the regulations operate with different geographical aspects. In contrast to the IMO Guidelines, which only cover parts of the Arctic, the IACS Unified Requirements are based on both Arctic and Antarctic conditions of navigation.[15] It must be added that the IMO Assembly recently decided to expand the Guidelines' geographical scope to also include Antarctica, south of 60°S, which was applied after January 1· 2011.[16]

In addition to the class regulations, the Unified Requirements address important aspects of construction for Polar Class ships. These rules are also considered as supplementary to the IMO Guidelines in technical matters with regard to hull and machinery standards.[17] The Unified Requirements form part of IACS member rules and apply to ships of member associations contracted for construction on and after 1 March 2008, but they are not mandatory.[18]

Having that said, the main purpose is to coordinate IMO/IACS standards for the operation and construction of vessels navigating in Arctic waters.[19] This process is not yet fully accomplished. However, when risk assessment is carried out, it is the responsibility of the ship owner to decide and select a Polar Class appropriate for the ship's intended service in ice-infested waters.[20] Moreover, all class societies have adopted these notations directly as their own regulations. Nonetheless, there are still different systems among individual classification societies (see Appendices 5.1 and 5.2). However, the main development today takes place within the Polar Class regulations.

5.1.1 National technical requirements

Technical requirements for navigation on the Northern Sea Route

Russia regulates shipping along the NSR on the basis of UNCLOS, Article 234 and domestic legislation (see Chapters 2 and 6). Regulations adopted in 1990 and 1996 allow navigation on the NSR on a non-discriminatory basis for ships of all nationalities based on *Guide for Navigation Through the NSR,* 1996 (NSR Guidelines).[21] This basic legal framework applicable to foreign shipping in the NSR has apparently not changed much until now. Indeed, many of these documents are made available by the Russian Ministry of Transport through their official webpage in exactly the same version as they were adopted in 1996.[22]

Technical requirements for ships passing through the NSR are provided in "Requirements for the Design, Equipment and Supply of Vessels Navigating the Northern Sea Route", a section included in the NSR Guidelines. These requirements apply to all ships with 300 gross tonnages (registered).[23] Additional requirements are provided with respect to the hull, which must be of a double bottom type that cannot be used for storage of petroleum products or other harmful substances.[24] Special requirements apply to the stability of the ship with reference to challenges of icing, e.g., potential ice accretion on horizontal and lateral surfaces of the ship, or when damaged.[25] The regulations also require a minimum of navigation and communications equipment, as well as certain provisions and emergency facilities.[26]

Ships navigating the NSR must satisfy the applicable Rules of the Russian Register of Shipping for vessels containing the following literal designations of ice resistance categories as part of the symbol of their ice class.[27] Ships are generally divided into a number of categories according to type and function. To ensure the safety of all types of vessels, each vessel is subject to various rules and standards regarding appropriate use of the vessel, the environment and other issues, as established by the classification societies and other organizations. The structural and machinery requirements for icebreaking vessels are established by several classification societies. An estimated comparison is presented in Table 5.1 above, of the literal designations of ice classes between those employed by the Russian and other classification societies.

In the *Ice Rules of the Russian Register* (1999), icebreakers are divided into four categories: LL6–LL9 (see Appendix 5.2). In addition, nine categories are introduced for ice-strengthened ships: LU1–LU9 (see Appendix 5.1).[28] The higher number of ice classes is a response to the described need to introduce higher ice classes for the more severe conditions, at the same time presenting equal steps between the ice classes. Statistical data on ice conditions in the various NSR parts are used to separate different seasons (winter–spring and summer–fall) as well as different severity in order to determine the required ice class for different times of the year (see Appendix 5.3). A vessel belonging to ice resistance category LU4, or equivalent ice class used by other classification societies, may be permitted to travel, under the control of icebreakers, along sections of the western part of the NSR up to 125°E, and along individual parts of the eastern part of the NSR during the summer period, if navigation conditions are favorable. Icebreakers are permitted to navigate the NSR under ice conditions that correspond to the designation of their respective ice class. On a case-by-case basis, the NSR Administration might permit the operation of an icebreaker under more severe ice conditions than those envisaged by its ice class. Such permission would be granted following a review of the appropriate documentation, provided by the owner of the icebreaker, confirming that the state of the hull, machinery, and systems of the icebreaker satisfy navigational safety requirements on the NSR and preclude the possibility of sea pollution. Class LU3 vessels might be permitted as an exception, upon special decision of the NSR Administration, in the summer navigation period in the western area of the NSR under favorable navigation conditions. Operation in ice of class LU3 vessels in the eastern area of the NSR is not permitted.

The operation of commercial vessels on the NSR is carried out in one of two modes. The first is unescorted mode, in which the vessel navigates in ice-infested waters without icebreaker assistance. When severe ice conditions are expected to lead to significant delay

Ship owner	Name of the ship (ice class)	Dwt	Departure and arrival (dates dd.mm.yy)	Cargo (tons)	Comments
Beluga Shipping Group	MV Foresight MV Fraternity	12,000 12,000	Vladivostok (21.08.09) Novy Port (07.09.09) – Murmansk	Heavy lift modules	First foreign ships in transit through the NSR Passage
Beluga Shipping Group	MV Houston	12,000	Norrkoping – Novy Port (26.07.10)	Heavy lift modules	Not a complete transit. Only destination-Arctic routed, whereas parts of the NSR was sailed
	MV Fortitude	20,000			
Murmansk Shipping Company	Indiga (1A super)	16,000	Murmansk (14.07.10) – Pevek (22.07.10)	Diesel fuel (13,000)	First ship to transit the NSR in 2010
	Varzuga (1A super)	16,000			
Sovcomflot	SCF Baltica (1A super)	117,000	Murmansk (14.08.10) – Pevek (25.08.10) – Ningbo (China) (06.09.10)	LNG (70,000)	The largest tanker ever to cross the NSR (as of 2010)
Nordic Bulk Carriers	MV Nordic Barents (1A)	41,000	Kirkenes (04.09.10) – China (30.09.10)	Iron ore	First foreign bulk ship ever to cross the NSR
Norilsk Nikkel	Monchegorsk (SA-15 class ship)	18,000	Murmansk (15.09.10) – Dudinka – Shanghai (17.10.10)	Iron ore	First ship ever to transit he the entire NSR without icebreaker assistance
Russian state owned	Georg Ots	12,600	Murmansk (09.09.10) – Anadyr (26.09.10) – Petropavlovsk – Vladivostok (10.10.10)	Passenger ship	First passenger ship ever to transit the NSR Without passengers

Source: The data has been retrieved from a number of various sources. All the sources have been double checked.

Table 5.2 An overview of transits on the NSR during summer season from 2009 to 2010.

in navigation, this mode is dropped in favor of the second mode: escorted navigation, or ice escort. During the escorting operation the icebreaker sails in front, opening a channel through the ice, and the commercial vessel follows behind. In severe ice conditions, the channel created by the icebreaker may close up before the escorted ship has passed through. In these cases the icebreaker may be obliged to tow the escorted vessel. In some cases, and when the ice and weather conditions are extremely severe, the vessels may be forced to perform a so-called close towing. During close-towing operations, the bow of the escorted ship is attached directly to a notch on the icebreaker, called the stern notch. Ships with bulbous bow are allowed to transit the NSR but, due to the bulb, it will not be possible to use close towing. On the other hand, the hull damage statistics on the NSR[29] it indicate that hull damages are most frequent when vessels are following an icebreaker and when close-towing operations are performed. This might be an issue in terms of the insurance premium.

Transit on the NEP

The interest in using the NEP for transit has increased significantly during the last years – from two vessels in 2009 and six vessels in 2010 to 34 vessels in the 2011 summer season. See Table 5.2 for more detailed information about the transits.[30]

Furthermore, according to the Russian Regional Minister of Property Relations Yuri Chuykov, the volume of goods shipped through the NSR was expected to almost double in 2012 (from 2011: 111,000 tons to 2012: 1.5 million tons).[31]

The last summer season (2011) had the highest number of vessels ever in transit through the NSR. According to Atomflot, the total cargo transported on the NSR was 820,000 tons. By comparison, in 2010 only six vessels used the NSR for transit and the total amount of cargo was 111,000 tons.[32] Fifteen of the 34 vessels transported liquid cargo (682,000 tons), three carried bulk (110,000 tons), four refrigerator ships transported fish (27,500 tons), two vessels transported general cargo and 10 vessels sailed in ballast.

The largest vessel ever, a 160,000 dwt Suezmax loaded with 120,000 mt gas condensate and the largest bulk carrier, 75,000 dwt Panmax vessel loaded with iron ore, used the NSR in 2011. A new speed record along the route was also set, with an average of 14 knots and eight days. In addition to condensate and ore, a seismic and international cruise vessel used the route.

Table 5.3 summarizes a number of the 34 registered transits that took place through the NSR during the 2011 summer season.

As reflected in Table 5.3, the authors have not been able to collect information about all the "34 registered transits" through the NEP for 2011.

Technical requirements for navigation on the Northwest Passage

Canadian maritime jurisdiction is divided into non-Arctic waters and Arctic waters. Shipping in the NWP is primarily governed by the *Arctic Waters Pollution Prevention Act*, 1970 (AWPPA) and other regulations under this Act, in particular the *Arctic Shipping Pollution Prevention Regulations*, 1978 (ASPPR) (see Chapters 1 and 6).[33] The ASPPR provides detailed requirements for ship design, construction and operation into the various zones. For example, ships over 100 gross tonnage and carrying oil in excess of 45 m^3 are not allowed

Ship owner	Name of the ship (ice class)	Dwt	Departure and arrival (dates dd.mm.yy)	Cargo (tons)	Comments
Neptune Maritime	Perseverance (1A)	70,000	Murmansk (29.06.11)– Ningbo China (20.07.11)	Oil condensate (60,000)	Route north of the New Siberian Islands. Did a return voyage with jet fuel – from east to west through the NSR
Marinvest	Marilee (1A)	74,000	Murmansk (02.08.11) – Ningbo, China (30.08.11)	Gas condensate (61,000)	Passed the Bering Strait 15.08.11
LR Ice Shipping Nine	Mariann (1A)	74 000	Murmansk (20.09.11) – Ulsan, South Korea	Gas condensate (60,000)	Route north of the New Siberian Islands
STI Heritage Shipping Company	STI Heritage (1A)	73,000	Vitino (19.07.11) – Map Ta Phut in Thailand (16.08.11)	Gas condensate (61, 000)	Route north of New Siberian Islands. Fastest transit through the NSR with an average speed of 14 knots. Eight days' transit of the NSR. Passed the Bering Strait 29.07.11
Sovcomflot	Vladimir Thikonov (1A)	162,000	Honningsvåg (20.08.11) – Map Ta Phut, Thailand (17.09.11)	Gas condensate (120,000)	Route north of New Siberian Islands. First supertanker (Suezmax) through the NSR. Seven-and-a-half days' transit
Neste Shipping	Stena Poseidon (1A)	75,000	Murmansk (30.08.11) – Incheon, South Korea	Gas condensate (57 000)	Route north of the New Siberian Islands. Passed the Bering Strait 08.09.11
Neste Shipping	Palva (1A)	75,000	Murmansk (11.09.11) – Huizhou, China	Gas condensate (60 000)	Route north of the New Siberian Islands. Passed the Bering Strait 21.09.11
Polarcus	Polarcus Alima (1A)	45,000	Hammerfest (15.09.11) – Cape Dezhnev (24.09.11)	Seismic vessel	This is the first known passage of a 3D seismic vessel along the NSR

Odyssey Bulk Ship Limited	Sanko Odyssey (1A)	75,000	Murmansk (31.08.11) – Jintang, China	Iron ore (72,000)	Passed the Bering Strait 10.09.11
Murmansk Shipping Company	Mikhail Kutuzov (1A)	23,000	Murmansk (13.07.11) – Jintang, China (10.08.11)	Iron ore (22, 000)	Escorted from the Kara Sea to Wrangel Islands. Passed the Bering Strait 25.07.11
Murmansk Shipping Company	Dmitry Pozharsky (Arc 4)	23,000	Murmansk (14.07.11) – Jintang, China (10.08.11)	Iron ore (22 000)	Escorted from the Kara Sea to Wrangel Islands (25.07.11)
Aurora Expeditions	Akademik Shokalskiy	1,700	Murmansk (10.08.11) – Anadyr (05.09.11)	Passenger ship	The first passenger ship with passengers to cross the NSR. Fifty passengers, 10 staff and 23 crew members on board
LLC Dalrifer	Nikolaev Communards	-	Bering Strait (4.8.11) – Kara gate (19.8.11)	Frozen fish products	Cargo - 4700 tons of frozen fish products
LLC Dalrifer	Capitan Spinner	-	Bering Strait (14.8.11) – Kara gate (23.8.11)	Frozen fish products	Cargo - 4700 tons of frozen fish products
Freeport Shipping	Reynfrost	-	Bering Strait (28.8.11) – Kara gate (6.9.11)	Frozen fish products	Cargo - 7000 tons of frozen fish
JSC ROMAN	The Coast of Hope	-	Bering Strait (30.8.11) – Kara gate (13.9.11)	Frozen fish products	Cargo – 8273 tons of frozen fish
Northern Shipping Company	Pioneer of Moldova	-	Kholmsk (19.8.12)	Equipment	The cargo - 545 tons of large equipment
Norilks Nickel	Polar (SA 15 class ship)	20 000	Murmansk (12.09.11) – Dudinka (21.09.11 – Shanghai	Iron Ore	Without Ice breaker assistance.

Source: The data have been retrieved from a number of various sources. All the sources have been double checked.

Table 5.3 An overview of transits on the NSR during summer season 2011.

to navigate in any zone unless they meet special construction standards set out in schedules to the regulations.[34] In addition, no tanker is allowed to navigate in any zone without the services of a qualified ice navigator.[35] However, Canada's ability to support any shipping activity on the NWP, whether it is community re-supply or transit assistance, is limited because of lack of funding for icebreaking services. This means that shipping on the NWP can be performed without icebreaker assistance.

Canada operates with a system dividing 16 shipping safety control zones (see Figure 5.2). The system assumes that the prevailing ice conditions in each zone are consistent from one year to another. The most difficult ice conditions are expected in Zone 1 and decrease progressively, with the lightest in Zone 16. The zones and dates are based on historical ice data and there have been no changes to the Zone System since its implementation in the 1970s,[36] despite changes in the sea ice regime.

Earliest and latest entry dates for each of the 16 shipping zones are set corresponding to the ice capability of 14 categories of ships. The categories include nine Arctic class ships with classifications based on the ice thickness that the vessel would have the power and strength to break,[37] and five ship types[38] (see Appendix 5.5). For example, the most powerful Arctic Class 10 vessel could operate year round in all the zones while the least ice-capable, Type E ship, would be excluded from entering the first six zones at any time of the year as those zones have the most severe ice conditions. The types are based on the Finnish-Swedish (Baltic) rules where Type A ships can operate in thick first-year ice and Type E ships are considered open water vessels with no ice strengthening.[39]

As previously discussed, there are seven potential routes through the NWP with various characteristics such as water depth, ice conditions, choke points, etc. (see Chapters 1 and

Source: Transport Canada

Figure 5.2 Shipping safety control zones.

4). Nevertheless, during a transit of the NWP, ships cross a number of individual zones. The combination of the Zone System, dates of entry and the main deep draft routes through the NWP will decide which Arctic Class will be chosen (see Appendix 5.4).

As an illustration, and for an east-to west-transit (see Figure 5.4), a shallow draft ship would typically transit Davies Strait and Baffin Bay on the Greenland side where ice is less prevalent. The ship then enters Zone 13 (Lancaster Sound) and proceeds into Zone 6 (Parry Channel). At this point, a shallow draft ship follows Peel Sound and enters Zone 7 (Queen Maud Gulf) northwest of King William Island. Zone 11 is entered as the ship approaches Dease Strait. Zone 12 covers the progress of the ship from the Amundsen Gulf to the Beaufort Sea. Meanwhile, a deep-draft ship would continue along Parry Channel, entering Zone 2 in Viscount Melville Sound, and thence southwest along Prince of Wales Strait, before entering Zone 12 (and rejoining the shallow-draft route) in the Amundsen Gulf.[40]

Variations in ice conditions can result in the Zone/Date System permitting the entry of a ship into a zone where known ice conditions are too harsh for the ship. This resulted in the creation of an *Arctic Ice Regime Shipping System* (AIRSS) in 1996, as a regulatory standard. AIRSS provides the master and ice navigator with a formula to calculate an Ice Numeral based on known ice conditions in the zone and the ice classification of the vessel. If the Ice Numeral is positive then the ship may proceed. If negative, entry is not permitted. AIRSS uses a ship classification system based on *Canadian Arctic Class* (CAC), in addition to the 14 categories of ships related to the various zones (see Appendix 5.4). Although AIRSS has the potential to replace the Zone/Date System, it is generally used to allow access to a zone outside of the proscribed entry dates. As an example, over the last summer seasons AIRSS was used to permit access of the tanker *Tuvaq* into Kugaaruk (Zone 5). Under the Zone/Date System a tanker of the *Tuvaq*'s ice class would not have been permitted at any time.[41]

However, it is a problem that there are multiple rules both for the NWP and also among the international classification societies.[42] Canada wants to take leadership in the future harmonization process and has recently adopted the IACS/IMO polar class rules into her own regulations. In addition, she encourages other countries to do the same. This means that the CAC system will be outdated in the future.[43]

5.2 Crew training and competence requirements in Arctic waters

A significant body of international public maritime law has established safety standards for seafarers. The IMO addresses mandatory competence and training requirements, primarily through the *International Convention on Standards of Training, Certification and Watchkeeping*, 1978 (STCW) and SOLAS. However, the contents of these conventions are not specific when it comes to requirements for navigators and crews operating in Arctic ice-covered waters. Nevertheless, seafarers operating in ice-covered waters face several challenges and risks that require specific training and experience. In that respect, the IMO Guidelines call for specialized training for seafarers in Arctic ice-covered waters.

The IMO Guidelines: Applicable to crews and navigators in Polar waters

Part C of the IMO Guidelines focuses on Operational Procedures, Crewing and Emergency Equipment. All ships operating in Arctic ice-covered waters are to carry an operating manual and a training manual for all ice navigators on board. Shipping companies are encouraged to develop such training manuals to include: "Summary of the IMO Guidelines; ice recognition; navigation in ice; and escorted operation."[44] Furthermore, the IMO Guidelines address specific fire safety, lifesaving, navigational, operational and crew training issues. This includes the development and inclusion of drills and emergency instructions, emphasizing changes to standard procedures made necessary for operations in ice-covered waters. These drills and emergency instructions would be incorporated into the routine vessel operational training.[45] The IMO Guidelines recommend that: "All of the ship's officers and crew should be made familiar with cold weather survival by training or self study of course material (...)" ,[46] addressing the safety measures mentioned above. In addition, it is recommended that: "As many as possible of the ship's deck and engine officers should be trained in ship operations in ice-covered waters."[47]

Moreover, the IMO Guidelines focus in particular on the role of the "*ice navigator*", which is defined as:

> (...) any individual who, in addition to being qualified under the STCW Convention, is specially trained and otherwise qualified to direct the movement of a ship in ice covered waters.[48]
>
> All ships in ice-covered waters should carry at least one certified ice navigator.[49]

In addition, the IMO Guidelines emphasize that:

> The Ice Navigator should have documentary evidence of having satisfactorily completed an approved training program in ice navigation. Such a training program should provide knowledge, understanding and proficiency required for operating a ship in Arctic ice-covered waters, including: Recognition of ice formation and characteristics; ice indications; ice manoeuvring; use of ice forecasts, atlases and codes; hull stress caused by ice; ice escort operations and; ice-breaking operations and effect of ice accretion on vessel stability.[50]

The IMO Guidelines: The complexity of the recommendations

Effective and safe sea transportation in the Arctic requires that the seafarers are skilled in different aspects of ice navigation. With the presumed growth in Arctic shipping, the need for qualified ice navigators and crew members will continue to increase. According to the ARCOP study, it is expected that there will be an estimated shortage of up to 4,000 seafarers with ice experience and training.[51] In addition, it is also important to note that ship owners, operators or others trading or expecting to trade in the Arctic develop an awareness and understanding of the challenges that Arctic navigation can pose for the seafarers working on ships. The IMO Guidelines reflect some measures of the standard setting approach of maritime safety regulation. However, although they include important recommendations for actors operating in Arctic waters, there are still certain complexities related to these recommendations.

Safe navigation in ice-covered waters depends much on the knowledge and skill of the ice navigator. The IMO Guidelines require that the ice navigator shall provide documentary evidence of having satisfactorily completed an approved training program in ice navigation. Yet, there exists no international recognized model course for ice navigators or qualification scheme for individuals who are to operate in ice-covered waters.[52] In addition, Section 14.2 of the IMO Guidelines is expressed in relatively broad terms, specifying only that a training program shall provide "(...) knowledge, understanding and proficiency required for operating a ship in Polar ice-covered waters". Although note is taken of the severe and unique circumstances faced by ship operators in ice-covered waters, such as specific fire safety, lifesaving, navigational, operational and crew training issues, provisions could have been made for a more detailed training program.[53]

With regards to crew, the IMO Guidelines provide little information on how to prevent, mitigate or avoid icing. This is a characteristic occurrence in the Arctic Ocean, when cold temperatures result in spray blown off the sea freezing immediately on contact with a vessel. If the ice is not frequently removed, it will build up on the ship's structure and could cause the vessel to destabilize or capsize. The Guidelines state that: "Ice accretion should be regularly removed from the lifeboats and launching equipment to ensure ease of launching when required [and that an] icing removal mallet should be available in the vicinity of the lifeboats".[54] This risk of ice-infested navigation could, however, have been regulated more extensively within the Guidelines. For instance, the Guidelines could have been more clear on how best to prevent, mitigate and avoid sea spray icing of vessels, for instance by referring to the environmental and vessel characteristics that determine the potential for such icing – like wind speed, air temperature and ship speed. Also, provision could have been made for alternative ice-removal equipment and how to protect vital components on deck.[55]

As previously mentioned, the IMO Guidelines recommend that "all of the ship's officers and crew should be made familiar with cold weather survival by training or self study of course material or publications (...)". [56] However, the term "*familiar*" in this respect does not necessarily require that crew must have appropriate education, for instance a certified training course in cold weather conditions, in order to operate in Arctic waters. Currently there are no cold weather survival course requirements. Furthermore, the IMO Guidelines recommend that "As many as possible of the ship's deck and engine officers should be trained in ship operations in ice-covered waters. "[57] Interestingly, the Guidelines do not recommend that "*all*" officers shall have the training necessary to operate under such conditions. According to the AMSA report, the IMO should provide specific, detailed and mandatory requirements for survival equipment (i.e., life boats, life rafts, immersion suits) and crew training. Internationally standardized crew training that includes compulsory education for ice navigation and emergency response in polar environments should exist. Crew training should include response knowledge for incidents likely to occur on specific types of ships. For example, cruise ship crew members will need more training on how to account for and direct numerous passengers during cold weather emergencies, while cargo ship crew members would benefit from salvage or other types of training. The requirements should apply to all vessels transiting the Arctic.[58]

On the other hand, the IMO Guidelines are the first international instrument to emphasize the need for specialized training of crew and ice navigators in respect of the Arctic. The IMO safety commission is currently in the process of improving the Guidelines. This is an important step towards the development of an international standard. However, in a non-binding form, the Guidelines' contribution to maritime safety in Polar ice-covered waters seems rather limited. The regulations impose no legal obligations upon the member governments. In consequence, no state has yet incorporated the Guidelines through national legislation.[59]

The International Northern Sea Route Programme (INSROP) goes one step further in identifying various navigational parameters that might be affected by seafarers operating in Arctic waters. The geographical areas of the NSR and the NWP include natural parameters (i.e., ice, climatic conditions, etc.) and societal parameters (i.e., legal and political parameters), acting together and making the navigation of ships a complex, multi-faceted challenge. Those areas are termed as the *hot spots* of the NSR and the NWP. These hot spots have their counterpart in the *cool spots,* which are areas that host just one single navigational parameter (such as ice) to be overcome without violating significant values like environmental interests or the wellbeing of indigenous people.[60] Moreover, there are certain societal and natural values/interests that seafarers should be aware of. These values/interests are clustering together into what is termed *socio-biodiversity,* which is a mix of nature/society – "all of which should be treated in a sustainable manner in order to reduce the likelihood of negative political reactions to increased navigation".[61] Furthermore, there are certain risks related to Arctic navigation. For instance, on the NSR and NWP there are several valued ecosystem components that are likely to be affected by shipping activities (i.e., emissions, oil spills, species disrupted by marine traffic, etc.). In addition, there are several indigenous groups living near the coast that might be affected by shipping activities on the NWP and NSR. The point is that seafarers must take these external safety parameters (natural and societal) into account, in addition to the IMO Guidelines procedures, in order to ensure sustainable shipping in the Arctic. Or as addressed by the INSROP: "We need a new breed of sea-ice captains and crews to meet the requirements of modern *multi-value navigation* in Arctic waters."[62]

STCW – Manila 2010

In 2010, the IMO Guidelines were further strengthened by the additional requirements of the STCW amendment, adopted at an IMO conference in Manila.[63]

The contents of the STCW amendments highlight the importance of competence needed by personnel operating in ice covered waters. However, this mainly applies to officers and not crew. The following sections are included in the amendments: ice characteristics – ice areas;

- ship's performance in ice and cold climate;
- voyage and passage planning for a ship in ice;
- operating and handling a ship in ice;
- regulations and recommendations;
- equipment limitations;
- safety precautions and emergency procedures;
- environmental considerations.

The amendment applied from 1 January 2012. As with the IMO Guidelines, these are also non-binding. Furthermore, the STCW code consists of two sections: (A) the binding part and (B) the non-binding part. The Manila amendment is covered in Section B, the non-binding part. [64]

This amendment and the expected development of an internationally recognized training program for navigation in ice-covered waters will be important for future work in the sense that it further highlights the need for improved skills and competence when operating in ice-covered waters.[65]

5.2.1 National crew training and competence requirements

Requirements for crews and ice navigators operating on the Northeast Passage

When applying for permission to enter the NSR, the Administration of the NSR requires a notification iin advance of the exact number of crew on board, and it is the Marine Operation Headquarters' (MOH) job to inspect the crew before they enter the NSR. The Russian legislation requires the following aspects:[66]

- The size of the crew for navigation in ice must be large enough to guarantee a three-shift watch.
- The ship master or a person substituting for him on a bridge watch must know the signals that are employed by icebreakers during leading through ice and presented in the Guide to Navigation.
- The crew of the vessel must be warned in advance regarding prohibitions against discharging pollutant substances and rubbish on the NSR as well as responsibility for any pollution of the sea and ice cover.
- The ship master or a person substituting for him on a bridge watch must possess the minimum level of knowledge of navigation in ice: experience of steering vessels under ice conditions along the NSR for not less than 15 days. In the absence of such experience, the presence of an ice navigator on board the vessel is compulsory.

In addition to these requirements, the regulations also emphasize that "(...) in cases where those persons have no such experience [at least 15 days' experience], the Administration/MOH may assign a State Pilot to the vessel to assist in guiding her through the NSR."[67] However, there are no specific requirements for certified training of ice navigators in addition to the 15-day rule of experience along the NSR. Thus, and in light of the legislation, it is unclear on what basis the MOH determines an ice navigator's qualification when considering his ability to navigate in the NSR. Nor is there any requirement for certified training for the crew on board ships that operate along the NSR beyond the requirements listed above.

If a ship does not completely satisfy the requirements, it may still be allowed to navigate on the NSR with icebreaker assistance or other special support for an additional expense. Furthermore, the ship can be expelled from the NSR. Article 10 emphasizes as follows:

> If a vessel navigating the NSR violates provisions of these Regulations, she can be removed from the Route. Direction in which a vessel will leave the Route shall be determined by Marine Operations Headquarters in consideration of the safety of the vessel, its crew and cargo, and necessary measures of nature protection.[68]

However, according to the Head of the Administration of the NSR, the current legal regime of Arctic marine shipping in the NSR is still based on these same 1990 Regulations and the 1996 Guide to Navigation. This basic legal framework applicable to foreign shipping on the NSR has apparently not changed much up until the present. Indeed, the Russian Ministry of Transport still makes many of these documents available through their official web pages in exactly the same version as they were adopted over 10 years ago.[69] In this case, it may be that the legislation requires a review of any amendment to accommodate the development of the technology and safety equipment available today.

Requirements for crews and ice navigators operating on the Northwest Passage

The ASPPR has also set out certain control requirements when it comes to crews and ice navigators. The owner or master of a ship proposing to navigate within any zone can apply for an *Arctic Pollution Prevention Certificate* from a Canadian marine inspector or from an examiner of a classification society outside of Canada, to indicate that the ship and the master comply with the regulations when operating in Canadian waters.[70] An inspector may issue an Arctic Pollution Prevention Certificate to the owner or master of a ship that is within Canadian waters. Where the owner or master of a ship fails to comply with the requirements "(...) without reasonable cause, the arctic pollution prevention certificate issued to him in respect of that ship is invalid if the ship navigates within a zone."[71]

Like the IMO Guidelines and the Russian Guide to Navigation through the NSR, the Canadian ASPPR also addresses the concept of ice navigators. Unlike the Canadian legislation, the Russian legislation and the IMO Guidelines do not require any certificate in order to get sailing permission. However, according to ASPPR, no tanker is allowed to navigate in any zone without the services of a qualified ice navigator. The ASPPR define a qualified ice navigator as having:

> (...) served on a ship in the capacity of master or person in charge of the deck watch for a total period of at least 50 days, of which 30 days must have been served in Arctic waters while the ship was in the conditions that required the ship to be assisted by an ice-breaker or to make maneuvers to avoid concentrations of ice (...)[72]

It is interesting to compare the Canadian definition of ice navigators with the Russian definition. The Russian definition requires experience of steering vessels in ice conditions explicitly on the NSR for not less than 15 days. Unlike the Russian requirements, the Canadian legislation does not require any explicit experience exclusively within the NWP. Furthermore, it refers to experience from ice maneuvring in Arctic conditions on a general basis. However, these issues are now raised among both the Canadian Coast Guard and Transport Canada, with the intention to state that ice maneuvring in the NWP and the NSR presents different challenges.

Furthermore, the complex archipelago, narrow straits, high tides, currents, variety of draft conditions in combination with a complex seasonal ice regime, represent major challenges for even the most experienced ice navigator.[73] For example, an ice navigator with extensive experience from the Baltic Sea is not necessarily well qualified to navigate along the NWP or NSR. In addition, an ice navigator with extensive experience from the NSR is not necessarily qualified for navigation on the NWP, since the two transport corridors represent different geographical areas in combination with different ice characteristics. This means that an ice navigator can meet unique risks and challenges in the different passages. One example that illustrates this complexity is the sinking of the ship *Explorer* in 2007. *Explorer* sank after hitting ice while sailing in Antarctica. Although this example is taken from Antarctic waters, it illustrates the different ice conditions met when navigating in various ice regimes, and thus it is interesting also in an Arctic perspective. According to the *Explorer* Investigation Report, the ice navigator observed multiyear ice, but the ice was much older than first assumed. In addition, the master of the *Explorer* was very experienced in Baltic waters, but was unfamiliar with the type of ice he encountered in Antarctic waters.[74]

This type of previous accident illustrates why explicit training on the NSR is necessary, such as required by the Russian regulations. It might also encourage the Canadian authorities to require training for masters in the NWP explicitly, in order to get permission to navigate. And not only referring to experience from Arctic waters in general, such as the legislation currently requires. Interestingly, Transport Canada pointed out that in cases where the captain or ice navigator is not familiar with navigation in a certain area of the NWP, a local adviser is often brought on board in order to make safe operation.[75] This emphasizes why specific NWP navigation training is needed.

Requirements for crews and ice navigators operating on the Transpolar Passage

When considering future possibilities for the Transpolar Passage, consistent safety provisions for ice navigators and crew, as well as adequate equipment and training provisions will be required. From an operational safety point of view, it is unlikely that any ship owner would cross the Arctic Ocean without an ice navigator and a crew trained for such conditions. In addition, ships crossing the Arctic Ocean will operate in a convoy, which means that there will be another ship to support in case of an emergency incident. Furthermore, the convoys must rely on self-support as the capability of any external rescue is limited. From a legal perspective, however, and the substantial shortcomings[76] in light of the current IMO Guidelines, this would be far from adequate for any possible and regular future marine traffic along the TPP. For example, low temperatures resulting in icing of vessels would be a constant problem. Moreover, cold weather course training would be an essential qualification to do a sufficient job on board any ship operating in the central Arctic Ocean. In addition, there are great challenges when it comes to emergency procedures, like rescue services for instance, due to the long distances to the mainland and operative ports offering rescue. If an accident should take place in these waters, it will take a long time for the rescue crew to arrive at the place.

5.2.2 Current practices on the Northern Sea Route, the Northwest Passage and the Transpolar Passage

The structure of any training is usually divided into four main categories:

- *Theoretical training*: Creates a general understanding of ice navigation, icing, icebreaking operations and escorting in ice.[77]
- *Practical training course*: Navigator is considered as a student or trainee on board a ship, in addition to the regular crew, in order to observe and learn from experienced officers.
- *On-the-job experience*: Inexperienced navigators are integrated as part of the regular crew and have to learn various ice operations from experienced colleagues.
- *Simulator training*: Enables navigators and crews to be trained properly for action in different ice conditions, including emergency situations. The main purpose of the simulator training course is to increase the professional level of the navigators appointed on large-capacity vessels or vessels with unusual maneuvrable characteristics.[79]

The training courses follow the requirements expressed in the STCW, the IMO Guidelines, the Canadian ASPPR, and the Russian NSR Guidelines.[79] Various maritime training institutions are developing or have developed ice navigation courses, employing full mission bridge simulators and associated software products. The IMO, for instance, has developed a program of model training to assist institutions developing ice navigation courses with an emphasis on meeting STCW requirements.

From a survey carried out during the ARCOP study, data were collected from well-known seafaring countries including Finland, Russia, Germany, Sweden, Canada, and the Netherlands. The survey showed that several theoretical courses were available for navigation in ice, lasting between 1 and 10 days, at a cost of between US$300 and 2000. The survey concluded that practical training courses are rarely offered. However, the ARCOP study found that the existing practical training course lasting between 3 and 30 days cost around US$2000.[80] The same survey discovered that most ship owners in North Europe do not train their personnel explicitly and extensively for navigation in ice-covered waters. Operators of icebreakers in North Europe do train their personnel, but training of inexperienced navigators is mainly performed by so-called "on-the-job experience". Theoretical training is mostly limited to nautical officers.[81] This statement was also supported by one representative from the Central Marine Research and Design Institute during one of the ARCOP workshops:

> There are no Russian requirements regarding the crew. There are the STCW requirements. In practice, the shipping company assesses the skills of the captain, and their assessment is trusted. This leads to the situation in which shipping companies have certified crews who have never seen real ice... Not all crew members or officers can have completed onboard practical training, because the demand for skilled officers is rising.[82]

With regard to the NWP, Canada is working to develop a model course using the simulator at the Marine Institute in St John's, Newfoundland.[83] However, there are currently

> (...) no real substantial qualifications when it comes to requirements for crew operating on the NWP. There is neither any formalized training at the moment (...) the CCG relies pretty much exclusively on 'On the Job Experience', achieved by retired CCG captains. Nevertheless, there should be more specified training required in order to enter the NWP.[84]

The most effective training is the so-called "structured training". This includes simulation with an icebreaker maneuvring in different types of ice and a commercial ship ice manoeuvring in combination with on-the-job experience. Currently, most training of seafarers has been accomplished in an ad hoc approach, with the instruction of persons new to vessel operations in ice-covered waters achieved by simple on-the-job experience and depending on knowledge transfer by experienced northern mariners. In other words, formal education requirements have been nonexistent,[85] despite various classification societies working to develop such certified training programs. The CCG recognizes that ships on voyages on the NWP have crew drawn from tropical countries possessing little knowledge of the details of ice navigation and little experience of living and working in cold climates:[86]

> Many of the chartered ships use crews typically from the Philippines because they are inexpensive (...) some of them have never operated in cold weather, they don't have the experience, the equipment (...) some of them neither have warm clothes.[87]

The general opinion is that theoretical training combined with practical training is essential for navigators of icebreakers, whereas theoretical training or on-the-job experience is considered sufficient for navigators of vessels that are assisted by icebreakers.[88] Opinions about simulator training are more divided. According to the AMSAbstudy, simulation training is now the most effective way of preparing crews for ice navigation.[89] Training centers in Russia, for instance, use TRANSAS equipment, e.g., Navy Trainer Professional with an ice navigation module, which simulates the vessel activity in different ice conditions. The training course "Preparation for navigation in ice conditions" developed at the Makarov Training Center in St. Petersburg consists of short specialized courses intended for captains, the chief mates and navigators. On the other hand, and according to the ARCOP study, the simulator training is not always found to be effective because reliable and efficient modeling techniques is currently unavailable, and because it is very difficult to create models for ships operating in ice.[90] Regarding the TRANSAS-based simulator, the ARCOP study argues that the functionality and realistic appearance of the ice field movements and the dynamic interaction between ship hull and ice are restricted to very simple scenarios such as navigation in level ice, navigation in an opened lead and ship handling during convoy operations. However, developments of the simulator training technology are in progress. For instance, the Ship Maneuvring Simulator Center (SMSC) in Norway has, in cooperation with Det Norske Veritas (DNV), developed advanced mathematical models for the realistic visualization of ice for real-time simulation. The purpose of this new ice maneuvring simulator includes ice management, different loading scenarios in drift ice and maneuvring in ice.

5.3 Search and rescue in Arctic waters

The International Convention on Maritime Search and Rescue (1979) provides a system for the rescue of persons at sea and cooperation among states, including rescue coordination centers, ship position reporting systems and advanced entry of rescue units into territorial waters of the states. Arctic state parties[91] of the Search and Rescue (SAR) Convention coordinate SAR incidents in their respective areas, make sure that rescue services are available during the shipping season and cooperate with each other.[92]

To facilitate maritime safety communications, the IMO adopted the *Global Maritime Distress and Safety System*. It is mandatory for all SOLAS Convention parties, cargo ships of 300 gross tons or greater and all passenger ships. NAVAREAs are *navigational areas* selected for the issue of navigational warnings, provide traffic services, broadcast meteorological forecasts and related maritime safety information (see Figure 5.5). Recently, the IMO's Committee on Communications and Search and Rescue approved the creation of Arctic NAVAREAs up to 90° North. Nonetheless, there are no coordination agreements any further north, which is important when considering any future TPPs. Some Arctic coastal states are responsible for coordinating one or more NAVAREAs within the Arctic. Arctic coastal states are responsible for the broadcast of maritime safety information in navigable waters within those areas. When it comes to service providers, Canada, Norway and Russia were identified, with the United States and Denmark agreeing to be Preparation Service providers for designated areas.[93]

As illustrated in Figure 5.5, Canada is responsible for providing meteorological information, traffic services, navigational warnings and coordination within the areas XVII and XVIII; Norway for the area XIX, the United States for areas VI and XII, and Russia for areas XX and XXI. The delimitation of such areas is not related to and shall not prejudice the delimitation of any boundaries between States. However, it is interesting that Canada is responsible for the coordination of the eastern part of the Bering Strait, the northern slope of Alaska, and the entire NWP. In other words, the NWP is a three-state route, and should in principle or practice be divided between the US, Canada and Denmark, but Canada is now

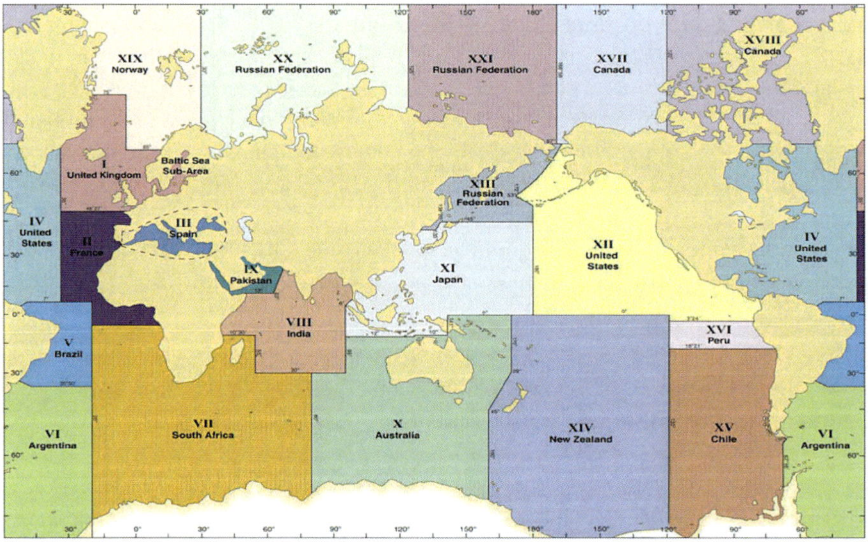

Source: ftp://ftp.wmo.int/Documents/SESSIONS/EC-PORS-1/Doc.4.3.pdf

Figure 5.3 Boundaries of Arctic NAVAREAs.

the coordinator responsible for the entire NWP in addition to America's northern slope of Alaska and Denmark's west coast of Greenland (see Chapter 1).

According to the NAVAREA (IV, XVIII), Canada and the US are SAR and meteorological information providers of Denmark's west coast of Greenland (see Figure 5.3). In addition, the Hudson Bay in Canada is also part of NAVAREA IV, for which the US acts as SAR Service Coordinator. In terms of Arctic sea routes, this area also supports the weather forecasts and SAR to the Arctic Bridge. However, US has not been issuing meteorological forecasts and preparation service for that area. Currently, Canada is providing such kind of service. Canada and the US should agree on their responsibilities, whether as issuing service or preparation service.[94]

Agreement on cooperation on aeronautical and maritime search and rescue in the Arctic

The first international agreement made exclusively for the Arctic region was signed at the ministerial meeting in Nuuk, on May 12, 2011. The SAR Agreement is the first legally binding agreement established under the auspices of the Arctic Council and covers search and rescue of aeronautical and maritime vessels and passengers.[95]

The Agreement was made in accordance with the 1979 International Convention on Maritime Search and Rescue (SAR) and the 1944 Convention on International Civil Aviation, both of which are established and widely recognized in international law. The objective of the Agreement is, however, to further strengthen aeronautical and maritime search and rescue cooperation and coordination in the Arctic. In order for that to be clear, each member state was given a particular Search and Rescue area which it is responsible for (see figure 5.4).

The main emphasis of the Agreement is to develop swift and efficient measures when accidents occur in the harsh Arctic region and to ensure, as much as possible, proper search and rescue operations. This is done through the clauses where the member states commit themselves to nominate certain national institutions from each state that will have full discretion in the field of search and rescue in the area. These national institutions are not only bound to take efficient measures, but also to notify other relevant national institutions when appropriate.

The most significant clause of the Agreement is Article 8, where member states obligate themselves to send a permission request to another member state when, for example, a rescue vessel enters the search and rescue area of another state. This is obviously not unusual, but the article also states that the receiving national institution "shall immediately confirm such receipt" and, as soon as possible, let the member state of the requesting vessel know if the request is permitted and under which conditions, if any.

The Agreement is only the beginning of a specific cooperation among the Arctic states and, as clearly stated in the Agreement, it will develop and be subject of amendments as search and rescue operations in the region become more numerous. The idea is then to make the Agreement more effective and cooperation more efficient. These objectives might be reached by the statements made in Article 9 of the Agreement, where signatories are obliged to maintain widespread cooperation on search and rescue. The most important cooperative measures of the member states is exchange of information on, for example, communication details, information about search and rescue facilities, lists of available airfields and ports and their refueling and resupply capabilities, knowledge of fueling, supply and medical facilities and information useful for training search and rescue personnel.

The Agreement will strengthen cooperation between the Arctic states and improve the way Arctic countries respond to emergency calls in the region. The necessity of such an agreement and cooperation is great, as accidents in the region rely as much on swift responses and efficient operations as on fully qualified rescue personnel and equipment.

5.3.1 Communication and support to navigation on the Northeast Passage

Marine communication and traffic services

As a coastal state, Norway has obligations under international law with regard to search and rescue services. A number of agreements on search and rescue have been concluded. For example, in 1995, Norway entered into an agreement with Russia on search and rescue cooperation in the Barents Sea. As a result of this agreement, Norway and Russia carry out joint annual sea rescue exercises.

Three satellite communication systems the IRIDIUM commercial system.[96] Communications using *Very High Frequency* radio (VHF), *Medium Frequency* radio (MF) and *High Frequency* radio (HF) as well as satellites are generally adequate for the NEP.

Vessel Traffic Management and Information Systems (VTMIS) have been used on the NSR for decades. Generally, a VTMIS gathers, evaluates and distributes vessel traffic and waterborne transport data to improve the safety and efficiency of transport.[97] The ARCOP study showed that there are not enough VTMIS along the NEP. One of their concluding proposals related to communication improvements was to define a vessel traffic service covering the total traffic volume of the NEP. In order to implement such a goal they found it necessary to improve the cooperation with local services along the route, such as port services and emergency services,[98] and also integrate new communication services to the system that are currently not part of the system. The work with the VTMIS showed that there are a number of Arctic information services that can be combined in the system. In the future, especially, the ice information should be part of the VTMIS system. Russia is planning to improve her existing maritime communication services during the next years with a 2020 timeframe, but it is still uncertain whether they will receive financial funding.

Although the ARCOP study considered the current Russian ice information system as an optimal ice data provider in order to support any future NEP navigation,[99] the AMSA study found that the existing charts insufficient to support current marine activities.[100] The Russian satellites are currently not able to provide Arctic navigation with reliable data due to technical problems. As a result, foreign satellite images are used. The national ice service in Russia (AARI) uses the American satellite NOAA observations in the visible range as well as the Canadian satellite RADARSAT. This is not consistent with ARCOP's recommendation that the Russian ice information system be considered as the standard platform for integration by other Arctic ice information systems. For instance, and to ensure sea trial of the leading containership of Norilsk Nickel of a highest ice class LU7 in March 2006, AARI ordered two radar images of the ice conditions in the Kara Sea from the Canadian satellite RADARSAT.[101] It is also interesting to note that the Beluga ships used their own

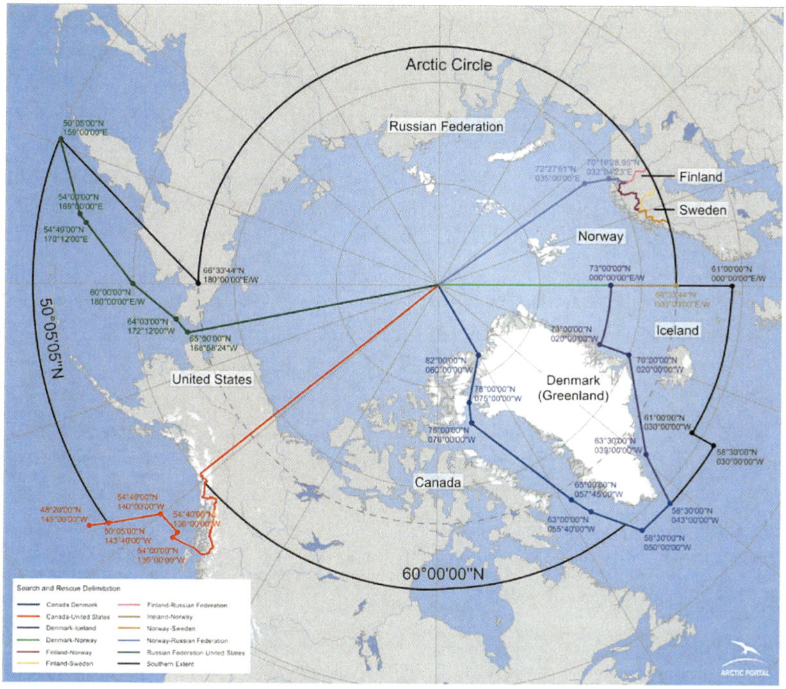

Source: http://library.arcticportal.org/1474/11/search-rescue_ensku_110524_minnka.jpg

Figure 5.4 The search and rescue regions relevant to the Agreement on Cooperation on Aeronautical and Maritime Search and Rescue in the Arctic (2011).

meteorologists' last summer, providing the ice navigator with up-to-date data in order that the ice information is as reliable as possible.

Aids to navigation and available icebreakers

The complex geography of the NSR in combination with marine traffic requires the installation of various navigational aids at important waypoints, including radio beacons, lighted marks or day beacons, radar beacons, radar reflectors and various buoys. However, the coast of the NSR is relatively low and flat, which makes it difficult to use radar. Thus, it has been necessary to equip the NSR with high-accuracy positioning systems.[102] Radio beacons are currently installed in 47 locations throughout the NSR, 17 of which are manned stations. Two types of radio beacon are in use, one with a range of 100 nm and the other with a range of 150 nm. At the same time, 30 locations – mostly in the estuaries – are equipped with radio beacons. A further 200 radar reflectors are also spread along the coast. The reflectors are installed because the low surrounding islands make radar navigation difficult. One thousand buoyant beacons are deployed along the NSR for operation during the summer navigation season. All aids are shown on nautical charts, and their characteristics are given in the NSR Guidelines.

The *Global Positioning System* (GPS; NAVSTAR satellite system) is a system for determining the precise location of objects using satellite data. Today, this information is used not only by ships but by aircraft and even by automobiles, GPS is an important system for ship positioning in the NSR as well as positioning in case of any marine accident or emergency situation. A similar system called GLONAS is used with Russian satellites, but in practice only Russian ships are equipped with the electronic devices necessary to receive GLONAS signals. The GPS system, which was originally developed for military applications, is provided to civilian users with the capability to determine positions within at least 100m with 95 percent accuracy. Nevertheless, fewer receivers are available and ships on the NSR are more likely to carry GPS. For higher accuracy with GPS, a *Differential Global Positioning System* (DGPS) has been introduced. For comparison, DGPS systems are not available in the NWP. Four DGPS stations are in operation in the Kara Sea. The plan is to cover all NSR coastal areas with DGPS systems.

The Russians have several plans to develop rescue service on the NSR. Currently, marine rescue coordination centers (located in Murmansk and Vladivostok) and Marine rescue sub-centers (located in Arkhangelsk, Petropavlovsk-Kamchatski) are in operation. These services are managed by the Main Marine Rescue Center of the State Marine Rescue Service of Russia jointly with the Department of icebreaker support (NSR Administration).[103]

The following services are available:

- multipurpose rescue ships,
- marine special units with oil spill response equipment,
- auxiliary vessels and salvage boats.

Regarding measures about the development of the rescue service, Russia has made structural plans along the NSR/NEP 11 years a head (until 2020). For example, and in a response by the Transport Ministry of Russia, installation of stations of emergency radio observation in Arctic points (total 12 units) will be implemented before 2020. Another example is the construction, by 2020, of rescue tugboats and marine diving vessels for the sea port of Pevek located in the East Siberian Sea, and the port of Naryan Mar located near the Pechora River. However, private financial support is needed as the Russian Government will cover only 60 percent of the expenses.[104]

The Russian icebreaker fleet is the world's strongest in both icebreaking capability and number of ships. Of the world's total number of 80 icebreakers and icebreaking supply vessels, 54 are operating in different parts of Russia.[105] However, this number includes both the Russian state-owned fleet and several icebreakers owned by private companies. Since 1959, Russia has built ten nuclear-powered icebreakers (see Table 5.4). Currently, Russia operates with five nuclear-powered icebreakers capable of year-round operations on the NSR. In addition, two nuclear-powered icebreakers are operational on rivers.

In 2008, the operative service of Russia's entire commercial icebreaker fleet was moved from the partially-privatized Murmansk Shipping Company (MSC) to the Russian state-owned company Atomflot.[106] Generally, the NSR icebreaker support operations are divided, west and east, between two organizations: the Atomflot headquartered in Murmansk and the Far East Shipping Company (FESCO), headquartered in Vladivostok. The nuclear icebreaker fleet services the western section of the NSR, extending from Murmansk to the

Ship name	Year	Operator	Class	Propulsion/ Shaft horse power	Operations/remarks
Arktika	1975	Atomflot	Arktika	Nuclear/75000	NSR/Out of service
Lenin	1959	Atomflot	Lenin	Nuclear/44000	NSR/Out of service
Sibir	1977	Atomflot	Arktika	Nuclear/75000	NSR/Out of service
Rossiya	1985	Atomflot	Arktika	Nuclear/75000	NSR/Port of Murmansk
Sovetskiy Soyuz	1990	Atomflot	Arktika	Nuclear/75000	NSR/ Arctic tourism/Port of Murmansk
Yamal	1993	Atomflot	Arktika	Nuclear/75000	NSR/ Arctic tourism/Port of Murmansk
50 Let Pobedy	2006	Atomflot	Arktika	Nuclear/75000	NSR/Port of Murmansk
Sevmorput	1988	Atomflot	Sevmorput	Nuclear/39400	NSR/Container ship/Port of Murmansk
Taymyr	1989	Atomflot	Taymyr	Nuclear/47600	River and coastal NSR operations
Vaygach	1990	Atomflot	Taymyr	Nuclear/47600	River and coastal NSR operations
Krasin	1976	FESCO	A1-Class	Diesel electric/ 36000	NSR (Far East)/Port of Vladivostok
Admiral Makarow	1975	FESCO	A1-Class	Diesel electric/ 36000	NSR (Far East)/Port of Vladivostok

Source: Brigham (1999), AMSA (2009), Bakharin (2006) and http://en.wikipedia.org/wiki/Nuclear_powered_icebreaker

Table 5.4 Russia's most powerful icebreakers.

River Lena as well as river ports on major Siberian rivers. The Arktika-class icebreakers can open passages through 1–2 m thick ice, which is sufficient to make possible year-round navigation in the region. The shallow water ships, the two Taimyr-class icebreakers and the Sevmorput-class cargo vessel, are designed to visit river ports and, generally, are not suitable for leading sea convoys. The newest nuclear-powered icebreaker has operated since 2006.[107] In addition, Russia holds several diesel-powered icebreakers operating on rivers and supports ships entering various ports along the NEP/NSR. For instance, Lukoil has two icebreakers to support its Varandey oil terminal in the Pechora Sea. Another example is Norilsk Nickel, which has built a fleet of six double-acting icebreaking bulkers that allow it to carry out year-round operations between Dudinka and Murmansk, independent of icebreaker support.[108] The eastern section of the route is serviced by the FESCO, which operates diesel icebreakers. When the ice is thick, the nuclear icebreakers help the Far Eastern Shipping Company keep the route open.[109]

Nevertheless, the Russian fleet is aging and requires investment and replacement during the coming years to maintain its effectiveness. The interest in using the NSR for transit sailing has also increased significantly during the last years (e.g., from two vessels in 2009 to 34 vessels in the 2011 summer season). If the trend of increasing transit shipping continues in the coming years, there will be a need for an improved fleet of icebreakers to escort ships through the NSR. According to experts in the nuclear shipbuilding industry, Russia needs at least three new icebreakers to sustain trade along the NSR for a sufficient period of time.[110] Russia is planning to complete the construction of a third generation nuclear-powered icebreaker by 2015. A federal program has allocated 17 billion roubles for the development.[111]

5.3.2 Communication and support to navigation on the Northwest Passage

Marine communication and traffic services

The *Canada Shipping Act, 2001* establishes a framework for SAR and communication operations in Canadian Arctic waters. The primary services provided in the Arctic are:[112]

- monitoring international marine radio distress frequencies;
- broadcasting ice and marine weather information, notices to shipping concerning hazards to navigation;
- screening ships entering Arctic waters to enhance the safe and efficient movement of maritime transportation and prevent pollution.

The CCG operates two seasonal Marine Communication and Traffic Services (MCTS) centers, one in Iqaluit and the other in Inuvik. The MCTS program provides marine safety communications and manages the movement of vessel traffic. Currently, MCTS are seasonal, opening at the end of May and closing at the end of November. In some areas of Canada, MCTS regulate the movement of commercial shipping through the mandatory participation of vessels in traffic control schemes. However, this is rather limited on the NWP. This means there is no single agency or Automatic Identification System that keeps up to date with vessel movements on the NWP, or is necessarily aware of all of the planned entries into those waters.[113]

A binding regulatory vessel traffic zone referred to as the Northern Canada Vessel Traffic Services Zone Regulations (NORDREG) has been adapted for the NWP. Coming into force on 1 July 2010, the Canadian Government, pursuant to the Shipping Act of 2001, passed the Northern Canada Vessel Traffic Services Zone Regulations (NORDREG),[114] covering Canada's coastal northern waters. More precisely, NORDREG covers the Shipping Safety Control Zones referred to above, but it also consists of additional waters (see Chapter 6).[115]

Under NORDREG, ships of 300 tons or more are encouraged to follow various reporting procedures including:

- the pre-entry provision of information about ice class;
- amount of oil on board (fuel and cargo);
- date of the Arctic Pollution Prevention Certificate if carried;
- name of the classification society.

It is also strongly advised to have an exit report.

Until 2010, the NORDREG was not mandatory, being a challenge to the sea safety on the NWP. As an example, in 2009 the CCG received reports of a total amount of 160 voyages in the Canadian Arctic. Even though the NORDREG is voluntary, ships regularly report their presence and positions in order to receive the benefit of services from the CCG (e.g., ice information, routing, icebreaker assistance, etc.). According to CCG, in 2009 between 98 and 99 percent of the total number of voyages reported their entering and exit.[116] According to Transport Canada, approximately 95 percent of the total number of voyages reported the same year[117] (i.e., the different sources refer to various percentages). Interestingly, CCG gave the following statement:

> (...) it occurs that cruise ship with up to 200 passengers don't report their entrance and/or position in the Canadian Arctic, because the NORDREG system doesn't require so. This is a serious issue of concern in terms of safe navigation. For example, if any passenger vessel accident should occur, it is essential that the CCG are informed in order to prepare SAR.[118]

This indicated, however, that the Canadian authorities did not have sufficient control of the current marine traffic movements on the NWP.

Aids to navigation and available icebreakers

The CCG responds to air and maritime SAR incidents on the NWP. Rescue centers are staffed by SAR coordinators and operate on a year-round basis. For most of NWP, the Canadian Forces provide fixed- and rotary-wing SAR aircraft from Trenton (see Figure 5.5), while the CCG relies primarily on its helicopters and icebreakers.[119] Regardless, it is a long way from Trenton to the NWP, which means it would take a long time to provide any rescue operation by air transport.

There are very limited modern aids to navigation along the NWP. Unlike southern Canada, for instance, the NWP does not have an available DGPS. There are some marine navigation aids on the NWP, maintained by the CCG, ranging from radar reflectors to buoys, lights, ranges and beacons. Most of these are for the benefit of community access, not for transits by deep draft vessels between the eastern and western NWP. The aids are generally set out from June by icebreakers. When the shipping season ends these aids are picked up and fixed aids are deactivated as the icebreakers leave the NWP towards the end of November.[120] This means that there are very limited marine support and salvage capabilities from the end of November to the end of May. The satellite coverage is considered as adequate. However, without base station reference points, the utility of GPS in close waters is limited. Position accuracy better than 30–100m cannot be expected along the NWP.[121]

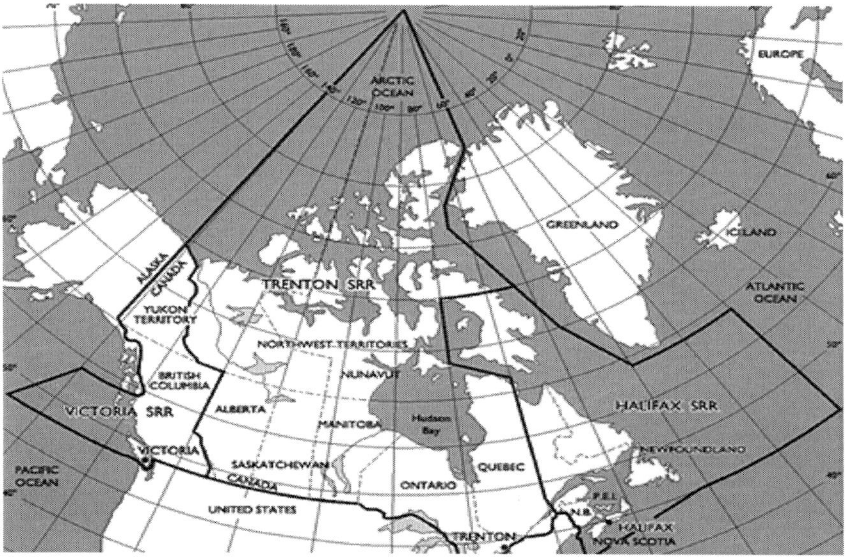

Figure 5.5 National defence SAR regions.[123]

The Canadian Hydrographical Service estimates that about 20 percent of its marine navigation charts relate to the NWP, and only 10 percent of these actually meet modern standards compared with 60 percent of southern charts. If the NWP is to be used by deep draft vessels on a regular basis, then an adequate suite of up-to-date charts is essential for all routes. These do not exist at present, and safe navigation by foreign vessels will not take place until there is confidence that a safe passage can be undertaken.[122]

Canada operates 21 icebreakers, 19 owned by the Canadian government and two by private firms (Table 5.5). The CCG icebreakers are classified as heavy (two vessels), medium (six vessels), and light (11 vessels). Every summer, the CCG icebreakers are deployed to the NWP at the end of June. At present, six icebreakers are deployed (two heavy, three medium and one light) to NWP operations. Anticipated longer active shipping season on the NWP raises a number of service level issues for the Government. One of them concerns the development of icebreaker services. The CCG is faced with an aging fleet and most of the ships date to the late 1970s/early 1980s. However, the current Government has committed 720 million Canadian dollars for a new Polar Icebreaker which is likely to be launched by 2015. However, Canada will need more icebreaker capacity in order to meet the anticipated increased NWP activity.[124]

5.3.3 Communication and support to navigation on the Transpolar Passage

When considering possibilities for the Transpolar Passage, there is currently no active vessel traffic service or other traffic management system in place in the waters of the Bering Strait. AIS capability and shore-based VHF communication services are also limited in the

Ship name	Louis S St. Laurent	Pierre Radisson	Des Groseilliers	Henry Larsen	Terry Fox	Sir Wilfred Laurier
Year	1969	1978	1982	1987	1983	1986
Status	Currently in service, 2010	Currently in service, 2010	Currently in service, 2010	Currently in service, 2010	Currently in service, 2010	Currently in service, 2010
Vessel type	Heavy Arctic icebreaker	Medium gulf/river icebreaker	Medium gulf/river icebreaker	Medium gulf/river icebreaker	Heavy Arctic icebreaker	Light icebreaker
Aircraft carried	2 helicopters	1 helicopter	1 helicopter	1 helicopter	1 helicopter	1 helicopter
Propulsion	Diesel electric	Diesel electric	Diesel electric	Diesel electric	Diesel with reduction gear box	Diesel electric
Ice class	100 A	100 A	100 A	Arctic Class 4	Arctic Class 4	Arctic Class 2
Operations /remarks	NWP/ Support the annual Arctic sealift to coastal communities	NWP/ Search and Rescue unit. Summer escort of commercial ships and start up and maintain navigation aids	NWP/ Participated in a number of science voyages	NWP/ Search and Rescue unit	NWP/ Support the eastern Arctic sealift to coastal communities in the summer	NWP/ Support scientific research, escort of ships occasionally

Source: www.ccg-gcc.gc.ca & http://en.wikipedia.org/wiki/Canadian_Coast_Guard

Table 5.5 Canadian icebreakers available on the NWP during the summer season.

Bering Strait region. Future sailings through the TPP will require significant improvements in marine communication technology. Communications using VHF, MF and HF as well as satellite are generally sufficient for the lower Arctic areas (parts of the NWP, southern Greenland waters and waters of the NSR). However, once the high Arctic is reached, voice and data transmission become problematic. Most modern ships are equipped with satellite digital communications equipment, not only for safety reasons but for the management and navigation of the ship. This equipment relies on geostationary INMARSAT satellites that do not provide service northward of about 80° N latitude, which makes it unsuitable for TPP. Other systems, such as the IRIDIUM constellation of 66 polar orbiting satellites, provide worldwide coverage, including the Arctic. The feasibility of communicating ice charts and satellite images to ships in the Arctic via the IRIDIUM system has been demonstrated but communications are limited. There is 100 percent coverage of the Bering Strait region from the Global Positioning System-Standard Positioning Service. However, the GPS constellation is not configured for optimal positioning in high latitudes, resulting in a potential degradation of position accuracy.[125] This can make marine transportation along various TPPs challenging in the future.

5.3.4 Communication and support to navigation on connecting sea routes

The Northern Pacific Corridor has limited capacity when it comes to SAR services.[126] For instance, the United States has no aids to navigation along the northern slope of Alaska (north of Kotzbue Sound), to support the NPC (see Chapter 1). Furthermore, there are only a few visual aids to navigation in the Bering Sea region, as well as limited capabilities to respond to an incident for lifesaving. Weather and oceanographic observations necessary to support SAR operations are also minimal. The current US short-range aids to navigation in the Arctic extend a short distance north of the Bering Strait where the largest zinc mine in the world (Cominco's Red Dog mine) near Kivalina receives ore carriers. North of the Aleutians along the coast of the Bering Sea, the US has some floating and fixed aids to navigation near the Pribilof Islands and Bristol Bay for tug, barge and fishing vessel traffic. In the Aleutian chain, there are several areas where navigational aids are maintained for local traffic, as well as for the high volume shipping traffic along the North Pacific Great Circle Route between North America and eastern Asia (see Chapter 1). However, the icebreaker capacity is currently limited. The US has three icebreakers, but not all of them are available. The *Polar Star* is currently out of service as it undergoes repairs. A smaller icebreaker, the *Healy*, is primarily used to aid scientific research. That leaves one ship operational right now as an icebreaker.[127]

Denmark has limited SAR capacity on the coast of Greenland but does have a permanent system of radio communication and radar beacons along the west coast of Greenland, supporting the Arctic Bridge (see Chapter 1) in the south.[128] Furthermore, there is a huge demand for SAR services along the western coast of Greenland.

In 2007, a Vessel Traffic Service for the coast of northern Norway was established in Vardø, operated by the Norwegian Coastal Administration. The service is designed to monitor and guide vessels, to promote safe and efficient navigation, and to protect the marine environment in the Barents Sea, and along the Norwegian coast, along the Northern Maritime Corridor (NMC). [129] Norway maintains aids to navigation along its entire coast, supporting parts of the NMC, and at Svalbard along the coast and in fjords. In addition, there are a number of fixed and floating aids to navigation in Svalbard internal waters. Based on analyses of both the changing traffic patterns and the utilization of better risk analysis methodology, it is expected that the requirement for aids to navigation in the Svalbard area will increase. This may partly involve SAR support for any future marine traffic through the Fram Corridor.

Iceland maintains a number of fixed and floating aids to navigation. The facilities include a DGPS system and radio beacons, in addition to a permanent system of radio com-

Figure 5.6 Ports and terminals on the NEP/NSR as of 2012.

munication for radio monitoring of its fishing fleet. Like Norway, Iceland has an advanced Automatic Identification System along its coast which covers the entire coastline. The maritime radio and satellite system has recently been renewed. The traffic monitoring is carried out by the Maritime Traffic Service in Reykjavik operated by the Icelandic Coast Guard. These systems will potentially support the Arctic Bridge in the south and the western leg of the NMC.

5.4 Arctic ports and terminals

5.4.1 Russian ports and terminals on the Northeast Passage/Northern Sea Route

There is uncertainty when it comes to the exact number of ports along the NSR, open to foreign vessels. The same uncertainty also applies to the operational status of the open ports. According to the Russian Government, 41 Arctic ports are open to foreign vessels and additional ports where regulated for visits of foreigners on board foreign cargo ships or tankers with Russian crew members are allowed (Figure 5.6).[130] More than half of the 50 ports on the NSR are today out of operation.[131]

The data on the different ports and terminals have been collected from a range of sources. Not all the NEP/NSR ports can be presented in the subsequent sections. Therefore, we have tried to focus on criteria such as: the port's operative status; openness to foreign vessels; available services; quality of services; control inspections; connecting infrastructure; and data on cargo turnover.[132] More details on the various ports and terminals are presented in Appendix 5, "Arctic ports and terminals".

Ports and terminals in the Barents Sea

A large percentage of today's intra-Arctic and destination-Arctic shipping activity using the NEP involves transportation through the Barents Sea. Development of port capacities in the Russian part of the Barents Region is directly connected to the increase of petroleum exports. In 2008, the ports/terminals of Varandey, Arkhangelsk, Vitino and Murmansk, directly or through the offshore terminals in the open areas of the Barents Sea, exported about 10 million tons of crude oil and petroleum products (in 2002 it was 4 million tons; and in 2004, the amount was almost 12 million tons). In 2009, the volume of oil exported to the western market through the Barents Sea was estimated to be approximately 15 million tons.[135]

The Port of Murmansk[134]

The Murmansk seaport is one of the largest transportation hubs in Russia and had a profit of €22.5 million in 2008.[135] Of all the western NEP ports, the sea port of Murmansk is the only ice-free port that operates on a year-round basis.[136] Large amounts of oil, coal, minerals, metals and other type of cargo are shipped in and out of the Murmansk port every year (see Chapter 3). In terms of Arctic marine traffic, the port is an important shipping destination along the NEP/NSR, and might be for any future traffic along the Arctic Bridge between Murmansk and Churchill (see Chapter 1). Murmansk NSR relevance is twofold. First, most deliveries to the NSR settlements during the summer season originate in Murmansk: Food, equipment and general cargo are loaded onto cargo ships in the port, and fuel products for the NSR are loaded onto tankers at the oil terminals close to Murmansk.[137] Second, Murmansk is the end terminal of the Dudinka-Murmansk line which is operated year-round by cargo vessels and ice-breakers, transporting oil and nonferrous ore from the Norilsk mines to the nonferrous metallurgical industry on the Kola Peninsula. The main limiting factor for the port of Murmansk is its inability to accommodate vessels larger than 45,000 dwt, most of the berths being designed for vessels around 10,000 dwt. Increasing the port's ability to accommodate larger ships is important also for the NSR, as vessels of approximately 50,000 dwt are considered optimal for NSR transit traffic, with potentially even larger vessels later if icebreaking technology improves and larger icebreakers enter operation.[138]

Moreover, in 2006 the design was accepted for a new oil complex and dry cargo called *Severny* (the northern) *Port*, located at the Pechanga Bay on the north coast of the Murmansk region. The project designer, Sevneft Company, intends that the complex will incorporate four terminals.[139] The planned total freight turnover was set to 65 million tons a year, with the prospect of increasing the capacity up to 200 million tons. Sevneft planned to start the construction in 2008, in order that the port could start operating in full scale by 2015. This project has not moved any further since 2006,[140] but in September 2009, the Murmansk governor Dmitry Dmitriyenko announced a 150 billion RUB (€3.4 billion) investment plan for Murmansk seaport.[141] As indicated, there are major development plans, but it remains to be seen whether the plans will become a reality in the future.

The Kola Bay terminals

In 2004, two new coastal oil terminals for offloading oil from railway tanks into sea tankers

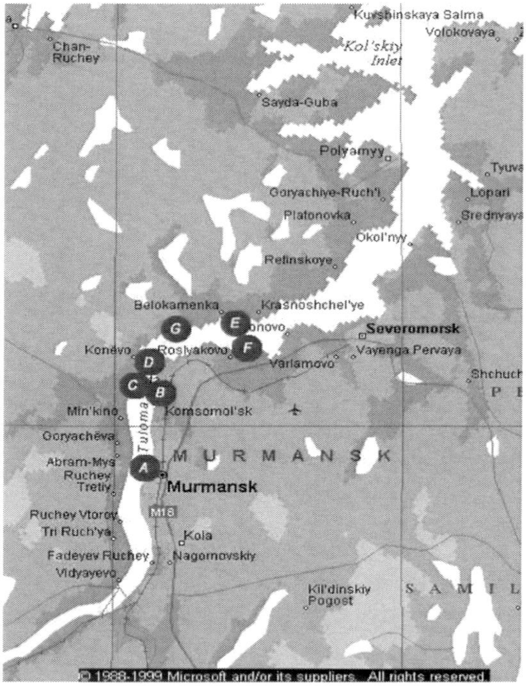

Source: Bambulyak and Frantzen (2005), p. 8

Figure 5.7 Existing and planned terminals in Murmansk and the Kola Bay.

were put in operation at the port of Murmansk, located at the Murmansk Sea Fishery Port (Figure 5.7, Location A) and at the former Ship Repair Factory (Figure 5.7, Location B). The same year, the two terminals received 3.7 million tons of export oil by rail and sent it to offshore transhipment terminals in the Kola Bay with the shuttle tankers.

In addition, there are four offshore transhipment terminals in the Kola Bay. The first offshore oil transhipment, terminal RPK-1 (Figure 5.7, Location C), was constructed by the Murmansk Shipping Company in 2002. Oil is delivered to RPK-1 from terminals in Varandey, Vitino, Murmansk Sea Fishery Port and Ship Repair Factory by shuttle tankers. The second offshore terminal, RPK-2 (Figure 5.7, Location D), in the Kola Bay was built by the White Sea Service Company, and the largest terminal, RPK-3 Belokamenka (Figure 5.7, Location E), was in operation as from March 2004. Belokamenka is the key unit of the northern oil export channel and receives oil from the terminals in Arkhangelsk, Varandey and the Ob Bay. In 2008, Belokamenka exported 3.7 million tons of crude oil, with almost half of it coming from the new Varandey terminal.[142] Currently, oil is transported by rail to a new terminal, RPK-4 Mokhnatkina Pakhta (Figure 5.7, Location F). In 2007, it was shipped 98 000 tons of heavy fuel oil for export.[143]

Ports and terminals in the White Sea

The Port of Arkhangelsk[144]

Archangelsk is a significant seaport in the White Sea. Traditionally it has been a central port for export of timber, but this export has been extensively reduced during the last 20 years.[145] Arkhangelsk was traditionally also the most important gateway for transport to the NSR and a main port for deliveries to NSR settlements. This position has now clearly been lost to Murmansk, even though considerable traffic still takes place during the summer season. One of the main drawbacks is that icebreaker escort is needed during large parts of the year, which is a serious shortcoming compared to Murmansk. Other shortcomings include the shallow harbor (maximum draft 9 m compared to Murmansk's 16 m). However, it was recently announced that a deep water port probably will be built at the island of Mudyug of the Dvina Gulf outside the town of Archangelsk. Construction of the new port is discussed in connection with the Belkomur Railway, a huge project that includes a railway connecting Arkhangelsk with Komi and Perm. This will shorten the distance between Siberia and the White Sea ports by 800 kilometers and open new efficient ways for cargo flows from the Urals and Siberia industrial zones to Europe (Figure 5.8).[146]

Since 2002, Rosneft, the owner of the Arkhangelsk regional petroleum storage depot in Talagi (see Figure 5.8), has also been involved in oil export. The same year, the terminal in Talagi shipped 1.9 million tons of oil and oil products for export. In 2007, Talagi sent 3.2 million tons of crude and petroleum products for export (2.4 million tons of crude via Belokamenka). Rosneft intends to increase the capacity of the Talagi export oil terminal to 12 million tons a year.[149] The *Onega Bay* of the White Sea (see Figure 5.8), established by the Volgotanker Company, was used for export oil transhipment operations during some months of the summer navigation in 2003 only. In 2008, Volgotanker applied for, but did not get, permission from the Russian authorities to restart the oil transhipment operations in the Onega Bay. Later, the same year, Volgotanker was declared bankrupt.[148]

In 2003, the Tatneft Company planned to build an oil export infrastructure in Severodvinsk (see Figure 5.8). The sea depth in the area allows the transport of 40,000 tons deadweight tankers. Tatneft planned to use the terminal for export of its own oil delivered to Severodvinsk by railway. However, the company later notified that they had changed their plans and started to investigate other possible locations.[149]

The Port of Vitino[150]

Vitino seaport (see Figure 5.8), in the Kandalaksha Gulf of the White Sea, was the first private seaport in Russia and it has been the major contributor to oil exports through the Barents Sea during the current years. The port specializes in storage and transhipment of such petroleum as black oil and gas condensate. Oil is also delivered to the port of Vitino by rail, mostly from the terminals in Yaroslavl where it arrives via trunk pipelines owned by Transneft. In Vitino, the oil is shipped to export directly or via the offshore transhipment complex in the Kola Bay. In 2007, Vitino exported 3.9 million tons of condensate and light oil products. And in 2008, 4.4 million tons (including 0.7 million tons of crude) were sent to big European ports directly or with transhipments in Norwegian Sarnesfjord in the Barents Sea.[151] Vitino has upgraded its infrastructure significantly during the last ten years.

Source: Bambulyak and Frantzen (2005)

Figure 5.8 The White Sea ports and terminals.

Still, one of the main shortcomings, similar to the port of Archangelsk, is that icebreaker assistance into the port is required during winter navigation.

Ports and terminals in the Pechora Sea

Varandey[152]

The oil loading terminal in Varandey (see Figure 5.9) was put into operation by the Murmansk Shipping Company in 2000. In 2004, the Varandey terminal shipped 560,000 tons of oil delivered by shuttle tankers to the transhipment terminal in the Kola Bay. The opening of a new Varandey terminal in 2008 was the most noteworthy event in Russian Arctic oil shipment to happen during the recent past.[153] The offshore terminal, operated by Lukoil, can alone double the annual volumes of oil shipped for export through the Barents Sea.[154] The same year, two icebreaking ships were built for Lukoil to support all-year-round oil shipment operation at the Varandey terminal. By 24 January 2010, 10 million tons of oil had been shipped through Varandey oil terminal in the eastern part of the Barents Sea since the start-up in 2008.[155] Varandey's shortcoming is its shallow waters, which only allow access to smaller vessels (maximum 60,000 dwt.)[156]

Kolguev[157]

Peschanoozerskoye oil and gas condensate field on Kolguev Island (see Figure 5.9) is the first production in the Nenets Autonomous Region. The first oil shipment from Kolguev took place in 1987. Since 2002, oil has been shipped by tankers for export, either intra-Arctic, by transhipment in the Kola Bay, or destination-Arctic shipping directly to Rotterdam.

Source: Bambulyak and Frantzen (2005)

Figure 5.9 The Pechora Sea and the oil terminals.

Shortcomings include the shallow harbor (maximum draft 10.5 m compared to Pirazlomnoye's 20.5 m).

Pirazlomnoye[158]

Pirazlomnoye (see Figure 5.9) is a large oil field among the proven oil reserves on the Pechora Sea shelf. Two icebreaking supply vessels were delivered to Sovcomflot in 2006, intended to support operations in Pirazlomnoye. There are major development plans and Gazprom plans to start oil production,[159] but it remains to be seen whether the plans will materialize.

Ports and terminals in the Kara Sea

Novy Port[160]

Novy Port (see Figure 5.7) has been developed in connection with the development of oil and gas fields on the Yamal Peninsula. Its importance relates mainly to the summer season (June–October). From 1979 until the early 1990s, it was a major port for import of large diameter pipes and equipment from Western Europe and Japan. Supply of oil and gas equipment to the Yamal Peninsula peaked in 1988 with 432,800 tons. The port may be used for transhipment between river barges from the Ob River and smaller seagoing vessels. Statistics is limited, but present port activity is believed to be very low. In 1998, the port had a recorded turnover of only 724 tons.[161] Total cargo turnover for the recent years are not available. In September 2009 the Beluga Group vessels sailing from South Korea via the NSR delivered heavy modules at this port. Shortcomings include the shallow harbor (maximum 7 m).

Ob Bay

In 1999, the RITEK Company made the first transhipment of export oil in the Ob Bay (see Figure 5.7). Since then, the oil extracted in the Western Siberia oil fields has been piped via local pipelines to the Ob River, and then delivered by Lenaneft river-sea tankers to the Ob Bay in the Kara Sea. From there, the oil is transhipped into shuttle tankers of 20,000 tons deadweight. Shuttle tankers transport oil intra-Arctic via the NSR to the offshore oil transhipment terminal Belokamenka in the Kola Bay.[162] In 2008, 238, 000 tons of crude were sent westwards by the NSR.[163]

Yamburg

Yamburg (see Figure 5.7) is located just south of Grdiny Point in the Ob Bay. A channel with a depth of 5.5 m leads to the port. The port is equipped with floating and motor cranes, and hospital facilities are available. The port mainly services the nearby oil and gas fields and a railway track connects the port with the giant gas fields of Yamburg and Noviy Urengoy. Yamburg is connected by a road and railroad to Novy Urengoi from where there are regular flights to Moscow. The port owner/operator is unknown; neither were any data of port turnover found. Nonetheless, the port was involved when, in September 2009, the Beluga ships sailed from South Korea via the NSR and delivered heavy modules in the Ob River.

Dikson[164]

Dikson Seaport is situated in the south-eastern Kara Sea near the entrance to the Gulf of Yenisei and is owned by the Norilsk Nickel Company (see Figure 5.7). The port specializes in handling river ships. The infrastructure includes an airport, hospital, radio relay line to Dudinka, radio aids to navigation and a rescue team. Furthermore, Dikson has the location to build up infrastructure for the development of oil and gas production in Taymyr and the Kara Sea. Due to its relatively deep harbor (15 m water depths are located at a reasonable distance offshore), Dikson has been mentioned as a suitable location for a new oil export terminal and for oil and gas condensate from the Vankor field. Rosneft had plans to construct the oil pipeline Vankor-Dudinka-Dikson and an oil loading terminal in Dikson. The capacity of the projected 710 km-long pipeline and the terminal was to be 18 million tons a year. Further, oil was supposed to be delivered intra-Arctic by shuttle tankers to transhipment terminals in the Barents Sea, and then by line tankers to the western market. However, in 2006, the Vankor oil transportation scheme was changed. Oil extract will instead be transported by the 550 km Vankor–Pur Pe pipeline.[165] The western Marine Operations Headquarters is also located at the port, performing control inspections for any foreign vessels sailing on the NSR from the west.

Dudinka[166]

The port of Dudinka (see Figure 5.7) is situated on the Yenisei River approximately 370 km from the mouth of the river and is operated by the Norilsk Nickel Company. Both a sea-vessel port and a river-vessel port are available. Dudinka has adequate connecting infrastructure, with an airport as well as the terminus of the Dudinka–Norilsk railroad. The port's main function is to export non-ferrous metals and ores from the Norilsk industrial complex located 50 km to the east, and to import supplies to the city. However, there are plans for future development of the port. When new oil and gas fields are developed in Taimyr and the Kara Sea, the plans for constructing oil-loading terminals in the ports of Dudinka and may come through. In 2002–2006, Dudinka shipped 20–40 thousand tons of oil a year for export via the Barents Sea.[167] Norilsk Nickel has built a fleet of six double-acting icebreaking bulkers that allow it to carry out year-round operations between Dudinka and Murmansk, independent of icebreaker support. The port is reported to be one of the few NSR ports in adequate technical condition.[168]

Igarka[169]

After the NSR was opened for foreign vessels and during the first half of the 1990s, Igarka was the only port on the NSR permanently open for foreign vessels, thus its historical importance.[170] Most of the export out of Igarka has traditionally taken place by ships of the Northern Shipping Company. The port specializes in the export of timber, and its main purpose has been to load and tranship timber and wood products both from the Igarka plant and from the upper reaches of the Yenisei River. Igarka was a significant port in the 1970s and 1980s, reaching a cargo turnover figure of approximately 800,000 tons in 1976. Foreign exports have also been considerable. However, traffic dropped sharply in the 1990s to an average of around 40–60,000 tons annually. Data from recent years are unavailable, neither are any data of the port capacity.[171] The port is open for navigation only during the summer season (June–October) and its shallow harbors only allow vessels with draft of up to 8m.

Ports and terminals along the Laptev Sea

Tiksi[172]

Tiksi is one of the principal ports for accessing the Laptev Sea. A large part of its activity consists of transhipment between sea vessels and river barges going up the Lena and Yana rivers. The Tiksi port has not shipped oil for export along the NSR since 2002. However, it is interesting due to its logistics, and as an example of how East Siberian oil can be delivered to the western markets. In 2001, the company Sakhaneftegaz in cooperation with the Murmansk Shipping Company started oil loading in Tiksi for shipping export via the NSR. The first 19,000 tons of crude from the Talakanskoye oil field in the Eastern Siberia were loaded to the tanker *Magas* and sent to the western market. The oil produced at the Talakanskoye field was delivered via a 110-km local oil pipeline to the oil refinery and the terminal in Vitim on the Lena River (see Figure 5.10). From there, it was transported via the Lena River on Lenaneft tankers belonging to the Lena River Shipping Company to oil storage facilities in the port of Tiksi. In Tiksi, oil was shipped to sea tankers and delivered to Rotterdam. Recently, oil from the Talakanskoye field has been piped in reverse mode by a launched 1100-km-long pipe to Tayshet. When the East Siberia Pacific Ocean oil pipeline

Source: http://en.wikipedia.org/wiki/Lena_River

Figure 5.10 Port of Tiksi and the Lena River.

is completed, Talakan oil will be piped eastwards.[173] However, the AMSA report stated that after completion of construction of the railway to Yakutsk on the Lena River, the role of the port of Tiksi on the NSR will most likely increase.[174] Recent data about the port are limited.

Ports in the East Siberian Sea

Pevek[175]

Pevek is one of the most important of the ports of the eastern NSR, both as a control headquarters for ships that enter the NSR from the east and as a center of NSR infrastructure because it has bunker, supply and repair facilities for NSR cargo vessels and icebreakers. The port is operational 150 days per year. However, all-year-round operation is possible with the assistance of icebreakers. Specialization is cargo handling for the extractive industry of the Chaunsky region and Pevek city, as well as cargo transhipment and storage. Pevek hosts the eastern Marine Operations Headquarters (MOH), operated by the Far Eastern Shipping Company, and located onboard one of FESCO's icebreakers stationed in

Pevek.[176] The port's authority stretches as far as 125° east, just west of the Lena River. Past this point, the western MOH, based in Dikson, has control. The headquarters is responsible for the management of all the NSR voyages from the east (i.e., convoys, provides up-to-date ice information, etc.). Despite its importance as the eastern MOH, the level of cargo handled in the port has been declining rapidly in recent years, although there is still gold mining around Bilibino and Leningradsky.[177] Port reconstruction and the construction of new objects of port services are required. Recent statistics from the port are limited.

Ports in the Bering Sea

Provideniya[178]

The port of Provideniya is situated in the deep Provideniya Bay on the Chukotka Peninsula. The overall depths in the bay are 30–35 m, and 9 m near the berths. The port is operational May to January with icebreaker assistance. Tug assistance is available and the Provideniya Hydrobase provides navigation information. The port has been considered the eastern gateway to the NSR, being the eastern start point for westbound convoys. The technical standard is reported to be insufficient and will require significant improvement in order to increase its relevance to the NSR.

5.4.2 Canadian ports on the Northwest Passage

In contrast to the NEP/NSR, there are very few ports along the NWP. There are essentially no adequate deepwater ports along the northern slope of Alaska or throughout the NWP. The lack of major ports and port facilities is also one considerable limiting factor in terms of increased shipping on the NWP. Furthermore, there are several obstacles related to trans-Arctic shipping in the near future (i.e., ice conditions, refuge, seasonality, geographical complexity, draft restrictions, choke points, insurance limitations and other costs) (see Chapters 1 and 4). Even so, most Arctic shipping scenarios indicate an increase in the years ahead, but the shipping will be destination-Arctic and intra-Arctic and mostly resource based. These expectations are also related to a not significant, but nonetheless increasing Arctic population and an increasing demand for consumer goods, community construction and industrial shipments. Based on these expectation significant port improvements will be required.

Port developments on the Northwest Passage

Nunavut (see Figure 5.11) is a region with a lot of opportunities but equally sizeable challenges. The natural resource potential is particularly important. Gold, diamonds, iron ore, uranium, rare earths, natural gas and oil are all present. However, most of these resources are land-locked in regions where transportation presents major physical and economic difficulties. Nunavut has no permanent road or rail links to the south or between communities and there is no deep-sea port. Moreover, shipping is seasonal with most communities normally receiving two supply calls a year. Air transportation also has constraints, with quality in many communities inadequate for a jet service. Of relevance to shipping, a community port programme has been initiated that is financially supported by both the federal and pro-

Source: http://www.casr.ca/id-arctic-empires-2.htm

Figure 5.11 Ongoing projects in the NWP.

vincial governments.[179] Recently, plans have been announced to build a deep-water port in Iqaluit. It will consist of a new single berth in the form of a 77m long caisson placed at the 11m depth mark and a new bridge that would connect the port facility to the mainland.[180] This port could be a main gateway to the NWP from the east. There are also plans to build a new deep-water port in Nanisivik. The existing port already has basic docking platforms and a fuel tank storage facility. Used by the occasional cruise ship and CCG vessels, the port is near Lancaster Sound, the eastern entrance to the NWP.[181] In addition, there are several resource-driven ongoing projects. Over the next 20 years, destination exports shipping will most likely include a port at the Mary River Iron Ore located at Baffin Island, which expects to produce annually 12.6 million tonnes of direct shipping iron ore per year and operate for at least 25 years.[182] European steel mills will be the primary destination for the ore. The project and its infrastructure will also provide a foundation for potential future production increases after 2014. Other ongoing development projects are:

- a port at Steensby Inlet at the Foxe Basin;
- Roche Bay magnetite port, near Igloolik at the Foxe Basin (possible operative within 2015);
- lake lead/zinc/copper concentrates shipping from either Gray's Bay or Bathurst Inlet (possible operative within 2015).

Destination import shipping will include logistics and fuel for the primary resource operations noted above as well as barge-mounted production modules for the Mackenzie pipeline and delivery of production modules to the Alberta Oil Sands.[183]

Churchill[184]

The port of Churchill is currently the only port that can be characterized as a "Canadian Arctic Seaport",[185] but it has not yet grown into any viable transportation link. As a result, there are not many economic activities in the community[186] and the population has been declining in recent years. The combination of port activity, some tourism, and government support should sustain the remaining population. Nonetheless, the port is a vital link in the transshipment of petroleum products and goods of all kinds to the communities of Nunavut. Vessel operators now provide scheduled services to the Nunavut region each season, providing a reliable supply of goods to communities in the Kivalliq region.[187] Previously, there have been discussions regarding the development of the "Churchill-Murmansk route", Arctic Bridge, as an alternative to shipping through the St Lawrence Seaway, which could shorten shipping time by some four days.[188] However, for this proposal to move forward, Churchill would require significant additional port and rail investments. The proposal is reportedly being studied at present by both Canada and Russia. Costs, cargoes and volume have not been determined (see Chapter 1).

5.5 The costs and bureaucratic procedures of Arctic shipping

5.5.1 Costs and fees on the Northern Sea Route

The main characteristic of the NSR is that parts of the route are covered with ice. This requires the assistance of icebreakers, for which the Russian government charges fees. The icebreaking fee plays the same role for the NSR as does the Suez Canal toll when it comes to the economics of using these routes. The administration of the NSR is rather bureaucratic and has very little transparency. Although the route in principle is open to international shipping, one needs to plan this a long time ahead, as applications for the passage must be filed months before.[189]

Administration and procedures

The overall responsibility for the NSR lies with the Administration of the NSR, with headquarters in Moscow. The practical administration, however, lies with two Marine Operations Headquarters (MOH), one on each side of the NSR. Their areas of responsibility are divided in the Laptev Sea at longitude 125°E. Ships planning to sail from west to east are referred to the MOH in Dikson, and those planning to sail from east to west must contact the MOH in Pevek.

The MOHs are responsible for scheduling, route assignment, navigational support and pilotage – in short, all operational aspects of a passage. The very first step in arranging for a NSR passage is to obtain official permission. The ship owner must submit an application to the Administration of the NSR in Moscow at least four months in advance, with a copy to the respective MOH, depending on the direction of the trip. For an additional fee, this time frame may be shortened to one month.[190] The application must contain the following elements:

- age and name of the vessel;
- flag of the vessel and the port of registry;
- address and communication numbers for the ship owner;
- gross and net tonnage and displacement of the vessel;
- principal dimensions of the ship (length, breadth, draft);
- engine output, the service speed and the design and propeller material;
- type of bow construction (bulbous or knife);
- ice class of the ship, name of classification society and date of last examination;
- presence of certificate of insurance or other financial security with respect to potential civil liability for environmental pollution damage;
- purpose of the voyage and the ports of origin and destination;
- preferred place for inspection by the administration's inspector;
- expected date of voyage.

Within 10 days after receiving the application, the NSR Administration will respond with their decision.[191] However, the NSR Administration is not always consistent. In 2008 and 2009, the Beluga Shipping Group applied for permission to sail through the NSR from South Korea. In both cases, the NSR Administration took several months to process the application.[192]

After preliminary approval has been obtained, a representative of the MOH must inspect the ship for ice navigation worthiness. The ship owner is expected to bear all costs associated with getting the ship to an agreed port for this inspection.

If the ship is found worthy, the ship is granted "Permission for leading through the seaway of the NSR". The MOH will then schedule the date, set up the specific route and provide the necessary icebreaker and pilotage support. Furthermore, if the ship is not in accordance with the requirements, she can still be guided through the NSR for an additional fee.[193]

Russian Northern Sea Route fees

UNCLOS Article 234 could be argued to define the basis for fees:

> Coastal States have the right to adopt and enforce non-discriminatory laws and regulations for the prevention, reduction and control of marine pollution from vessels in ice-covered areas within the limits of the exclusive economic zone, where particularly severe climatic conditions and the presence of ice covering such areas for most of the year create obstructions or excep-

> tional hazards to navigation, and pollution of the marine environment could cause major harm to or irreversible disturbance of the ecological balance. Such laws and regulations shall have due regard to navigation and the protection and preservation of the marine environment based on the best available scientific evidence.

The fee system is meant to finance several aspects of public services to make the fairways open to traffic and to protect the environment. The main cost elements would be:

- icebreaker support, including maintenance and renewal of the icebreaking fleet;
- maintenance costs on fairway infrastructure;
- ice–air reconnaissance flights;
- hydro-meteorological services;
- satellite communication;
- pilotage.

The philosophy of the Russian government has always been that the tariffs should be set so that the actual traffic finances the cost elements listed above. This has led to a rather strange development of the fees. It is difficult to talk about an average fee, as the fee will be dependent on a number of different factors, like:

- type of ship/cargo;
- size of ship (larger ships have lower fees per ton);
- ice class of ship;
- ice conditions (determines how much icebreaker assistance is needed);
- competences of captain and crew (inexperienced crew will require more pilotage/helmsmen);
- the actual route.

The NSR is divided into three different regions:

A: Novaya Zemlya to Severnaya Zemlya (60°E to 90°E)
B: Severnaya Zemlya to the Bering Strait (90°E to 169°W)
C: All areas north of the 78°N parallel

The fees in region A are 70 percent of those in region C, while region B's fees are 80 percent of those in region C. The region C tariff will be applied for any full transit voyage or one that traverses at least two of these regions.

In the mid 1980s, some 6–7 million tons of cargo were carried along the NSR and the icebreaker fees were around US$3 per ton. In the mid 1990s, the traffic had been reduced to around 2.2–2.4 million tons, so the fee was increased to US$7.50 per ton in 1995. The traffic continued to decline to less than 2 million tons in 2000, and the resulting loss of revenue meant that the state had to subsidise the icebreaker operations quite substantially. In 2003, the Ministry of Economical Development and Trade issued a decree called "About Changing Rates for Ice-breaking Fleet Services on the NSR", which led to a substantial

Type of cargo	Fee US$/ton
Mechanical engineering products	74.60
Nonferrous metal	51.90
Container cargo	31.70
Liquid cargo	16.00
Northern delivery	7.80
Timber cargo	4.50

Table 5.6 Icebreaking fees for major cargo types 2003.

increase in the icebreaker fee to an average of US$23 per ton. The tariff is differentiated according to type of cargo as given in Table 5.6.

If we assume that one container corresponds to 24 tons in its loaded condition, a fee of US$31.70 per ton would amount to US$760. The current freight rate for North Europe to Asia traffic is around US$800/TEU. Using the NSR for a container vessel would effectively double the freight rate, so this fee system would definitely not attract any container traffic on a regular basis. To put this into perspective further, a 4300 TEU ship would have to pay US$240,800 for a Suez passage. Assuming a 70 percent load factor, that means that the fee per loaded TEU is US$80 or only 10.5 percent of the NSR fee.

In 2009, the Murmansk Shipping Company was approached in order to obtain information about the current fee structure.[194] The request went unanswered, but in a telephonic follow-up call a fee of US$40 was mentioned per ton of container cargo. This level of fee seems quite unrealistically high, and it is probably fair to conclude that fees are negotiable. As mentioned in Section 6.6.1, the Beluga ships paid only €60,000 in icebreaker fees, which corresponds to only US$2.25 per dwt for high-value project cargo.

Other expenses will normally also occur when using the NSR, some of which are:

- communication services (US$2.20 per minute for telefax and US$4.50 per minute for telephone;
- Russian helmsman (if needed) will typically cost US$33 per da;
- maps and various guidebooks will typically cost US$700–900;
- bunker filling fee is US$6.30 per ton and the oil is at world market prices;
- supply of fresh water can vary from US$1 up to US$12 depending on place and quality.

5.5.2 Costs and fees on the Northwest Passage

Essentially, the NWP is open to navigation, and there is no fee system for this passage.[195] The Canadian government monitors the fairway, but there are examples of ships having navigated the NWP without the knowledge of the authorities.[196] The NWP is a dangerous route to navigate, so without a formal system for transit passage, one can only classify a transit here as hazardous. If a ship ends up in a distressed situation, the Joint Rescue Coordination Centers will no doubt respond to any SAR incidents. It is highly unclear, however, what the bill for such an operation will amount to. It is unlikely that the Canadian government will introduce any icebreaker fees in the future. Currently, it is very costly to ship cargo on the NWP, and the Canadian government wants to encourage the economic viability for any future transportation.[197]

5.6 Marine insurance in Arctic waters

Marine insurance is an essential service to international shipping. It is generally agreed that, without marine insurance, international commercial navigation in the Arctic would not be economically viable.[198] Unlike most aspects of marine transportation, there is no international convention on marine insurance.[199] Moreover, insurance practices are driven by insurance markets. Marine insurance allows ship owners and carriers generally to take on the risk of trading on Arctic routes and liabilities imposed by statute. Shippers and consignees[200] would also want to protect their cargoes. Today, even though many of the risks associated with shipping are well known and understood by insurers, there are still risks related to Arctic navigation that need to be identified. Underwriters normally base their underwriting premiums on a historical loss record, such as statistics, and the frequency of accidents is a key element in the evaluation of risk when navigating in Arctic waters. Knowledge of accident rates can help determine insurance rates. However, lack of Arctic empirical data and significant statistics makes it difficult for insurers to compose an overall risk assessment related to the risks of the various Arctic shipping routes. With the exception of the Russian NSR experience and the research that was provided through the INSROP and the ARCOP, there is only limited knowledge about the marine insurance aspects of Arctic shipping. As a result, the provision of insurance for Arctic shipping tends to be on a case-by-case basis,[201] expensive and also requiring self insurance.[202]

Of particular importance for Arctic shipping is Protection and Indemnity (P&I), offered through so-called P&I Clubs, and represented by the International Group of P&I Clubs.[203] Each club is an independent, non-profit-making mutual insurance association, providing cover for its ship owner and charterer members against third party liabilities linked to the use and operation of ships. Clubs cover a wide range of liabilities, including personal injury to crew, passengers and others on board, cargo loss and damage, oil pollution, wreck removal and dock damage.[204] In addition, underwriters normally charge a surcharge with respect to Hull and Machinery (H&M) and Cargo Insurance. However, at this time, there do not appear to be evidential insurance market patterns for Arctic shipping, partly because of the rarity or specialized activities to date.[205]

5.6.1 Marine insurance aspects of the Northern Sea Route/Northeast Passage

An important initiative in the marine insurance aspects of the NSR was the INSROP (1993–1998). According to the Russian legislation, a compulsory insurance is required for foreign ships trading in the NSR.[206] Furthermore, it appeared unlikely that the shipping

Russian Arctic seas	**Average intensity of navigation**	**Percentage of damages N=800**
Kara Sea	Highest	40%
Laptev Sea	Unknown	20%
East Siberian Sea	Unknown	21%
Chukchi Sea	Lowest	14%

Source: The data is taken from Østreng (1999a), pp. 10–11

Table 5.7 Ice damage to ships on the NSR (1954–1990).[215]

industry would develop an interest in the NSR without the necessary economic incentives or much better information about the route.[207] The substance of these aspects was not fully recognized when INSROP was initially designed. During INSROP it became clear "... that the Russian as well as the global marine insurance market was willing and able to underwrite NSR navigational and related risks, provided that appropriate and adequate information structures of the risks involved could be presented to underwriters."[208] Although the NSR had been used for many years this had been confined to specialized, mainly Soviet shipping. In that respect, shipping tended to be insured under state schemes. During the 1990s, the state monopoly in the Russian insurance sector ceased to exist, and the Russian insurance sector was now a proper marine insurance market with many companies competing for business for various types of risk coverage, assisted by a modernized legal system. In other words, the Russian marine insurance market is today able to underwrite NSR coverage specifically.[209]

INSROP recognized that risks requiring coverage were H&M, cargo insurance, P&I and a number of miscellaneous risks.[210] P&I insurance, including cargo insurance, depends on ship size and type of cargo, and H&M insurance compiles with the ship price, taking account for the estimated damage rate. Furthermore, those H&M policies do not cover navigation in ice-infested waters and require special arrangements with the insurer. For instance, and although double hulled vessels have been operating in Arctic waters for a long period, a double hull could involve higher repair and maintenance costs and, thus, higher insurance premiums. Regarding P&I Clubs, INSROP found that some of the Clubs had experience of Russian operations. However, claims on the NSR would be higher than those experienced elsewhere. In particular, wreck removal, pollution salvage and towage, cargo, and crew claims were of particular concern.[211] INSROP estimates showed that the total insurance rate would be expensive for the ship owners and almost twice the comparable insurance rate prevailing in shipping via the Suez route.[212] Overall, there was a significant lack of reliable statistical data and the insurers suggested that initial premiums would be conservative until a better data base is provided. Russia was not willing to share much of its official data during INSROP, for instance on the rate of damages. However, some statistics of damages along the NSR were presented. It was stated that during the period 1954–1990 a total of 800 damages (22 a year) were reported on the NSR (see Table 5.7).

Mode of operation	Average frequency
Independent movement	5–10%
Movement behind icebreaker	50–60%
Close towing	15–20%
Breaking down of ice near the ship, compacting	5–10%
Ground touching in shallow water	1–2%

Source: Grishchenko (1999), p. 88

Table 5.8 Hull damage frequency in different modes of operation (1954–1990).

The largest number of accidents were found in August–September during the easiest ice conditions. Moreover, it was stated that 82.5 percent of all ship damages were caused to the hull. The rest was caused to propellers, shafts and steering gear.[213] Relatively many old ships of L1 class had been involved in the damages. L1 class ships, as the lowest rated ice capable class, do not have appropriate strengthening and are sufficient for navigating in summer seasons in very light ice conditions. The most hull damages have occurred in the eastern NSR, East Siberian and the Chukchi seas, due to their more difficult ice conditions principally.[214]

Of note, is that approximately 10 percent of ice damage to ships occurs when navigating independently; 60 percent of damage occurs when following an icebreaker; and the remaining 30 percent results from towing, grounding and other operations in ice (see Table 5.8).

On the other hand, the potential insurers would need a more reliable and extensive database in order to calculate the risk. That was the INSROP conclusion drawn through further contacts with the major marine insurance actors. One of the main problems was lack of significant statistics showing the relative variations. Insurers are not comfortable in responding to hypothetical or theoretical proposals. However, INSROP stated that scattered data about damages to vessels caused by ice do exist, but a more analytical approach and description discussing the damages caused to vessels of higher ice class in different ice conditions would be needed. The actual damages should also be compared with theoretical estimates and class regulations. More detailed descriptions of both the ice conditions and the actual damages would make it possible to apply these data on other vessels, i.e., more detailed statistical data on the operations along the NSR must be presented.[216] In the next stage, these data should support the different risk estimation procedures of future operators.

ARCOP was a three-year (2003–2006) research project that concentrated specifically on accessing the oil and gas resources of the Arctic regions of Russia. The main purposes were to determine what specific changes had taken place since INSROP and what risk factors would likely affect the insurance.[217] ARCOP faced complexities similar to that experienced during INSROP. This was due to the fact that the marine insurance industry was generally unwilling to provide information on risk coverage based on hypothetical or theoretical questions. However, it was again stated that if marine underwriters are presented with a realistic situation, i.e., full information on a named vessel, information on the route to be taken, the cargo to be carried, the duration of the voyage, and such other information that a practical insurer may require, it appeared to be certain that a realistic insurance rate would be quoted by the insurance negotiator involved.[218] This was basically the same conclusion drawn from the INSROP project, namely that marine insurers appear to be willing to underwrite the various risks involved in NSR shipping operations. However, this could not be fully confirmed until more practical evidence of the actual planned shipping services was being provided.[219]

Non-Russian commercial transits of the Northern Sea Route/Northeast Passage

What came out the ARCOP and the INSROP studies was that marine insurers are capable of responding to new risk coverage demands in many innovative ways, but such risks must be based on practical experiences. Given the organizational structure of the insurance industry and how it works, it seemed clear that the most effective way of evaluating the future of insurance for NSR shipping is to perform experimental voyages on the NSR with non-Russian ships. According to ARCOP, there appeared to be little new evidence that cargo exporters and importers had expressed further interest in the use of the NSR and there was less concern with actual trans-Arctic shipping. Neither INSROP nor ARCOP research had been able to present marine insurers with significant data.

The *Kandalaksha* (1995) experimental voyage, using the NSR/NEP from Yokohama to Kirkenes, carried out by INSROP,[220] had been more a test of propulsion technology, communications and navigation systems, ice strength, ship hull design, etc., than a commercial venture. Although the experiment was important and successful, the project group concluded that further experimental voyages were needed in order to test the commercial aspects, including marine insurance.[221]

The Arctic Demonstration and Exploratory Voyage (ARCDEV) of April/May 1998, was carried out to demonstrate the technical feasibility of transporting oil/gas condensate from the Russian Arctic to Western Europe by icebreaking tanker and to evaluate the economical situation of such transport. The Finnish tanker *Uikku* was used to find out the objectives of creating basic material on economics of marine transportation including marine insurance cost. According to the Russian NSR Guidelines, a mandatory insurance is required and financial security issued for potential liability in case of marine environmental pollution. The fact that nuclear-powered icebreakers are widely used in the NSR created a specific problem related to insurance cover. The marine insurance underwriters excluded insurance coverage for any kind of consequences of damages from radiation of nuclear materials. On the other hand, and in accordance with the Russian regulations, the Russian nuclear-powered icebreaker operators do not have any liability insurance which would cover for any kind of consequence of nuclear-related incidents. This was a problem in the case of *Uikku*. According to the project group, it implied that the ship owners are provided services which remove their insurance cover and they expose themselves to risks that are not accountable.[222] The following response was presented:

> It is common knowledge that nuclear related risks are beyond the scope of cover under all marine insurance policies. Thus any kind of nuclear related incident or liability would be sorted out in a way that bona fide ship owners would be held harmless. Uninsured nuclear risks are

> alien to western operators and an obstacle to make reliable risk assessment for investors. In a logical sense those who gain profits from using optional nuclear fuel would solve the problems with additional insurance costs and liabilities in case of incidents.[223]

Moreover, their concern was related to the fact that the assisted vessel cannot choose what kind of fuels used on icebreakers. In addition, the project group disclosed that H&M coverage operates with a much higher premium than normal. The higher premium was made due to the fact that there has not been any major hull damage in past operations. Again, this is similar with INSROP and ARCOP findings. Even if this was a practical case for the insurance industry, it reacted with a high risk assessment due to lack of available statistics.

Recently, the transits of German company Beluga Shipping of the NEP during August and September 2009 was one practical experiment that would be of interest to insurance companies relative to their need for reliable data in their risk analysis. Two ships delivered various heavy lift modules from Ulsan, South Korea, to Novy Port/Yamburg at the River Ob, via the NEP/NSR by passing the Novaya Zemlya and continuing to Arkhangelsk, where ships loaded steel pipelines for delivery to Nigeria. The exact marine insurance coverage and cost is currently not available, but according to one representative from Beluga, the marine insurance aspect of the NEP/NSR transits was not a significant barrier when considering the economic profitability.

The INSROP and ARCOP marine insurance projects focused on responding to some specific questions in order to determine whether risk coverage for the NSR navigation was realistic. One of them related to certain time- and cost-saving advantages in using the NSR over the Suez Canal routes. However, the exact economic advantages are not known. This was also due to the lack of detailed knowledge related to costs for icebreaker assistance and ice navigators, real-time ice data, reliable weather forecasts and administrative procedures and acceptance in the Russian bureaucracy. However, the Beluga case gave some indications in that respect. For instance, by using the NSR/NEP over the Suez Channel, Beluga made financial savings of about US$100,000 alone for bunker costs plus US$20,000 daily for each day traveling the NEP whiched shortened the usual voyage time.[224] Overall, about US$300,000 per vessel were saved by transiting along the NSR instead of using the Suez Channel. Another factor that must be added to the calculation is the cost related to icebreaker escort and navigation. According to Beluga Shipping representatives the two Beluga ships had Russian Ice Navigators steering the ships through the passage. In addition, two nuclear-powered icebreakers were hired during the transits. The icebreaker hire was handled as a regular service component included in the "Charterers Agreement", and it was arranged that the client should cover the costs for icebreaker assistance. According to Beluga, the cost of hiring icebreaker escort was around €60,000, not a very significant amount. Furthermore, the icebreaker assistance for parts of the voyage was something that both the client and the insurers paid great attention to. "However, nobody insisted, since icebreaker assistance is a regular part and integrated into the NSR regulations anyway. You need to pay for them no matter whether you use them or not."[225] According to Beluga, special meetings were held between their experts and representatives of the insurers. The details of Beluga's planning and preparation processes, the routing proposal, technical details of the vessels, etc., had been presented to the underwriters who then decided on the extra premium applicable for these particular cases., Beluga was not willing to give details about the extra premium, since that was part of private negotiations. Still, Beluga was willing to share some of the insurers' risk concerns, including:

- vessels may encounter unknown, unexpected coastlines or areas of shallow water;
- vessels may encounter sea areas strongly covered with broken ice;
- vessels may get stuck in ice;
- vessels may lose propulsion due to damage or lost variable pitch;
- less functionality of radar/nautical equipment.

Another aspect that would reduce the risk and high insurance premium of transiting the NEP is how the project was organized and prepared in advance. In that respect, Beluga had gone through detailed preparations and procedures for the voyages, i.e., modern ice-hardened vessels, experienced seafarers, a committed department for Health, Safety, Security & Environment, a future-oriented department for Research & Innovation and it has engaged chartering experts. In addition, Beluga had three full-time employed meteorologists with access to up-to-date satellite pictures and analyses of the ice situation on a daily basis. During the transits, they provided the captains with daily routing suggestions avoiding ice contact as best as possible to minimize the risk.

It might only be a speculation as the insurance cost in the case of Beluga is unknown, but if the financial savings are as significant as suggested, it is also possible that the company was willing to pay a relative higher insurance premium, as opposed to the ARCDEV project in which the insurance was regarded as an economic obstacle.

However, Beluga was quite restrictive regarding the insurance issues. According to Beluga: "(...) the risk issues are seen as an experimental advantage and marks a main difference between our approach and others who wants to sail the NSR, but don't know how to deal with the concerns"[226] In other words, Beluga claimed that they knew how to satisfy and reassure the insurers, "(...) but we are not revealing that to everybody for quite obvious reasons."[227]

5.6.2 Marine insurance aspects of the Northwest Passage

There is limited knowledge and research on the marine insurance aspects of the NWP, in contrast to the NSR research as previously presented. In light of any expected increased marine transportation in the years ahead, there are several risk aspects which could imply that the insurance premium will be high to cover H&M and P&I. For instance, the combination of draft and severe ice conditions along the NWP would be significant risk factors which would affect the insurance rate. The standard route through the NWP, for instance, is draft limited to 10 meters. This has the potential to exclude many of the standard types of ship unless they are very lightly loaded. The deep draft routes via the Prince of Wales Strait (or M'Clure Strait) would be suitable in this sense, but have the most extreme ice conditions in the Canadian Arctic. It is plausible that if sea ice melt in the central Arctic Ocean continues, as many climate models indicate, there is potential for more mobile multiyear sea ice to be swept southward through the NWP. In terms of marine transportation in the NWP, ice conditions are likely to get worse rather than better (see Chapters 1 and

4). The observations also show a very high year-to-year variability of sea ice coverage on the NWP, an important factor of uncertainty when considering marine insurance. The severe ice conditions would potentially damage hulls, which means that additional H&M insurance would be required in order to cover the risk. The P&I cost would most likely be very high for any ship owner that enters the NWP. The Canadians have a very strict regime when it comes to shipping regulations, operating with control zones regulated by the ASPPR. One example is the sinking of the *Arctic Ublureak* in 1983, when the Canadian government required that the vessel should be removed the following summer. The cost of wreck removal was probably borne by the P&I Club involved, not by the hull insurers, because the vessel was a constructive total loss and the cost of removing the vessel exceeded its value when recovered.[228]

When considering the availability of icebreaker assistance it seems also quite limited. Currently there are six icebreakers available in the Canadian Arctic including the NWP, but only during the summer season (from June to November). However, two of the icebreakers are assigned to research projects. In addition, these are all diesel powered and far from being as powerful as the Russian nuclear icebreaker fleet. Altogether, in light of any increased destination-Arctic shipping activity, as anticipated,[229] it is quite uncertain to what extent the assistance capabilities are. However, this will most likely exceed the insurance cost.

5.6.3 Marine insurance aspects of the Transpolar Passage

Regarding aids to navigation, it is quite limited on the NWP. For instance, there is no DGPS system in contrast to the southern part of Canada, which is an important safety capacity for ships operating in the NWP.

Currently, there has been no commercial voyage using the central Arctic Ocean as a shortcut. However, if one considers the possibility of any future TPP, the insurance rate is likely to be as high, if not higher, than the risk factors already discussed for the NWP and NSR/NEP. Higher risks for ice damage to ships and potential damage to cargoes in extreme cold temperatures, and the nonexistent maritime infrastructure in the Central Arctic Ocean (such as salvage, ports and emergency response), will most likely be factors in determining future TPP insurance rates. If an accident should occur hundreds of nautical miles in ice and outside any reasonable distance for any salvage of SAR operation, it is likely that the premium cost would be so high that it would not be economically profitable for ship owners.

5.7 Information needed in order to make further risk assessments

The "Arctic Infrastructure" chapter and the main aspects that have been discussed, reflect the major challenges and uncertainties insurers face when the criteria for premium will be decided for vessels operating in Arctic waters.

Technical requirements

Arctic shipping includes sailing in both the most severe ice conditions and open water.

This will differ with respect to season, but also in relation to various Arctic geographical areas. Given that there are yet no international mandatory requirements regarding ice class, it is left to the ship owner to decide the operational elements. However, the insurer will most likely base the insurance rate on a combination of the operational area, ice conditions and the vessels' technical standard. If the ship is sailing in waters where ice conditions are unpredictable, the underwriters will most likely require that the vessel has an adequate ice class. Although it is not a direct requirement in the regulations, this will still be an indirect requirements from the underwriters. Data on the possible accidents and statistical accident rates for the NSR, NWP and TPP would be valuable when preparing preventive actions and estimating the additional costs caused by these accidents.

Crew training and competence requirements

According to *Assuranceforeningen Gard*, a well-prepared and trained crew and an experienced and trained ice navigator are two of the most important aspects to consider when insurance deals are negotiated. Important factors include how many times the ice navigator has made this route before, and whether crews are prepared for winter navigation. However, previous discussions in this chapter have shown that the international and national requirements for crews and ice navigators in the NSR and NWP are very limited, as are the ship owners' requirements. This could mean that large parts of crews operating in the Arctic have limited training. On the other hand, it would be difficult for underwriters to examine the crew's operational basis, since thereis as yet no internationally recognized educational certificate.

SAR

As discussed in this chapter, available SAR facilities diverge between different Arctic areas. Based on previous discussions in this chapter, it is reasonable to conclude that the SAR facilities generally are more accessible on the NSR than on the NWP. One of the issues ship owners are facing is whether the vessel will sail on the NSR independently or with the icebreaker assistance. One might think that any use of icebreaker assistance would reduce the risk of accidents significantly. However, the available statistics of hull damage average frequency on the NSR show that approximately 10 percent of ice damage to ships occurs when navigating independently; 60 percent of damage occurs when following an icebreaker; and the remaining 30 percent results from towing. Furthermore, it is still unclear whether these statistics are valid and significant. It does not state, for example, how much of the operations have taken place in the winter and/or summer season, or the specific areas operated along the NSR. Other SAR factors of importance for underwriters would be the availability of electronic maps, real-time ice information, weather forecasts and how many days ahead the weather forecast would be reliable. Moreover, are any SAR units available within a reasonable distance? And how much time does it take for any rescue unit to arrive in terms of the sea route's total extent?

Arctic ports

As previously discussed, availability of Arctic ports and port facilities differs between the NSR and the NWP. There are very few ports along the NSR that meet modern southern port standards. The underwriters would base their risk assessments on many of the criteria discussed in this chapter. Typical issues would relate to available facilities such as are there ice free ports available? Is there available icebreaker assistance and/or tugs when vessels enter the port? Are there adequate loading/unloading facilities? Any transhipment challenges? One of the main obstacles identified along the NSR, for instance, is lack of adequate water depths suitable for potential vessels trading into the ports. These are all factors that would be important when underwriters consider the likelihood for any damage to cargo and vessels when using the ports.

5.8 Some concluding remarks

This chapter has examined six aspects of the available infrastructure on the NSR, the NWP and various TPPs. A few temporary concluding remarks are in order – temporary because they are likely to go through a process of change to emerge in a somewhat different form in subsequent chapters of this book. Moreover, in Chapter 7 we will analyze and compare the three transport passages in terms of shipping and Arctic infrastructure.

- The IMO Guidelines is an incomplete operational framework for crews and navigators operating in Arctic ice-covered waters. The Guidelines need considerable improvement in order to provide adequate safety measures. Moreover, there is an increasing demand for competent seafarers and it is expected that there will be an estimated shortage of 4,000 seafarers with appropriate experience and training. Currently, there is no international recognized education for ice navigators or qualification scheme for individuals operating in Arctic waters. In addition, there are examples of ship owners operating in the Arctic whose crews are insufficiently trained for winter navigation. Training courses are expensive and the most common type of practice today is the so called "on-the-job experience", which means that inexperienced and unqualified seafarers are integrated as part of the regular crew and have to learn various ice operations from experienced colleagues.
- There are some 50 ports on the NEP/NSR, but only a very few meet the standards required by the international shipping industry. The majority of ports are ice covered most of the year and cargo vessels need icebreaker assistance in order to enter. The main problems are poor technical standards, lack of water depth, and lack of sufficient port facilities and connecting infrastructure. Recent investments and developments of NEP/NSR ports are primarily related to current or planned oil and gas activities. There are currently no adequate ports on the NWP.
- The current situation for the NSR is that icebreaker fees are negotiable, and that the fee system in itself is highly non-transparent. Until the NSR is established as a regular transit route, the actual fees one will have to pay will be open to negotiation. If the Russian government wants to attract transit traffic through the NSR, the fee levels must be commercially reasonable. The NWP has currently no fee system and the Canadian government will most likely not introduce any fees as it will discourage the economic viability for any transportation on the passage.

- IMO has divided Arctic waters into SAR coordination and responsibility areas. Canada is responsible for providing meteorological information, traffic services and coordination of the eastern part of the Bering Strait, the northern slope of Alaska and the NWP. Russia is responsible for the NEP, from the Barents Sea in the west to the Bering Sea in the east. The first international agreement made exclusively for the Arctic region was signed at the ministerial meeting in Nuuk on 12 May 2011. The SAR Agreement is the first legally binding agreement established under the auspices of the Arctic Council and covers search and rescue of aeronautical and maritime vessels and passengers. There are limited aids to navigation and traffic service systems in order to monitor ships entering Arctic waters. GPS and DGPS are adequate for some areas, but in most areas it is not available. In addition, Canada operates a mandatory reporting system, NORDREG, for vessels that operate on the NWP. On the other hand, the Russian Administration regulates navigation on the NSR through the NSR Guidelines. Before any vessel can enter the NSR, the Marine Operational Headquarters requires a control inspection in west or east, depending on which direction the vessel is sailing from. Russia has a considerably larger and stronger icebreaker fleet to support operations on the NSR than Canada has for the NWP. Russia has five nuclear icebreakers available for NSR support, in addition to several river and diesel icebreakers operating on a year-round basis. Canada operates six icebreakers (two heavy, three medium and one light) for seasonal (May–November) NWP operations. Both fleets are aging and need improvements in light of any increased ship traffic in the future.
- Previous studies show that the international insurance market would be willing and able to underwrite NSR risks. Nevertheless, lack of empirical data and significant statistics make it difficult for marine insurance underwriters to compose an overall risk assessment related to the risks of the various Arctic shipping routes. As a result, the provision of insurance for Arctic shipping tends to be on a case-by-case basis. Some statistics are available, based on Russian marine traffic (1954–1990). Approximately 10 percent of ice damage to ships occurs when navigating independently; 60 percent of damage occurs when following an icebreaker; and the remaining 30 percent results from towing, grounding and other operations in ice. However, further research is required in order to examine the risk.

Notes

1 Such as the *Convention on the International Regulations for Preventing Collisions at Sea*, 1972 (COLREGS), the *International Convention on Load Lines*, 1966 (Load Lines Convention), the *International Convention on Standards of Training, Certification and Watchkeeping for Seafarers*, 1978 (STCW), and the *International Convention on Maritime Search and Rescue*, 1979 (SAR Convention)

2 SOLAS (1974), Chapter 1, and Van der Zwaag et Al. (2008), p. 14

3 SOLAS (1974), "Technical Provisions"

4 Jensen (2007), pp. 8, 11

5 Jensen (2007), p.17

6 Ibid p. 17

7 Ibid, and Van der Zwaag et al. (2009), p.310

8 Such as the IACS *Unified Requirements for Polar Class Ships*

9 IMO Guidelines (2002)
10 As defined in IMO Guidelines (2002), paragraph G-3.2
11 IMO Guidelines (2002)
12 Jensen (2008), p.109
13 IMO Guidelines (2002), Section G-3.18
14 Van der Zwaag et al. (2008), p. 70
15 Jensen (2007), p.14
16 IMO Guidelines (2002), Resolution A.1024 (26)
17 Jensen (2007), p.14
18 Van der Zwaag et al. (2009), p. 18
19 Jensen (2007), p.18
20 Jensen (2007), p.18
21 The following sections are included in the NSR Guidelines: "Regulation for Navigation on the Seaways of the NSR (1991)"; "Regulations for Icebreaker-Assisted Pilotage of Vessels on the NSR (1996B)"; "Requirements for Design, Equipment and Supply of Vessels Navigation the NSR (1996C)"
22 www.mintrans.ru
23 Requirements for Design, Equipment and Supply of Vessels Navigation the NSR (1996C), Art.2.10 and 11
24 Ibid, Art.3.1
25 Ibid, Art. 6.2 to 6.5 .
26 Ibid, 7.1 to 7.3 and 8.1 to 8.3
27 Ibid, Art.2.2 and ARCOP D2.2.1 (2004)
28 ARCOP Workshop 1 (2003), p.18
29 See Section 5.6.1 of this chapter
30 http://www.barentsobserver.com/34-vessels-in-transit-on-northern-sea-route.4991248.html and
31 http://rus-shipping.ru/ru/stats/?id=53
32 http://www.barentsobserver.com/shipping-to-double-on-northern-sea-route.5018647.html
33 http://ria.ru/arctic_news/20111128/500547773.html
34 AWPPA (1970), Chapter A-12
35 ASPPR (1978), pp. 3 and 6
36 ASPPR (1978), p. 26
37 http://www.tc.gc.ca/marinesafety/debs/arctic/acts-regulations/images/zones2.JPG
38 10, 8, 7, 6, 4, 3, 2, 1A and 1
39 A, B, C, D and E
40 Van der Zwaag et Al. (2009), p. 53
41 CASA (2007), p. 124
42 CASA (2007), p. 124
43 Interview with Transport Canada, 15.01.10
44 Interview with Transport Canada, 15.01.10
45 IMO Guidelines (2002), Section 13.3.3
46 AMSA (2008), p. 190
47 IMO Guidelines (2002), Section 14.1.3
48 IMO Guidelines (2002), Section 14.1.4
49 IMO Guidelines (2002), Section G-3.10
50 IMO Guidelines (2002), Section 1.2.1
51 IMO Guidelines (2002), Section 14.2
52 ARCOP Workshop 6 (2006), p. 15

53 AMSA (2008), p. 217
54 Jensen (2008), p. 110
55 IMO Guidelines (2002), Section 11.5.3
56 Jensen (2007), p.16
57 IMO Guidelines (2002), Section 14.1.3
58 IMO Guidelines (2002), Section 14.1.4
59 AMSA (2008), p. 217
60 Jensen (2007), p. 23
61 Østreng (1999b), pp. 365–366
62 Østreng (1999b), p. 366
63 Østreng (1999b), p. 425
64 Section B – V/g "Guidance regarding training of masters and offcers for ships operating in polar waters", Manila 2010
65 Annex 2, Resolution 11 refers to IMO Guidelines (2009)
66 Eger and Kristiansen (2011)
67 See the Regulations for Design, Equipment and Supply of Vessels Navigation the NSR (1996C), Art. 91 to 9.5
68 Guide for Navigation through the Northern Sea Route, (1996a), Art. 4
69 Regulations for Icebreaker-Assisted Pilotage of Vessels on the NSR (1996b), Art. 2.17,
70 www.mintrans.ru
71 ASPPR (1978), Chapter 12
72 ASPPR (1978), Chapter 18
73 ASPPR (1978), Chapter 26 .
74 Based on an interview with the Canadian Coast Guard, 14.01.10
75 http://en.wikipedia.org/wiki/MS_Explorer
76 Based on an interview with Transport Canada, 15.01.10
77 See section: "The IMO Guidelines: The complexity of the recommendations"
78 ARCOP D3.7.1 (2004), p. 6
79 Tankers, Gas Tankers, Chemical Tankers, Bulk Carriers, Containerships, RO-RO, Barges and Passenger Vessels AMSA (2008), p. 204
80 AMSA (2008), pp. 201–204
81 ARCOP D3.7.2 (2005)
82 ARCOP D3.7.1 (2002), p. 4
83 ARCOP Workshop.9 (2006), p. 129
84 AMSA (2008), p. 191
85 Based on an interview with the Canadian Coast Guard, 14·01.10
86 Based on an interview with the Canadian Coast Guard, 14·01.10
87 Based on an interview with the Canadian Coast Guard, 14·01.10
88 Based on an interview with the Canadian Coast Guard, 14·01.10
89 ARCOP D3.7.1 (2002), p. 6
90 AMSA (2008), p. 203 and ARCOP Workshop 9 (2006), p. 17
91 ARCOP Workshop 9 (2006), p. 17
92 All the Arctic states have ratified the Agreement
93 Van der Zwaag et al. (2009), p. 22
94 IMO COMSAR (2006)
95 http://www.weatheroffice.gc.ca/marine/index_e.html
96 The Arctic Council (2011), The Agreement on Cooperation on Aeronautical and Maritime Search and Rescue in the Arctic. Ministerial meeting in Nuuk, May 12, 2011 http://library.arcticportal.org/1474/1/Arctic_SAR_Agreement_EN_FINAL_for_signature_21-Apr-2011.pdf

97 Grishchenko (1999), p. 80
98 ARCOP WP D3.6.1, (2004), p. 12
99 ARCOP WP D3.6.1, (2004), p. 36
100 ARCOP WP D1.1 (2005a), p. 16
101 AMSA (2008), p. 208
102 AMSA (2008), p. 194
103 Grishchenko (1999), p. 78
104 AMSA (2008), p. 195–196
105 AMSA (2008), p.196
106 http://portal.fma.fi/portal/page/portal/fma_fi/merenkulun_palvelut/talvimerenkulku/icebreakerfleets.pdf. The Russian state-owned fleet plus several privately-owned companies' icebreakers are included in these numbers
107 Atomflot is part of the state nuclear corporation Rosatom
108 http://www.bellona.org/articles/articles_2009/atomflot_prays_for_sun
109 http://www.barentsobserver.com/norilsk-nickel-boosts-arctic-independence.4647401-116320.html
110 Bakharin (2006), p. 26
111 http://www.bellona.org/articles/articles_2009/atomflot_prays_for_sun
112 http://www.freerepublic.com/focus/f-news/2234272/posts
113 CASA (2005), p. 22
114 CASA (2005), p. 29
115 Northern Canada Vessel Traffic Services Zone Regulations (SOR/2010-127) http://www.gazette.gc.ca/rp-pr/p2/2010/2010-06-23/html/sor-dors127-eng.html
116 The waters of Ungava Bay, Hudson Bay and Kugmallit Bay that are not in a shipping safety control zone; the waters of James Bay; the waters of the Koksoak River from Ungava Bay to Kuujjuaq; the waters of Feuilles Bay from Ungava Bay to Tasiujaq; the waters of Chesterfield Inlet that are not within a shipping safety control zone, and the waters of Baker Lake; and the waters of the Moose River from James Bay to Moosonee
117 Based on an interview with the Canadian Coast Guard, 14.01.10
118 Based on an interview with Transport Canada, 15.01.10
119 Based on an interview with the Canadian Coast Guard, 14.01.10
120 Rompkey (2009b), p. 43
121 AMSA (2008), p. 176
122 CASA (2007), p.149
123 CASA (2007), p.149
124 Rompkey (2009b)
125 AMSA (2008), p. 149–150
126 AMSA (2009), p. 109
127 AMSA (2008), p. 176
128 http://arcticfocus.com/2009/07/13/coast-guard-needs-more-icebreakers-for-arctic/
129 AMSA (2009), p. 163
130 AMSA (2009), p. 164
131 Russian Federation (2006)
132 Interview with former Head of NSR Administration Vladimir Mikhailichenko in:
133 https://www.riverlake.ch/news.php?newsid=87
134 Often, information from the different sources has been inconsistent; in these cases, the author has tried to apply the best of his judgment when choosing which data to emphasize
135 Bambulyak and Frantzen (2009), p. 31
136 See Appendix 5.6 for detailed information of the Port of Murmansk

137 http://www.barentsobserver.com/murmansk-port-and-sovcomflot-on-privatisation-list.4656621-116320.html
138 AMSA (2008), p. 201 and CNIIMF (2004), p. 6
139 Ragner (2000b), p. 86
140 Ragner (2000b), p..81
141 AMSA (2008), p. 205
142 Bambulyak and Frantzen (2009), p. 60
143 http://www.barentsobserver.com/murmansk-port-and-sovcomflot-on-privatisation-list.4656621-116320.html
144 Bambulyak and Frantzen (2009), p. 59
145 Bambulyak and Frantzen (2009), p. 57
146 See Appendix 5.7 for detailed information about the Port of Archangelsk
147 http://www.ascp.ru/en_htm/9.htm and Ragner (2000b), p. 80
148 http://www.barentsobserver.com/one-step-closer-to-deep-water-port-in-arkhangelsk.4609737.html
149 Bambulyak and Frantzen (2009), p. 50
150 Bambulyak and Frantzen (2009), p. 51
151 Bambulyak and Frantzen (2009), p. 51
152 See Appendix 5.8 for detailed information about the Port of Vitino
153 Bambulyak and Frantzen (2009), p. 53
154 See Appendix 5.9 for detailed information about the Varandey Terminal
155 Bambulyak and Frantzen (2009), p. 53
156 Bambulyak and Frantzen (2009), p. 42
157 http://www.barentsobserver.com/10-million-tons-shipped-from-varandey-oil-terminal.4725304-116321.html
158 Ragner (2000b), p. 88
159 See Appendix 5.10 for detailed information about the Kolguev Terminal
160 See Appendix 5.11for detailed information about the Pirazlomnoye Terminal
161 Bambulyak and Frantzen (2009), p. 46
162 See Appendix 5.12 for detailed information about Novy Port
163 Ragner (2000b), p. 82
164 Bambulyak and Frantzen (2005), p. 5
165 Bambulyak and Frantzen (2009), p. 41
166 See Appendix 5.13 for detailed information about the port of Dikson
167 Bambulyak and Frantzen (2009), pp. 40–41
168 See Appendix 5.14 for detailed information about the port of Dudinka
169 Bambulyak and Frantzen (2009), p. 41
170 Ragner (2000b), p. 83
171 See Appendix 5.15 for detailed information about the port of Igarka
172 Van der Zwaag et al. (2009), p. 67
173 Ragner (2000b), p. 84
174 See Appendix 5.16 for detailed information about the port of Tiksi
175 Bambulyak and Frantzen (2009), pp. 39–40
176 AMSA (2008), p. 207
177 See Appendix 5.16 for detailed information about the port of Tiksi
178 Ragner (2000b), p. 87
179 http://en.wikipedia.org/wiki/Pevek#Port
180 See Appendix 5.18 for detailed information about the port of Provideniya
181 CASA (2007), pp. 36 and 49

182 http://www.casr.ca/id-iqaluitport2.htm
183 http://www.cbc.ca/canada/story/2007/08/10/port-north.html
184 www.baffinland.com
185 AMSA (2008), p. 151
186 See Appendix 5.19 for detailed information about Churchill Port
187 Interview with Transport Canada 15.01.10 and Canadian Coast Guard 14.01.10
188 CASA (2007), p. 54
189 http://www.portofchurchill.ca/
190 CASA (2007), p. 55
191 The information basis for this section is ARCOP Workshop 7 (2006), ARCOP Final Report (2006), Liu & Kronbak (2009), and the NSR Guidelines (1996)
192 Regulations for Icebreaker-Assisted Pilotage of Vessels on the NSR (1996b), Arts.2.1 and 2.2
193 Regulations for Icebreaker-Assisted Pilotage of Vessels on the NSR (1996b), Art. 2.3
194 The information was received from the Beluga Shipping Group.
195 NSR Guidelines (1996), Art. 2.5.
196 Liu & Kronbak, 2009
197 According to Transport Canada and the Canadian Coast Guard
198 See Section 5.3.2 of this chapter
199 Interview with Transport Canada, 15.01.10
200 ARCOP D2.4.2 (2006)
201 Van der Zwaag et al. (2009), p. 40
202 A consignee is the person to whom the shipment is to be delivered
203 Gold (1999), p. 257
204 Self insurance is a risk management method in which a calculated amount of money is set aside to compensate for the potential future loss
205 The major clubs are based in Japan, Norway, Sweden, the United Kingdom and the United States
206 http://www.igpandi.org/
207 ARCOP D2.4.2 (2006c)
208 Regulations for Navigation on the Seaways of the NSR, 1991
209 Gold (1999), p. 253
210 Gold (1999), p. 255
211 Gold (2000), p. 116
212 ARCOP Workshop 7 (2006), p. 55
213 INSROP WP 28 (1996a), pp. 5–7
214 INSROP WP 95 (1996) and Kitagawa (2001), p. 97
215 INSROP WP 28 (1996a), p. 24
216 INSROP WP 28 (1996a), p. 24
217 Kitagawa (2001), p. 156
218 See ARCOP D.2.4.1 (2004b) and Gold (2009), pp. 307–310
219 ARCOP Workshop 7 (2006)
220 Gold (1999), p. 256
221 Gold (1999), p. 256
222 Gold (1999), p. 255
223 ARCDEV (1998), p. 26
224 ARCDEV (1998), p. 24
225 Beluga Group Report (2009)
226 Conversation with Beluga representative
227 Conversation with Beluga representative

228 Conversation with Beluga representative
229 Van der Zwaag et al. (2009), p. 41
230 AMSA (2009)

6 Ocean law, coastal waters, jurisdictional controversies – indigenous peoples' rights*

This chapter starts with an overview of the parts of international ocean law laid down in the third UN Law of the Sea Convention (UNCLOS), assumed to be of particular relevance when navigating along coastal sea routes in the Arctic. These parts of international law relate more precisely to the coastal states' regulatory rights in their maritime zones which ships navigating in such waters have to obey. As jurisdictional controversies could cause uncertainty in terms of navigation, disagreement on the legal status of parts of the Northeast Passage/the Northern Sea Route and the Northwest Passage respectively will be dealt with as well. Before going more specifically into these sea routes, the chapter will give an overview of IMO Guidelines relevant for the Arctic, as well as discussing what can be defined as corporate governance in these waters. As there is a growing interest about the Arctic from the international community, the chapter also provides some insight into new treaties and political initiatives that will extend cooperation in the Arctic in relation to issues as well as to parties.

Shipping in the Arctic implies sailing in areas inhabited by various groups of indigenous peoples that for ages have been using Arctic lands, waters and ice to sustain their livelihood and their culture. The second part of the chapter deals with indigenous peoples' rights to self determination laid down in international conventions and declarations as well as in national legislation, claims agreements and settlements with their respective Arctic mother states. These are rights that will have to be taken into consideration by all parties operating in the Arctic.

*Section 6.1– 6.4 is written by Brit Fløistad, Lars Lothe, Ocean Futures. Section 6.2.3 is written by Willy Østreng, Ocean Futures and Section 6.5 is written by Brit Fløistad.

6.1 International ocean law, IMO Guidelines, "Soft Law" including corporate governance and Arctic initiatives

6.1.1 International ocean law: Coastal maritime zones

The Arctic Ocean is, like all other oceans, subject to international ocean law laid down in UNCLOS. Of particular relevance in regard to navigating along Arctic coastal sea routes is, as indicated above, coastal states' maritime zones such as internal waters, territorial sea, contiguous zones and Exclusive Economic Zone (EEZ), all of which can be found in the waters of the five states bordering the Arctic Ocean. Navigating along the Northeast Passage/the Northern Sea Route and the Northwest Passage thus means sailing in waters which are under some form of coastal state jurisdiction.

The point of departure in determining the breadth of these coastal waters is the baseline defined as "the low water line along the coast".[1] Where the coastline is indented or cut off, or when there is a fringe of islands in the immediate vicinity of the coast, a straight baseline, from appropriate points,[2] can be drawn, not departing to any appreciable extent from the general direction of the coast. In regard to bays, a baseline closing the bay can be drawn if the distance between the entrance points of a bay does not exceed 24 nm.[3] Here, account can be taken of the reality and importance of economic interests of long usage to the region concerned,[4] and the 24 nm limit could also be exceeded in so-called "historical bays". [5] The waters between the coast and the baseline will form part of the coastal state's internal waters. In these waters, a coastal state's sovereignty equals that of its land territory. That is to say, a foreign ship will need permission from the coastal state to enter these waters, the exception being a customary law duty for coastal states to give refuge to ships in need of assistance. Beyond its internal waters, the coastal state has a territorial sea, not to exceed 12 nm. Here, foreign ships have the right of innocent passage.[6] Passage here means that a ship traverses such waters, continuously and expeditiously, without entering into internal waters or calling at a port. Foreign ships can stop and anchor only if this is incidental to ordinary navigation, necessary because of *force majeure,* or for rendering assistance to other ships or persons. Submarines entering coastal states' territorial waters are required to surface and show their flag. For the passage to be innocent it must not be "prejudicial to the peace, good order or security of the coastal state", and must take place in conformity with international law. The coastal state can adopt certain laws and regulations relating to innocent passage through its territorial waters, to which foreign ships will have to comply. Such regulation can apply to navigation safety, to protecting facilities and installations, cables and pipelines, and to the preservation of living resources and the environment. The coastal state is also entitled to prescribe stricter rules in regard to innocent passage through its territorial sea[7] as long as such rules do not "apply to the design, construction, manning or equipment for foreign ships unless they are giving effect to generally accepted international rules or standards." [8] When necessary for safe navigation, the coastal state can also designate or prescribe certain sea lanes for foreign ships exercising their right to innocent passage, particularly when such ships are carrying nuclear or other dangerous or noxious substances. The coastal state has a duty to indicate clearly such lanes on a chart.[9]

While sovereignty in the territorial sea follows from sovereignty over the land mass, the second set of maritime zones has to be specifically proclaimed. More precisely, a coastal state can proclaim a contiguous zone, which is a zone contiguous to, and extending another 12 nm beyond, the territorial sea, that is 24 nm from the baseline. In such a zone, the coastal state can exercise control over foreign ships in order to prevent infringement of its customs, fiscal, immigration or sanitary laws.

Pursuant to UNCLOS, coastal states have the right to proclaim an Exclusive Economic Zone.[10] The breadth of such a zone may not exceed 200 nm from the baseline. Within its EEZ, coastal states have "sovereign rights for the purpose of exploring and exploiting, conserving and managing the natural resources, whether living or non-living ...".[11] That is to say that the coastal states have an exclusive right to resources as well as a duty to manage such resources in a sustainable way. With regard to the right of shipping in an EEZ all states, whether coastal or land-locked, enjoy "the freedoms referred to in UNCLOS, Article 87"; that is to say, the freedoms of the High Seas regime, which includes the right to navigation.[12] The coastal state may lay down certain rules and regulations in its EEZ to which foreign ships will have to comply. In so doing, the coastal state shall, however, "have due regard to the rights and duties of other coastal states" and also "act in a manner compatible with the provisions of the Convention". National regulations within an EEZ must thus be in accordance with international law, implying the possibility of detecting the limits of coastal states' lawmaking.

In addition to the coastal state regimes mentioned above, UNCLOS contains a so-called Arctic clause – article 234 – which gives coastal states bordering the Arctic Ocean certain rights regarding lawmaking within their EEZ. Constituting a special regime for ice-covered areas, the article reads as follows:

> Coastal States have the right to adopt and enforce non-discriminatory laws and regulations for the prevention, reduction and control of marine pollution from vessels in ice-covered areas within the limits of the exclusive economic zone, where particularly severe climatic conditions and the presence of ice covering such areas for most of the year create obstructions or exceptional hazards to navigation, and pollution of the marine environment could cause major harm to or irreversible disturbance of the ecological balance. Such laws and regulations shall have due regard to navigation and the protection and preservation of the marine environment based on the best available scientific evidence.

The precise interpretation of article 234 could be disputed, raising questions such as: what is to be considered "particularly severe climate conditions"; what is meant by ice covering areas "most of the year"; and when can the presence of ice be said to create "exceptional hazards to navigation"? [13] What seems clear, however, is that the purpose of article 234 is to prevent, reduce and control pollution from vessels navigating in what is considered particularly vulnerable waters and to achieve this by giving coastal states a broader right to control shipping in their EEZs than what is otherwise inherent in this regime. Such laws and regulations must, however, be non-discriminatory, i.e., they apply to national as well as foreign ships. They also have to be based on the best available scientific data, which means that a coastal state may not implement laws and measures solely on its own judgment of potential hazards and harm to the environment. Ships enjoying sovereign immunity, such as warships and non-commercial governmental ships are exempted from article 234.[14] As the Arctic clause only applies within a coastal state's EEZ, it has no bearing on shipping on the High Seas outside such zones.

Canada was the driving force for article 234 to be included in UNCLOS, the reason being the country's own Arctic Water Pollution Prevention Act of 1970 which by this inclusion would be regarded as being consistent with international law.[15]

As for the continental shelf, i.e., the sea-bed and subsoil throughout the natural prolongation of the land territory, the coastal state exercises sovereign rights for the purpose of exploring and exploiting the natural resources on its shelf.[16] Coastal states' rights in regard to their continental shelves do not affect the legal status of the superjacent waters,[17] and has thus no bearing on navigation. Coastal states' oil and gas production on their continental shelves may, however, have a certain impact on shipping as installations and structures, including safety zones, will have to be respected by ships navigating in the vicinity.[18] However, such structures and zones must not be established when they interfere with the use of recognized sea routes.

6.1.2 The Polar Code: IMO Guidelines

The International Maritime Organization (IMO) is a specialized agency for the United Nations with 169 member states. The IMO's main task is to develop and maintain a comprehensive regulatory framework for shipping addressing issues within "safety, environmental concerns, legal matters, technical cooperation, maritime security and the efficiency of shipping".[19] The IMO must be characterized as the central and representative forum for all issues of concern to the shipping industry internationally.

The core legal instruments for safety and environmental protection at sea have been, and still are, the SOLAS convention (Safety of Life at Sea) , the STCW convention (The International Convention on Standards of Training, Certification and Watchkeeping for Seafarers 1978/1995), and MARPOL (International Convention for the Prevention of Pollution from Ships), all ratified by every Arctic state. These conventions were adopted within the framework of IMO, they apply globally and, of course, also to the Arctic Ocean. Specific rules for the Arctic were originally suggested by Germany as an amendment to the SOLAS convention.[20] Development of safety for vessels operating in polar waters began in 1993 within the framework of the IMO. In 2002, the IMO Guidelines for Ships[21] (the IMO Guidelines) operating in Arctic ice-covered waters were adopted,[22] applicable to all parts of the Arctic Ocean. The IMO-Guidelines are therefore a supplement aimed at the special circumstances in the Arctic. This is precisely mentioned in the preamble:

> ACKNOWLEDGING that the polar environment imposes additional demands on ship systems beyond the existing requirements of the International Convention for the Safety of Life at Sea (SOLAS), 1974 and the International Convention for the Prevention of Pollution from Ships, 1973, as modified by the 1978 Protocol relating thereto (MARPOL 73/78), as amended.[23]

For a more detailed analysis of the IMO Guidelines' material contents see Chapter 5. Some main characteristics discussed in this chapter relate to the issues dealt with by the IMO Guidelines: construction, equipment, operational, and environmental protection and damage control. First, we will discuss the meaning of the current IMO Guidelines being "non-binding" and quickly analyze the desirability and impact of a binding regime.

The IMO Guidelines are non-mandatory, being merely recommendations. For the IMO Guidelines to become legally binding thus depends either on the individual states implementing the regulations in their national legislation,[24] or their adoption as a binding treaty, maybe in the form of an amendment to the SOLAS convention. The IMO have been trying since 2008 to reach agreement on legally binding the Guidelines. Here, it may be of interest to note that in his foreign policy statement to the Storting on 10 February 2009, Minister of Foreign Affairs Jonas Gahr Støre pointed to the importance of not only completing the amendment of the IMO Guidelines but to make them mandatory. Currently, (March 2012), preparatory work is being done to produce a Polar Code which will be binding by way of reference from the existing SOLAS and MARPOL conventions. The latest estimates by IMO indicate the Polar Code will become binding from 2016. Even though this work is uncompleted, the IMO Guidelines already have major implications in their current form and status.

Firstly, the IMO Guidelines are adopted by a very representative and authoritative international body reflecting the intentions of all IMO member states. In its introductory remarks, the IMO Assembly suggests that member states should give considerable weight to the Guidelines:

> INVITES all Governments concerned to take appropriate steps to give effect to the annexed Guidelines for ships constructed on or after 1 January 2011; ENCOURAGES all Governments concerned to take appropriate steps to give effect to the annexed Guidelines for ships constructed before 1 January 2011 as far as is reasonable and practicable; RECOMMENDS Governments to bring the annexed Guidelines to the attention of ship owners, ship operators, ship designers, shipbuilders, ship repairers, equipment manufacturers and installers and all other parties concerned with the operation of ships in polar waters.[25]

Further, the IMO Guidelines are already of considerable practical significance through their actual use by the shipping industry and classification societies by adopting the IACS Unified Requirements for member societies (see Chapter 5). The main reason may very well be simply that the Guidelines provide the actors with a useful tool. Although not tried before any court, it is a reasonable assumption that compliance with the Guidelines will be given substantial weight in any case where negligence on the part of the ship-owner or operator becomes an issue. As most liability, e.g., for oil spill, will be strict liability, the question of negligence will most likely emerge if the insurer seeks recourse from the insured after first having paid for the loss incurred. By the adoption of the IACS requirements which are linked to the IMO Guidelines by reference, adherence and compliance will also have an impact on the possibilities of getting insurance at all. It should here be mentioned that availability and cost of marine insurance is a restraint to Arctic shipping. The insurance industry has been willing to provide insurance on a case-by-case basis, but the lack of actuarial experience makes it very difficult to assess the risks involved.[26]

Making the Guidelines mandatory will still make a substantial difference. We will see later in this chapter, and we have seen in Chapter 5, that the Arctic states have passed national legislation within the areas covered by the IMO Guidelines, most importantly Russia and Canada. However, legally-binding IMO Guidelines (in the form of a treaty or amendment to the SOLAS convention) will provide an obligation for the Arctic states to adopt legislation with standards at least as ambitious as those contained in the Guidelines. This provides potential for harmonization of the requirements that would be beneficial to the shipping industry. However, the Arctic states have very different legal and legislative traditions and they would in any case be free to adopt regulation with higher standards than the Guidelines. One should therefore not exaggerate the harmonization effect. Although the Arctic states can enforce the regulations within their national waters, it would add substantial value if all members of the IMO joined in adopting them, not least so that non-Arctic flag states become bound by the rules. Further, adoption of mandatory rules will certainly add to existing regulations in areas beyond national jurisdiction, such as the TPP.

As for the material contents of the Guidelines, we will mention the following with reference to the more detailed text in Chapter 5.

The unconditional requirement in the definition of "ice-covered waters" is a sea-ice construction of 1/10 coverage or greater, which poses a structural risk to ships. There is no requirement, as in article 234, for "ice most of the year". Thus the IMO Guidelines for Ships apply to the entire Arctic Ocean with the explicit exception of some areas including waters adjacent to the Norwegian mainland coast and the waters adjacent to the Kola Peninsula (see the Introduction and Figure 3).

The point of departure for the IMO Guidelines is that the Arctic represents an environment which imposes additional demands on ship systems and that safe operation in ice-covered areas requires special attention.[27] The IMO Guidelines thus define additional special measures for safety of life and protection of natural environment deemed necessary for the Arctic Ocean, containing specific guidance concerning equipment for fire safety, lifesaving and navigation, and paying special attention to operation procedures, crewing and emergency equipment. According to the Guidelines, all ships operating in Arctic ice-covered waters are to carry an operating manual and a training manual for all ice navigators on board.[28] The ship's operating manual shall also contain special procedures for protecting the environment.

The IMO Guidelines introduce a system for differentiating between polar and non-polar ships, as it is recognized that not all ships will be able to navigate safely in all areas of the Arctic at all times of the year.[29] As for shortcomings, it has been pointed out that there is no recognized internationally-approved training course for ice navigators, and that the stipulation in the Guidelines that the ice navigator should document that they have completed satisfactory training is phrased in rather broad terms.[30] The IMO Guidelines also lack a requirement as to documented navigation service in Arctic ice conditions.[31] It is also maintained that the serious hazard of ice-infested navigation should have been regulated more extensively.[32]

6.1.3 "Soft law", including corporate governance and social responsibility

The legally binding instruments applying to states and private subjects in the Arctic are treaties, customary international law, national acts and regulations, and regulation adopted by local authorities (such as autonomous republics of the Russian Federeation).

However, in the interface between formally-adopted laws and regulations on the one hand and politics on the other, there are norms that probably have an influence on how the involved parties act, and which also influence political and legislative developments and the deliberations by competent courts of law on a running basis. Certainly, these non-binding norms will form both a backdrop and, to some degree, a basis for policy-making and the adoption of rules both internationally and within the Arctic Counsil, and it is therefore in place to review them briefly.

Closest to binding, formal law (and many would argue a part of binding law) are those concepts and norms that are normally termed "general legal principles". In international environmental law, one example is the principle of awareness, which reinforces the importance of taking due regard to nature even if the detrimental effect of practices are not proven, but only appears as a more or less likely risk.

Further, over the last decades, certain policy concepts have developed and, again, it is easiest to find examples within the environmental area. The most well-known example is that of "sustainable development", first coined by the Brundtland Commission. Another is "environmental security". More recently, the concept of "stewardship" has been launched – not specifically for the Arctic, but as a general policy and governing principle. It is certainly more all-embracing than the other concepts described. Good stewardship with regard to the Arctic must, as a practice, apply both to policymaking, governing and enforcement, the point being that all interests involved (environmental, indigenous people's rights etc.) as well as the need for economical and industrial development are taken into due consideration on all levels.

While, e.g., the IMO Guidelines have been developed through cooperation between states and could become binding public rules, steering tools in the form of policies and internal rules on corporate governance and corporate social responsibility have been developed by business itself. Over the last decades, focus has been steadily growing on corporate governance and corporate social responsibility. These are not topics specifically aimed at the Arctic, but in order to complete the picture of influences on corporate behavior they are well worth noticing, especially in relation to indigenous people's rights, environmental issues, and human safety. This section is by no way meant to reduce the importance of mandatory rules and government control, but it is a relevant point that what could be termed "self-restraint in self-interest" has been coming more and more to the forefront – at least for major businesses. Further, the growing focus on this issue makes commercial actors much more competent when it comes to compliance with regulations applicable to their activity.

Any serious industrial player in the shipping industry today will have a set of self-imposed rules that typically address topics such as:

- company ethics and codes of conduct;
- compliance with laws both domestic and wherever operations take place. For a shipping company operating in the Arctic, a typical corporate steering document would be a manual listing the requirements to be met;
- health, environment and safety (HES): most detailed manuals that address these topics will be part of the corporate steering documents.

Certainly, implementation in the corporate steering documents of the IMO Guidelines would be an example of good corporate governance.

In general, it can be assumed that the implementation of good corporate governance and corporate social responsibility today are central elements in building a good reputation, and thus has become a central part of business itself. Further, a growing recognition has emerged within businesses that the adverse effects of non-compliance represent a major risk, both in form of legal and financial consequences and in damaged reputation, and this has contributed to development in the field. An extensive discussion of this is not warranted in our context, but it should be mentioned that an extensive literature exists in the field. It is also true that the development of internal control system partly is driven by external expectations from authorities, media, NGO's etc. Good examples of corporate governance and corporate social responsibility can be found on the homepages of major companies, see for example www.statoil.com.

6.1.4 An Arctic treaty: Still of interest and relevance?

One characteristic of international law is that it is often described as a dynamic legal treaty or by customary international law being established through state general practice and being defined as "accepted law". Another characteristic is that international law could be general in character but also that it is geographically limited to a specific area. An example of the latter is the Svalbard Treaty where a need was seen for a particular treaty and where the geographical application is defined to the islands of Svalbard and its territorial waters.

The Arctic has certain special features: it is partly ice covered, it is environmentally vulnerable and it is subject to rapid climate changes. It therefore might not come as a surprise that, in an area like the Arctic in which a good many parties are showing their interest, and which is assumed to contain valuable resources and is a particularly vulnerable environment, the question would be raised of a special Arctic treaty to reflect the peculiarities of the area. Article 234 of UNCLOS and the IMO Guidelines both indicate the need for special legal instruments in the Arctic, but they are at the same time examples of such instruments being established within already-existing legal frameworks. With regard to an Arctic Treaty, readers are reminded that the UN Convention on the Law of the Sea has been described as "possibly the greatest legal development ever to take place".[33] It thus seems relevant to assume that there will be ways to relate to maritime matters in the Arctic within the legal framework of this Convention. It is also worth noticing that the IMO Guidelines for ships operating in Arctic ice-covered waters now are moving towards becoming legally binding and, if so, this will be a major part of what should be comprised by an Arctic treaty as such.

Concrete initiatives for a specific Arctic treaty have not been many. The most comprehensive suggestions have been generated by the World Wildlife Foundation (WWF).[34]

The core players in the Arctic – namely Canada, the USA, Russia, Norway and Denmark – have expressed as their opinion that the present legal framework should be built on rather than a special treaty being created for the Arctic. They have the view that the current international law "provides a solid foundation for responsible management by the five coastal States and other users of this Ocean through national implementation and application of relevant provisions."[35] These states thus[36] see no need for developing "a new comprehensive international legal regime to govern the Arctic Ocean".[37] Additional reference was made to the five Arctic states "cooperating closely with each other and with other interested parties", and to a will "to strengthen this cooperation based on mutual trust and transparency."[38] This was reiterated in 2009 by Russia, Norway and the EU at a conference

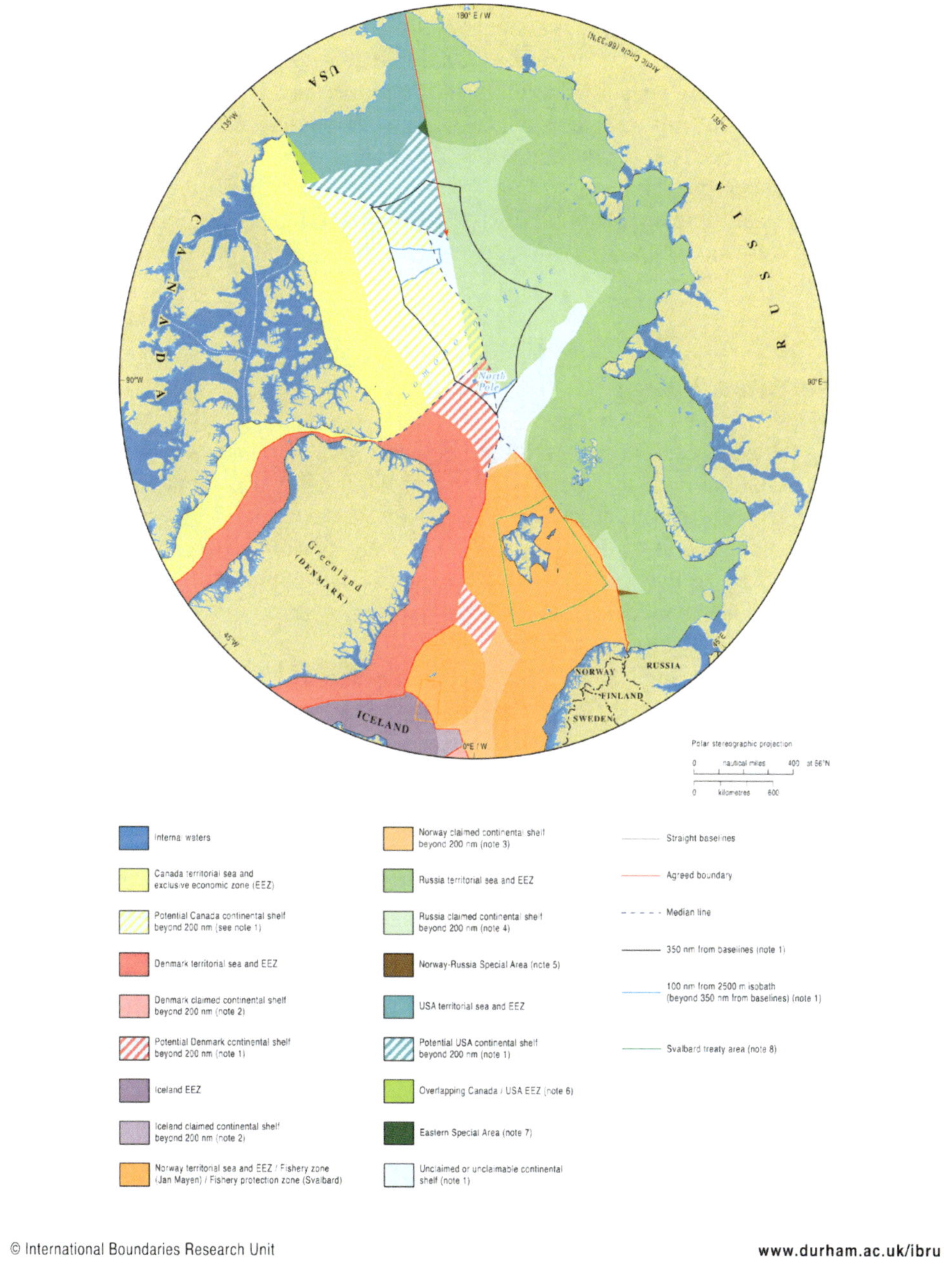

Source: http://www.dur.ac.uk/resources/ibru/arctic.pdf

Figure 6.1 Jurisdiction in the Arctic.

entitled Arctic Frontiers[39] where these parties stated that UNCLOS, the IMO and the UN Commission on the Limits of the Continental Shelf are sufficient international instruments for managing the Arctic and the expected results of climate change.

The Arctic states make special mention of the Arctic's characteristics, and the need to protect the environment, ensure free passage, strengthen the rescue capabilities and to take indigenous people's needs into consideration.

Against this background, an Arctic treaty seems to be something for the future, if it becomes a reality at all. Implementation of existing regimes, on the other hand, will remain the focus for governance of the Arctic.

6.2 The Northeast Passage

Sailing along the Northeast Passage and the Northern Sea Route means passing through the waters between the northern coastline of mainland Norway and Svalbard. However, predominantly it means sailing along the coast of the Russian Federation, which means having to relate to laws and regulations laid down by Russian authorities in these waters. It also means facing jurisdictional controversy regarding the definition of the Northern Sea Route as well as an unsolved delimitation line in the Bering Strait.

6.2.1 Northern Norway and Svalbard: Coastal waters

Norway extended its territorial sea from 4 to 12 nm in 2003, and at the same time a contiguous zone out to 24 nm was established.[40] The extension to 12 nm territorial sea also applies to Svalbard and Jan Mayen. A 200 nm EEZ was introduced in the Norwegian legislation in 1976,[41] and established outside mainland Norway as of 1 January 1977. In June of the same year, a 200 nm fishery protection zone was established outside Svalbard,[42]and a fisheries zone outside Jan Mayen. The Norwegian government gave two reasons for not implementing an EEZ outside Svalbard.[43] Firstly, it was stated that fishery protection would be satisfactorily achieved by a fishery protection zone. Secondly, it was acknowledged that parties contracting to the Svalbard Treaty might challenge a Norwegian economic zone outside Svalbard,.[44] leading to potential confrontations, which was seen as not being in the interest of Norway. As for shipping in the 200 nm zone outside Svalbard, this falls under the governance of the High Seas regardless of whether the zone is a fishery protection zone, a Norwegian EEZ or an EEZ under the regime of the Svalbard Treaty. Regulatory measures will in all three cases be carried out by the Norwegian Government.

The most important Norwegian act relevant to Arctic shipping is the Shipping Safety Act of 2007.[45] The Act deals with all issues related to safety at sea. Geographically, the Act covers ships sailing in Norwegian territorial waters, including those surrounding Svalbard, the EEZ outside mainland Norway and the Norwegian continental shelf. The Act may, however, be extended to include foreign ships outside these geographical areas, as far as is compatible with international law. The Act does not contain specific rules for Arctic waters, but has detailed provisions for technical and operational safety, human safety, environmental safety, etc. Of specific interest is Chapter 9 of the Act, which makes a safety certificate mandatory for ship-owners, and imposes a duty to establish adequate internal safety management rules. There are no provisions for clearance before entering Norwegian zones, as is the case in Canadian and Russian waters, but there are provisions giving the Norwegian Maritime Directorate extensive powers for inspections and the possibility of imposing orders. Further, the Act contains both civil and criminal liability provisions.

Subject to any restrictions deriving from international law, the Pollution Act of 1981[46] also applies to the Norwegian EEZ if the source of pollution is a Norwegian vessel or "otherwise decided by the King". Pursuant to the Pollution Act of 1981, regulations were passed by the King in 1997[47] that apply not only to all ships sailing in the Norwegian EEZ, including the waters outside Svalbard, but also to ships on the High Sea in the vicinity of the Norwegian coast if particularly harmful pollution from these ships threaten the Norwegian coast. According to this Act, intervention against acute pollution or the risk of acute pollution on the open sea and in outer Norwegian territorial waters shall take place in accordance with international agreements to which Norway is a party. Svalbard – its land territory and territorial waters – is explicitly exempted in the Pollution Act, as a special Act on the protection of the environment of Svalbard and its territorial waters has been passed.[48] According to this law, the release, dumping or incineration of waste or other material from ships or vessels is prohibited. When significant environmental damage is caused, the Governor of Svalbard can order the party to pay environmental compensation to the Svalbard Environmental Protection Fund.

Not surprisingly, Norway has passed substantial legislation pertaining to all aspects of offshore petroleum operations, with the Petroleum Act of 1996[49] as the cornerstone. The Act does not directly affect navigating the High Sea, but it might be worth mentioning that it makes it clear that unauthorized vessels must not be present in safety zones which are required around oil and gas installations and facilities. Vessels taking part in petroleum production on the continental shelf will have to conduct their operations in conformity with standards set by the Norwegian Petroleum Directorate.

6.2.2 Russian Federation: Coastal waters

In Russian legislation the Northern Sea Route is described as "the national unified transport line of communication of the Russian Federation in the Arctic, including the Vilkitskii, Sho-kalskiy, Dmitry Laptev and Sannikov straits".[50] A 12 nm territorial sea appeared in Soviet legislation in 1960, at a time when the outer limit of the territorial sea, according to international maritime law, was 3 nm. Regarding the Northern Sea Route, Russian authorities have made it clear that navigating along these waterways must be conducted in accordance with provisions laid down in Federal Russian law. Legislation establishing the status of internal sea waters, territorial sea and contiguous zone outside the Russian Federation is stated in a Federal Act of Act of 17 July 1998,[51] while a 200 nm EEZ was established in a Federal Act of 2 December 1998.[52] The point of departure for determining the outer limit of these waters is the baseline, described in the Federal Act of July 1998 as the low-water line along the Russian coast. In places where the coastline is deeply indented and cut into, a straight baseline is drawn joining islands, reefs and rocks, as well as across a bay when the breadth does not exceed 24 nm. When belonging historically to the Russian Federation, a straight baseline can also be drawn when the mouths of the bay are broader than 24 nm.[53]

The Act defines the territorial sea to 12 nm beyond the internal maritime waters. It confirms the right to innocent passage through these waters, but also lays down certain regulations such as safety of navigation (including the use of sea lanes), conservation of living resources, prevention of infringement on fisheries and preservation of the environment. With regard to the environment, the Act of July 1998 makes it clear that the dumping of wastes and the discharge of harmful substances in the territorial waters are prohibited, with conditional exceptions for dumping and disposal when incidental to or derived from the normal operations of ships. The Act also makes it clear that wilful and serious pollution of the environment, contrary to the requirements of the laws of the Russian Federation, shall be considered to be prejudicial to the peace, good order or security of the Federation.[54] In order to ensure safety of navigation, to safeguard the State's interests and to protect the environment, areas where navigation is prohibited and which are temporarily dangerous for navigation may be established.[55] The Act also gives the Federal Executive Body for Defence the authority to temporarily suspend the right of innocent passage in specific areas of the territorial sea for the purpose of conducting training with weapons.[56] A contiguous zone, extending 24 nm from the baseline, has been established in which the Russian Federation shall take necessary measures in order to prevent infringements of its customs, fiscal, immigration or sanitary regulations.[57]

As indicated above, an Exclusive Economic Zone was established in the Federal Act of 2 December 1998. Measures, norms and control of the marine environment laid down in the Federal Act for internal waters and the territorial sea are to apply also in the EEZ, taking into account international rules and standards. The right for all states to enjoy the freedom of navigation is stated; as is the competence of the Russian Federation within its EEZ to establish a legal regime in areas of environmental emergencies and disasters and to ensure immediate actions to eliminate the consequences of accidents involving pollution by oil or other substances. Further provisions protecting the marine environment in the EEZ of the Russian Federation is led down in a special chapter of the Act,[58] containing provisions dealing specifically with the protection and preservation of ice-covered areas with a wording almost identical with UNCLOS article 234.[59] In order to prevent pollution from vessels by oil, toxic liquid and refuse, the chapter also contains a provision for adopting special mandatory measures for individual areas in the EEZ.[60]

July 1991 marked the opening of the Northern Sea Route to foreign vessels; at the same time, the Regulations for Navigation on the Seaways of the Northern Sea Route, approved by the USSR Minister of Merchant Marine in September 1990, were enacted. The Regulations lay down provisions for navigation through the NSR on the basis of non-discrimination and for the purpose of ensuring safe navigation and preserving, reducing and keeping under control marine environment pollution from vessels. The Regulations cover three main areas. The first relates to the actual navigating through the NSR, demanding that the owner or the master of the vessel submit a notification and request to Marine Operation Headquarters and also information about guarantee of payment for icebreaker dues. Having been admitted, the vessel is obliged to follow assigned sea lanes. A certificate of due financial security with respect to the civil liability of the owner for damage inflicted by polluting the marine environment is required. Compulsory icebreaker-assisted pilotage is established in certain areas. The second part of the Regulations defines the procedure for submitting requests, organization of the pilotage as well as obligations and responsibilities, making it clear that navigating behind an icebreaker does not relieve the master of the vessel from control over the safety of navigating his vessel. The third part of the Regulations lays down specific requirements regarding design, equipment and supply of vessels navigating the NSR. More precisely, the requirements apply to hull, machinery installations, systems and arrangements, stability and watertight integrity, navigational and communication facilities, supplies and emergency outfit and to manning.

Other regulatory measures laid down by the Russian Federation with which foreign vessels sailing along the Northern Sea Route have to comply are: Guide to Navigating through the Northern Sea Route, Regulations for Icebreaker and Pilot Guiding of Vessels through the Northern Sea Route, and Requirements for the Design, Equipment, and Supplies of Vessels Navigating the Northern Sea Route, all adopted in 1996. The Guide is a comprehensive document consisting of three parts, where the second and most substantial part gives a detailed navigational description of the Kara, Laptev, East Siberian, and Chucki Seas with their straits and islands.[61] The Guide also contains the full text of the Regulations of 1990 mentioned above, as well as the Regulations of 1996. According to the latter, vessels that intend to sail along the Northern Sea Route have to submit a request to the Administration of the NSR at least four months in advance. It is important that the request satisfies the Requirements of 1996. If the response is positive, an inspection takes place at the expense of the owner. Information has to be given to the Administration about time of arrival and, when entering the NSR, the vessel must be brought under the control of the West or East Marine Operations Headquarters for icebreaking support and organization.[62] The third part of the Guide addresses ice navigation under different conditions, with or without the assistance of icebreakers, as well as salvage and rescue support. Finally, it provides the full text of the Requirements of 1996,[63] the main purpose of which is to secure safe navigation and the protection of the Arctic marine environment. Reference is here made to vessels having to possess at least the Russian Federation Register of Shipping ice classes of L1, UL or ULA, with additional requirements related to such aspects as the hull, the machinery plant and propeller blades, stability under ice conditions, navigation and communication equipment, treatment of waste, emergency facilities and, not least, to the crew.[64] A document worth mentioning is Federal Rates Service dated 26 July 2005. Referring to the Standing on Federal Rates Service ratified by the Government of the Russian Federation, the document contains rates on icebreaker fleet services along the Northern Sea Route.[65] It is predicted that the listed rates will increase.[66]

6.2.3 Historical sectorial controversies and the legal uncertainties of sea ice: Still relevant?

For decades, the legal status of the NSR has been a fairly low key, but nevertheless contentious, political issue in US–Soviet/Russian Arctic relations. Russia claims the ice-covered straits of her Arctic archipelagos as internal waters under several theories, including those of historic bays and closed by straight baselines. In Russian legislation, internal waters also include the waters of the White Sea, Cheshkaya and Baridaratskaya bays, as well as the waters separating the mainland from the archipelagos. In general, all bays and gulfs where the entrance does not exceed 24 nm are regarded as part of the internal waters of the Federation.

The USA is the only country that through declarations and navigation has explicitly challenged the Russian regime. The US declarations include protests to the establishment of Arctic baselines and claims that Russian Arctic straits are international and subject to transit passage. The US accepts more extensive coastal waters rights with respect to prescriptive and enforcement competence, but holds the question open with regard to ice covered straits.

Possible jurisdictional controversies

With regard to the Northern Sea Route there are some uncertainties related to the definition of Russia Arctic waters (see Chapter 1), related more specifically to the implication of the Sector principle laid down in a Russian Decree of 1926. In this section, we discuss the Sector principle and its eventual impact on the legal status of sea ice and the freedom of the High Seas of the Arctic Ocean.

The Sector theory

The Sector theory warrants claims to sovereignty over sectors of the Earth's surface, as measured by meridians of longitude. Applied to the Arctic, the sector is defined by two meridians of longitude extending from the North Pole south to the most easterly and westerly points on the Arctic Circle pierced by the state. Under certain interpretations of this theory, nations possessing territory bordering on the Arctic Ocean have a rightful claim to all islands and even sea ice and ocean areas lying to their north within the sector.[67] The principle is ambiguous in that two different interpretations exist about its legitimacy, function and applicability in international law:

- Some regard it as a special version of the *contiguity principle,* i.e., the right to claim sovereignty over areas based on the propinquity of the claimant state to the territories being claimed. In this interpretation, it will provide individual Arctic rim states complete control over all activities occurring in "pie-shaped slices of the region".[68]
- A practical form of delimiting geographical areas whose sovereignty is secured by other means of sovereignty claims, such as *effective occupation*, etc.[69] In this interpretation, the "sector idea may be unobjectionable, if other elements in the situation are satisfactory".[70] As pointed out by Castel, "whatever its role, the sector principle must be viewed in the perspective of the total ... claim – a claim with broad foundations, ..., which is consistent with the principle of international law".[71]

On 15 April 1926, the Soviet Presidium of the Central Executive Committee adopted a decree based on the Sector theory, declaring all lands and islands situated in the Arctic Ocean, between the coastline of the USSR and the North Pole and between sector lines drawn from the easternmost and westernmost points of the coast up to the North pole, the territory of the Union (see Figure 6.2). Islands already under the sovereignty of other countries, such as the archipelago of Svalbard, were explicitly exempt from the claim. According to the Soviet lawyer, V.L. Lakthine, this decree was prompted by the need of the Soviet Union to protect its northern coasts from enemy attacks and foreign infiltration as well as to safeguard economic national interests within the sector. Among other things, the sector was to prevent other states from establishing airbases within the Soviet "gravitation area" of the

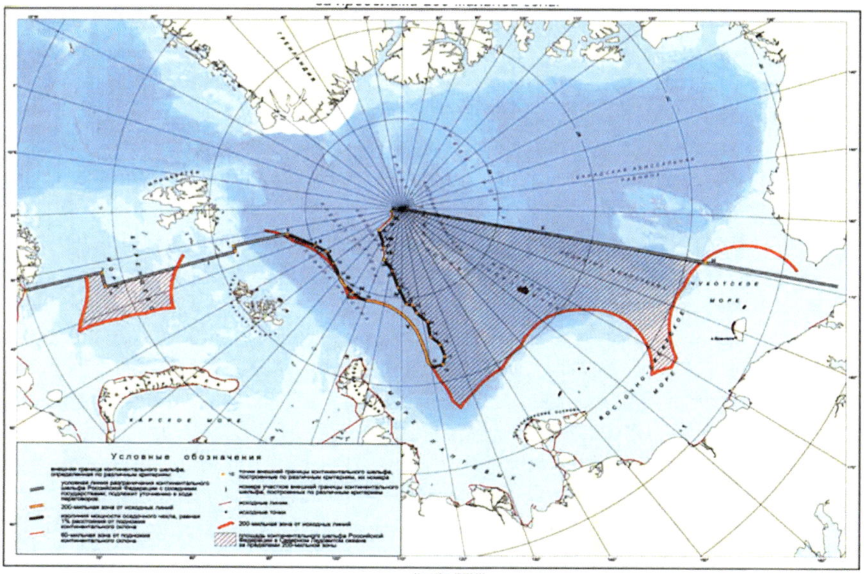

Source: http://www.un.org/depts/los/clcs_new/submissions_files/rus01/RUS_CLCS_01_2001_LOS_2.jpg

Figure 6.2 The Soviet sector claim and 200 nm economic zone.

Arctic Ocean, and to preserve the Northern Sea Route. It was a self-protective move to prevent "future imperialist interventions in the Soviet gravitation sector".[72]

The decree refers exclusively to "lands and islands" and not to any other features of the ocean space contained within the sector. The Sector theory thus places land territory in the legal possession of a claimant state even before it has been discovered. In this respect it deviates from international law, posing the question of how a state can claim sovereignty over areas of which she knows absolutely nothing, not even whether it exists. Although the Sector theory is not part of accepted international law – and as such is a most controversial principle – the Soviet claim to territorial sovereignty over Arctic islands has not been seriously challenged by other states. This implies that the claim is regarded as a practical means of delimiting an area of sovereignty which is secured by other means accepted by international law (see point 2 above). What has caused the most controversy is that the Sector theory has been claimed to apply also to the ocean areas within the sector.

The policy of not invoking the Sector principle in support of sovereignty claims over maritime areas has never officially been altered, either by the Soviet Union or by the Russian Federation. What is more, there are no legislative acts on the part of the Russian Federation claiming any special rights in the Arctic sector outside of the 200-mile EEZ (with the possible exception of the High Latitude and Near-Pole routes discussed in Chapter 1).[73] However, during the Stalin era leading Soviet experts on international ocean law maintained in writing that grounds existed for extending the sector claim to include: (a) the waters between the island; (b) the waters between the island and the mainland; (c) the pack-ice and ice-islands; (d) the remainder of the area within the sector; (e) the air space over the entire sector; and (f) the seabed within the sector.[74] According to Lakthine, the spirit of the decree was to prevent and control foreign intrusions into the Soviet sector. In order to make this come through, the concept of "land and islands" should be extended to comprise also sea ice and adjacent waters and the air space above the sector. If not, the sector would be regarded as High Seas "with all its associated consequences".[75] In Lakthine's mind, the decree was in accordance with existing international ocean law at the time.[76]

The Soviet state-controlled press followed suit, first in 1928 and again in 1950, advancing the Sector claim to apply to the open ice-infested waters, including drifting sea ice. These authors based their views to a large extent on the Marxist–Leninist theory of the class struggle and what they regarded as the expansionist character of imperialism. Accordingly, Soviet maps were subsequently published depicting the sector lines as boundaries for "Soviet Arctic areas", comprising 43% of the Arctic Ocean including a substantial part of the Central Arctic Basin (see Chapter 1). This move found support in the analysis of E.B. Pashukanis – a highly respected Soviet legal expert – who in 1935 declared the 1926 Soviet Decree to be lawful "with the aim to form the northern borders of the USSR legally and precisely".[77] A. Sabanin – legal advisor to the People's Commissariat for Foreign Affairs – took issue with this stand and argued that the demarcation lines did not mean borders for the polar sector.[78] In so doing, he reiterated the official Soviet position.

In the 1960s, 1970s and early 1980s, Soviet lawyers promoted the view that all Arctic states were entitled to their own sector in the Arctic Ocean, but they disagreed on what features of the sectors could be claimed as the territory of the coastal state. One group[79] argued that national sovereignty could only be claimed over islands, whereas others argued that such claims could be extended to both land and water (sea ice included).[80]

In the period 1962–68, the US government, who challenged the legality of the Sector principle, put the Soviet position to test, dispatching US Coast Guard vessels to conduct hydrographic research in what international ocean law regarded as High Seas in the Chukchi, East Siberian, Laptev, Kara and Barents Seas. These missions were openly to assert the High Sea status of these waters. Soviet authorities on their part made it plain that the Laptev and Sannikov straits could not be transnavigated by US vessels because these straits were historically part of Soviet internal waters, where Soviet sovereignty applied without any curtailment. The Coast Guard vessels did not transit those straits but, this fact notwithstanding, Soviet authorities considered the US presence in the High Seas of these waters so provocative that in 1966 they publicly declared them to be of an "unfriendly nature".[81] This diplomatic reaction implied that even waters defined as international by international ocean law were not quite so in the official mind of Soviet authorities. According to recent Russian sources the reaction was due to the fact that the "Americans were making use of warships for reconnaissance and military applied research on the seaways of the NSR". The USSR policy was to deter "...the NATO naval forces from the Soviet Arctic".[82]

In 1966 the Military Publishing house of the Ministry of Defence of the USSR published *A Manual of International Maritime Law*, stating that: "The sovereign rights of the USSR in the Arctic sector stem from the enormously effective economic, organizational and scientific research activities of the USSR in mastering the Northern Sea Route, from the historical fact of discovery, from research on the polar seas and islands by Russian navigators, and also from the special geographic and climatic conditions of this region."[83] No specification was made of what the sovereign rights were and where they applied, but the argument covered both discovery, activities, land, sea ice, geography and climatic conditions.

Twenty years later – in 1985 – the Soviet Navy published an international law of the sea manual stating that the lateral lines of the Arctic sector converging at the North Pole should *not* constitute state boundaries. This seemingly categorical statement was, however, immediately qualified in the next sentence which said that: "...the special character and importance of the Arctic seas for the coastal States give ground to consider the polar sectors as zones of their economic and defence interests, and to use the appropriate meridians for the delimitation."[84] This was a reiteration of what Lakhtine wrote in 1928 that strategic defence and economic considerations played a prime role in the adoption of the 1926 Decree.[85]

In the post-Cold War period, the sector principle has neither been officially rejected nor reaffirmed nor reconsidered. There are still a few analysts within Russia arguing that the Arctic sector should be defined as an integral part of Russia with a vertex at the North Pole. On the other hand, prominent voices within the State Duma contend that real arguments in support of the Soviet Sector concept simply do not exist any more, stating that the concept of the Arctic sector as part of the state territory has collapsed as a result of the adoption of the UN Law of the Sea Convention of 1982.[86] As observed by Leonid Timtchenko, there is a "...lack of clarity in Soviet policy; even today, after the collapse of the USSR, the official position of its successor – the Russian Federation – relating to the sector concept is still not clear."[87] Erik Franckx concludes, "Although there still existed a great margin of flexibility as to [the Sector theory's] exact field of application in doctrinal writings, it can be concluded that this theory has been transformed into a "fall-back" notion, in order to sustain more limited claims if need may arise."[88] The uncertainty and confusion involved is what other scholars have labeled *creative ambiguity*, feeling it to be deliberate and planned.[89] In light of this ambiguity, it is interesting to note that Russian analysts like B.M. Klimenko argue that, although the norms and laws of UNCLOS undoubtedly apply to the Arctic, it is only the islands within the sector that belong to the coastal state, but these waters: "...may be modified according to the special conditions of the region" (see discussion of special Arctic treaties above).[90] In line with this, V.N. Kulebyakin argues that sector states have sovereign rights over both land and sea territories in the Arctic with a reference to Article 234 of UNCLOS, which gives coastal states the right to adopt and enforce non-discriminatory laws and regulations for the prevention, reduction, and control of marine pollution from vessels in ice-covered areas within the limits of their exclusive economic zones (EEZs). Although, Article 234 confers extensive authority to the coastal state in ice-covered areas,[91] most international lawyers would probably agree with Leonid Timtchenko "that it is doubtful that this article and the (UNCLOS) as a whole contain legal grounds for the sectoral concept".[92]

The legal and political challenges relate partly to the inherent imprecision of the terms "Arctic sovereignty" and "polar areas". The question of sovereignty in the Arctic Ocean is a very special one, and to some extent is incapable of exact definition, owing primarily to the undefined status of sea ice in international ocean law. Only in the polar areas does sea ice exist in sufficient quantities and with sufficient stability (irrespective of the effects of global warming) to support permanent structures. The Eskimos use the sea ice as a substratum for hunting, trapping and transport, whereas states use it as platforms for scientific research and landing of aircrafts. This permits sea ice to be employed as land and contrasted with water, posing the question: Should sea ice be regarded as "Arctic ocean water" (which in fact is what it is), or as "Arctic ocean territory" (which is what it is from any functional practical point of view). At the outset, sea ice could be regarded as both land and water. The challenge of the Arctic Ocean, in terms of international law, is that: "...the Arctic Ocean has the characteristics of both land and sea, and yet it is totally unlike both." According to US writers, it is the "unusual nature of the Arctic Ocean that makes it difficult to resolve questions of exactly where 'Arctic waters' end and 'Arctic territories' begin, and whether the concept of an 'open sea' is applicable in the Arctic Basin."[93] On the basis of the blurry complexity of this problem, two schools or theories of law have developed: *the ice-is-water theory* and *the ice-is-land theory*.

The ice theories

The *ice-is-water theory* recognizes that the sea ice could be occupied and used for multiple human purposes, but contends that such occupation is temporary. The constantly-changing appearance and position of sea ice in the Arctic Ocean imply that it lacks the stability required by international law to be the object of permanent sovereign possession. It is illogical to claim national sovereignty over an object that could melt and disappear from the face of the Earth. Ice is water and will never be anything but water. Water expanses are already regulated by international law.[94]

The proponents of the *ice-is-land theory* argue that ice is not merely a form of water, but a solid substance, claiming that laws "...evolved for a liquid (and the ... arguments in favour of freedom of the seas all rest on the liquid quality of the sea) do not necessarily apply to a solid. Ice can be settled upon; ice is a barrier to navigation; ice possesses definite limits. There is little about ice to invite a comparison with water."[95] Some 40 years ago, Gordon W. Smith stated that the problem of sea ice "cannot be dismissed simply by assuming that the absence of a particular law is in itself a proof that another law exists".

There is no international law of sea ice.[96] This implies that the ambiguities of the interpretation of the Sector principle also introduce uncertainties to the regime of the NSR – the most likely highway for Arctic transportation of mineral resources in the short and medium term. The idea has been launched that a Sector claim to achieve sovereignty over floating ice "must evidently claim what might be termed *ice space* rather than the ice itself, in the same manner as a state claims the airspace above its soil rather than the air".[97]

Ice-covered space and Russian legal state behavior in the Arctic

As discussed in Chapter 1, the NSR is a series of different sailings lanes, and ice conditions at any one place and time will decide the sailing course to be set. At times, due to the accumulations of pack ice in the straits, surface vessels operating in convoys will be forced to proceed north of the large island masses to find suitable routes through the ice – the corridor feature of Arctic navigation. In such cases, the navigation lanes used are legally claimed to be national and under full Russian control and jurisdiction. However, uncertainties exist about the northernmost limits of the NSR.

In the *Regulations for Navigation on the Sea Ways of the Northern Sea Route*, published by the Head Department of Navigation and Oceanography, USSR Ministry of Defence, 1990 (Decree 1990) article 1,2. the NSR is defined to be "...situated within its inland seas, territorial sea (territorial waters), or exclusive economic zone adjacent to the USSR Northern Coast and include sea ways suitable for leading ships in ice, ...".[98] Apparently, the NSR is here partly defined officially as a *coastal route* within the internal and territorial waters of the Russian Federation and/or partly as a *high latitudinal route* located within the Russian 200 nautical miles EEZ, which belong to the High Seas (see Chapter 1).

Prominent Soviet/Russian lawyers, A. L. Kolodkin, former President of the Soviet Association of Maritime Law, and Dr M.E. Volosov, former Chief Secretary of the Executive Committee of the same Association, provide an extended explanation. In an article in *Marine Policy*, March 1990, they argued that the distinctive *climatic* and *hydrographical factors* of the region should be taken into account when discussing the legal status of the NSR, and that "the States of the region must possess certain privileges and prerogatives, above all with respect to regulating the access of users, including foreign users, and also with comprehensive control over all types of activity carried out in the region".[99] This also affects the definition of the northern limits of the NSR: "...under any circumstances *a significant part of the Northern Sea Route* lies within the Soviet economic zone, or the territorial and even internal waters of the USSR."[100] Indirectly, the authors here indicate that a *small part of the NSR* may lie outside of the EEZ in international waters. This is so because the "integrated nature of the Northern Sea Route as a transport route is not affected by the fact that individual portions of it, at one time and another, may pass outside of the aforesaid boundaries where the USSR exercises its sovereign rights or sovereignty in full (i.e. it may pass into the high seas)".[101] Thus, as long as part of a voyage includes waters under undisputed Russian jurisdiction, Kolodkin and Kolosov define the NSR to include sea lanes running beyond its own economic zones in high latitudes. Thus, in principle, all conceivable sea lanes south of the North Pole, and even across the Pole itself, could be part of the NSR as long as the voyage in question passes through North Russian coastal waters (see Chapter 1). This interpretation seems to be confirmed in a recent Russian–Norwegian study on remote sensing of sea ice along the NSR. In this study, Russian researchers without any further ado state that voyages along the NSR are carried out along four different routes: a coastal route, mid-route, high-latitudinal route and a near-the pole-route (see Chapter 1). According to the authors, the latter route passes the large circle across the North Pole because the most important factor in planning shipping routes in ice "appears to be the optimal use of the zones with the easiest ice conditions, regardless of whether they are encountered on coastal or high-latitudinal routes".[102] Two factors attract interest in this respect: First, the NSR is depicted as starting in Murmansk whereas the official definition of the western point of the NSR is Novaya Zemlja. Second, the near-the pole-route as depicted in Chapter 1 does not necessarily involve North Russian coastal waters in terms of internal waters, territorial waters and the EEZ. This deviates from the reasoning of Kolodkin and Kolosov, and since Ivan Ye Frolov as Director of the state-owned Arctic and Antarctic Institute (AARI) in St Petersburg is a central player in the Russian Arctic, these deviations add to the ambiguity of the Russian position.

The rights of innocent passage through the NSR is being curtailed by "an aggregate of legal and material circumstances" which, according to Kolodkin and Kolosov, make the

Figure 6.3 The Arctic Ocean sea floor.

NSR a *national transport route* in which the regulation of navigation is "the prerogative of the USSR as a coastal State". In support of this claim, the Norwegian Indreleja case ruled by the International Court of Justice (ICJ) in the Anglo-Norwegian fisheries case in December 1951 is invoked. The Court ruled that the entire sea route from Varangerfjorden to the Porsangerfjord, exploited and equipped exclusively by Norway, is under the complete control and administration of Norway irrespective of whether parts of it are within internal and territorial waters.[103] The political implication of this is that under certain specific circumstances the NSR regime in principle comprises more than 75% of the ocean space within the sector defined by the sector principle, providing jurisdictional power to the coastal state over a fairly big part of the Central Arctic Ocean between the Russian EEZ and the North Pole.

The legal reasoning of Kolodkin and Kolosov implies that a transpolar shipping route can run in international waters between the Russian territorial sea and the North Pole and still be under the jurisdiction of the Russian Federation. It is interesting to note, however, that Kolodkin and Kolosov restrict the jurisdiction they claim for the Russian Federation in international waters to sailings which also include portions of the NSR/NEP. In the sea belt between the seaward boundary of the territorial sea and the seaward boundary of the EEZ, and on the High Seas beyond the economic zone, they claim the freedom of navigation is exercised without limitations and in accordance with UNCLOS.[104]

Ten years later, Brubaker reasoned that the NSR in some cases could be extended beyond the EEZ. This reasoning is based on the fact that all the articles of the 1990 Decree, except for Article 13, contain the term "marine areas adjacent", and Article 1 of the Decree does not clarify whether the provisions are also applicable beyond the EEZ. On the grounds that the "mandatory nature of the measures, ... , to be taken by the authorities ... may seemingly be argued to encompass the high seas, if these are understood as an 'adjacent area' under the legislation and claims indicated",[105] the terms "ice-covered areas" and "adjacent areas" remain unclear in Russian legislation, opening up for extended geographical interpretations of where the sea lanes of the NSR runs.

In August 2007, a privately-funded submarine expedition planted the Russian flag under the North Pole on the Lomonosov Ridge at the depth of more than 4000 meters. On board was Dr Arthur Chilingarov, a noted polar researcher and a Hero of the Soviet Union. Dr Chilingarov previously was Deputy Chairman of the Russian Duma and a close associate of President Vladimir Putin, who prior to the dive appointed Chilingarov as Head of the Russian scientific program for the International Polar Year of 2007–2008. Immediately after the successful dive, Chilingarov claimed that the purpose of the expedition was to prove that the Arctic Ocean was Russian territory, and that the resources therein belonged to Russia.[106] This statement, in combination with Chilingarov's prominence in Russian politics and Arctic science, gave him broad coverage in world media. A few months after the dive, Chilingarov was appointed Hero of the Russian Federation. This incident gave rise to heavy protests, not least from Denmark and Canada, both of which claim the Ridge as part of their respective continental shelves (see Figure 6.3). The dive and flag planting had no legal foundation in UNCLOS. Article 77 (3) of the Convcention reads: "The rights of the continental shelf do not depend on occupation, effective or notional, or any express proclamation." These rights are secured by the Convention itself and find global manifestations. As pointed out by the Norwegian Foreign Minister: "A flag no longer creates rights. To plant a flag on the sea-bed of the North Pole makes the North Pole no more Russian than [Roald] Amundsen's flag on the South Pole made the [South] Pole Norwegian."[107] This being the case, speculations nevertheless flourished in the media: Did the dive serve political purposes? Was it a trial balloon on the part of Russia to claim more of the seabed within the Russian sector than warranted in UNCLOS? Was the dive a reawakening of the Sector principle in new clothing?

In the aftermath of the flag planting the Russian Foreign Minister, Sergej Lavrov, has made strenuous efforts to counteract the impression that this was an official act on the part of the Russian government. In his version, this was nothing but "a brave stunt" undertaken by a private group of Russian citizens.[108] In multiple later statements Lavrov underscores that "Russia strictly abides by the norms and principles of international law and is firmly determined to act within existing international agreements and mechanisms", and that "... despite alarmist forecasts, Arctic disputes, when they arise, are likely to be matters of law and diplomacy rather than conflict and stalemate".[109] This complies with the Ilulissat declaration of 28 May 2008 signed by all the Arctic five, including the Russian Federation: "... the law of the sea provide for rights and obligations concerning the delineation of the outer limits of the continental shelf ... we remain committed to that framework."[110] Although Russia continues to gather bathymetrixc data in the Central Arctic Ocean, presumably to support her claim to the Lomonosov Ridge, she at the same time exchanges bathymetric data on continental shelf demarcations with both Denmark and Canada.[111] In this respect, Baker concludes that the regional "...cooperation is in stark contrast to media assertions of the potential for conflict in the Arctic and to other speculations, often focusing unduly on the Russian Federation, that an unregulated race for the North Pole is underway. In fact, Russia was the first State, Arctic and otherwise, to submit information under internationally agreed procedures to the Commission on the Limits of the Continental Shelf".[112] This attitude is also reflected in the concept of the Russian Arctic sector as defined in the *Fundamentals of Russian Arctic Policy of 2009* which only includes the internal waters, land and islands, territorial sea, the exclusive economic zone and the continental shelf of Russia, within which "Russia possesses sovereign rights and jurisdiction in accordance with international law" (see Chapter 1).[113]

The willingness to implement UNCLOS was also put to a practical test in the negotiations on the delimitation of the Barents Sea between Norway and the Russian Federation. In an early phase of these negotiations, the Soviets claimed the sector line to be the ocean boundary between the two countries, whereas Norway argued the median line to be a more reasonable line of division. Thus, the disputed area amounted to 176,000 square kilometers.[114] Actually, the Soviet Union took the stand that there was "no disputed area east of the sector line",[115] implying that this was Soviet "ocean territory". The Sector Decree had, according to Soviet negotiators, "...gained such an administrative prescriptive title in Soviet practice that it carried a special psychological and political significance".[116] The formal negotiations went on for 36 years, and in late April 2010 the parties agreed to divide the disputed area in two equal parts. In this agreement, Russia accepted a fairly radical deviation from the sector line, "losing" nearly 80,000 square km of ocean territory to Norway (see Figure 6.4). As of the present, the western sector line is nothing but theoretical history. By applying existing international ocean law, the Russian government, in most instances, gets sufficient coastal state control over Arctic maritime spaces to play down the role of the Sector principle. However, as long as the theory is not officially denounced, creative ambiguity by playing on hydrographic and climatic factors could be applied by Russia to claim sovereignty over the high-latitude and near-the-Pole routes in the High Seas of the Arctic Ocean. This is not an easy option on the part of Russia since it most likely will create serious conflicts within the group of Arctic eight, or at least this is what the Russian government has to take into consideration. According to Russian analysts, should the nationalistic sentiments in Russian policy gain in popular support, the likelihood is for a more ambitious advancement of the Sector principle in which the sector may be depicted as part of the territory of Russia. If, however, democratic values and institutions get strengthened, the likelihood is for Russia to secure her economic interests by elaborating a new approach towards an international regime and play down the impacts of the Sector claim.[117]

Unsolved delimitation line

In 1990, the then Soviet Union and the United States signed a Maritime Boundary Agreement delimiting their borders in the Bering Sea, Arctic Ocean and northern Pacific Ocean. The agreement referred to the cession of the Alaska Treaty of 1867 dividing the Bering Sea along a south-westerly axis. But without a map attached to the Treaty at the time of its signing, the superpowers disagree over the legal status of an area of approximately 50,000 square kilometers of economic, environmental and strategic importance. The agreement was entered into by the then Soviet Union, and has not been ratified by the Russian Federation, which has only agreed to abide by it on a provisional basis. The United States

Senate gave its advice and consent to ratification as early as 16 September 1991, and has no intentions to reopen the issue.

6.3 The Northwest Passage

When navigating along the various routes of the Northwest Passage, ships will pass through the national waters of Greenland/Denmark, northern Territories of Canada and Alaska/USA. The coastal zones established by these states, and the national legislation introduced in these respective waters, will form the legal framework for sailing the NWP. Sailing along this Passage also means facing jurisdictional controversies still pending, such as the drawing of the Canadian baseline and thus the definition of Canada's internal waters, as well as two unsolved delimitation lines between Canada and Denmark/Greenland and Canada and the United States.

6.3.1 Greenland: Coastal waters

Denmark extended its territorial sea from 3 to 12 nm in 1999, while the territorial sea of Greenland remains 3 nm. A 200 nm EEZ was established outside mainland Denmark in May 1996[118] and outside the coast of Greenland in October 2004,[119] including maritime delimitations towards Canada's EEZ, with later amendment indicating the delimitation line towards the fisheries protection zone outside Svalbard.[120]

The Danish Act on Protection of the Marine Environment of 1993[121], which as of 2004[122] also applies to the waters outside Greenland, aims to prevent and reduce pollution of the marine environment from ships, aircraft and floating and fixed platforms. The Act contains rules that owners, users, masters, and pilots of ships have to report immediately to the Minister for the Environment any discharge or dumping incidents in contravention of the Act. An authorized body has the right, without a court order, to examine a ship in order to prevent pollution if discharge has taken place or is likely to take place.

The Self-Government of Greenland is responsible for the marine environment in Greenland's territorial sea, while the responsibility in the EEZ outside Greenland's territorial waters is with the Danish Government or, more precisely, with a special unit[123] under the Ministry of Defence targeted for and located in Greenland. The intention seems, however, to be that the Self-Government will take over responsibility also in the EEZ.[124]

Denmark's Act on Protection of the Marine Environment of 2009[125] does not at present apply to Greenland, but it is inherent in the Act that this may be proclaimed through a Royal Decree.

6.3.2 Canada: Coastal waters

Canada passed legislation extending its territorial sea limit from 3 to 12 nm in 1970.[126] Linking this to the Northwest Passage, it was stated that by this bill Canada had undisputed control over two of the gateways of the Northwest Passage.[127] However, control here did not exclude innocent passage, as the waters were defined as territorial rather than internal. In 1973, Canada claimed these waters to be internal, which today can be seen in the

Canadian Oceans Act of 1996.[128] The straight baseline method applied by the Canadian Government in defining its internal waters has caused controversy, which is dealt with in more detail bellow.

As for maritime zones, including the territorial sea, definitions are laid down in the Oceans Act of 1996. Apart from defining the above-mentioned baseline and the extension of territorial waters measured to 12 nm from this baseline, the Act also establishes a contiguous zone of 24 nm from this baseline in which Canadian authorities have jurisdiction to prevent the entry of a ship when there are reasonable grounds to believe that an offence under fiscal, immigration or sanitary federal law might be committed. Arresting a foreign ship for breaking these laws will, however, require the consent of the Attorney General of Canada.

Canada's 200 nm Exclusive Economic Zone was introduced when the Oceans Act came into force in 1997, stating the rights and duties of a coastal state in such a zone laid down in UNCLOS. The Ocean Act also refers to fishing zones[129] established within Canada's EEZ, with the more specific regulations laid down in the Fisheries Act of 1985,[130] with the purpose of protecting fish and fish habitat in Canadian waters. These regulations are not related particularly to shipping except by referring to another Act that could be of relevance when sailing in Canadian waters, namely the Canada Shipping Act of 2001.[131] This Act, which is one of many Canadian Acts related to shipping, applies not only in respect to Canadian vessels everywhere but also "in respect to foreign vessels in Canadian waters".[132] The purpose of this Act is to protect individuals who participate in marine transportation and commerce, to promote safety in marine transportation, and to protect the marine environment from damage due to navigation and shipping activities. It thus contains regulatory measures on registration, listing and recording of vessels, safety, navigation services, incidents, accidents and casualties, pollution prevention and measures related to pleasure craft. Also the Marine Transportation Security Act of 1994 applies to foreign ships in Canadian waters, with regulations[133] requiring foreign ships to report to the Canadian Coast Guard upon entering these waters.

Of relevance to shipping is, not least, the current Arctic Waters Pollution Prevention Act introduced in 1970.[134] This was initiated as a response to the US tanker *Manhattan* having sailed though the Northwest Passage without permission from the Canadian Government, causing concern about what such tankers might imply in Canadian waters. The Act applies to what is defined as Arctic waters, "except where otherwise provided", with Arctic waters being defined as Canada's internal waters, territorial waters and EEZ.[135]

According to the Act, no person or ship shall deposit or permit the deposit of waste in the Arctic waters or in any place where the waste could make its way into the Arctic waters. Undertakings that are in danger of causing deposit of waste, or waste having been deposited by way of accident shall immediately be reported to Canadian authorities. The responsibility of masters of ships in this respect is particularly mentioned. So too is the joint liability of the owner of cargo and of ships navigating within Arctic waters for costs, expenses and loss or damage caused by deposit of waste in these waters. Liability is absolute and does not depend on proof of fault or negligence, and claims can be sued for in any court of competent jurisdiction in Canada.

Specifically with regard to shipping, the Act gives Canadian authorities the right to prescribe certain areas of what is defined as Arctic waters as shipping safety control zones and to implement specific regulatory measures in such zones.[136] The actual control zones are laid down in a Shipping Safety Control Zone Order[137] describing their latitudes and the longitudes. Canadian authorities can implement regulations that apply to ships of any class entering the zone, about aspects such as: hull and fuel-tank construction (including the strength of material used); quantities of fuel, water and other supplies; navigational aids and equipment; manning (including the number of qualified navigating and look-out personnel); and type of cargo with its maximum quantity and method of storing. The Canadian authorities can exempt ships of foreign powers from such regulations when it is evident that appropriate measures and reasonable precautions have been taken, but the authorities could also make regulations whereby the owner or master of the ship entering a shipping safety control zone will have to provide a certificate as evidence of the compliance with the standards prescribed in the zone, as well as evidence of financial responsibility in the form of insurance or any other satisfactory form. A specific feature when it comes to Canadian authorities' jurisdiction within the shipping safety control zones is thus that it encompass regulatory measures normally to be decided by the Flag state only.

Through an amendment of December 2009, the Shipping Safety Control Zone was extended from 100 nm to the outer limit of the Canadian EEZ.[138] This is in line with Arctic Waters Pollution Prevention Act extending the limit to up to 200 nautical miles offshore.[139] When extending the control zones, reference was made to Article 234 of UNCLOS giving coastal states "the right to adopt and enforce non-discriminatory laws and regulations for the prevention, reduction and control of marine pollution from vessels in ice-covered areas within the limits of the exclusive economic zone". The amendments were thus considered to be consistent and compatible with international law.

Coming into force on 1 July 2010, the Canadian Government, pursuant to the Shipping Act of 2001, passed the Northern Canada Vessel Traffic Services Zone Regulations (NORDREG),[140] covering Canada's coastal northern waters. More precisely, NORDREG covers the Shipping Safety Control Zones referred to above, but it also consists of additional waters prescribed in article 2.[141] NORDREG is motivated by the need to exercise effective control in the Arctic, especially since shipping activities may increase. It replaces the former voluntary reporting system. The Regulation requires that foreign and domestic vessels of a certain size report to the Canadian Coast Guard if traveling through Canada's Arctic waters. This new mandatory requirement will ensure vessels report information such as identity, position and destination to the Canadian Coast Guard. Pursuant to the Regulations, clearance must be obtained before entering the zone and the Canadian authorities could require extensive reporting and conduct close monitoring of the activities.

Prior to the new mandatory Regulations being passed, the Canadian government expressed that "in case of non-compliance, the vessel would be contacted to require that it comply with the Regulations. Where necessary, enforcement action may include follow-up communications with the flag state of a foreign vessel, notification through Port State Control procedures, and possible prosecution in accordance with the CSA [Canadian Shipping Act] 2001 and consistent with international law".[142]

Worth mentioning also is that Canada has adopted very strict liability and compensation regulations laid down in the Marine Liability Act,[143] establishing liability for ship owners when it comes to damages caused by oil spills.

6.3.3 United States – Coastal waters

The Northwest Passage runs along the coast of Alaska through the Bering Strait (separating Russian and Alaska) through several waterways before going through the Canadian coastal waters.

The United States' jurisdictional claims in the Arctic Ocean are essentially no different than those applied elsewhere of its coasts. In 1988, the country extended its 3 nm territorial sea to 12 nm, while a 200 nm EEZ had been established off the coast of the United States on 10 March 1983. Even though not having ratified UNCLOS, the proclamation of an EEZ refers to international law and defines the rights of the United States within its economic zone to be the same as those laid down in UNCLOS.

The United States has consistently held the position that the Northwest Passage is an international strait in its entirety. The dispute with Canada on this point is elaborated in Section 6.3.5 below. The obvious and explicitly stated reason is the United States' strategic and security interest of free access.

We have not found legislation on a national (federal) level applicable to the coastal waters along Alaska comparable to the regulatory regime of, e.g., Canada, and regulations on the state level (Alaska) are not elaborated on in this book. It should, however, be noted that, as far as accidents and oil spills/other pollution are concerned, a strict liability regime applies.

6.3.4 Jurisdictional controversy

The jurisdictional controversy with regard to the Northwest Passage is, as indicated earlier, the Canadian straight baseline and thus what constitutes Canadian internal waters as well as Canada's still-pending delimitation lines between her own coastal waters and that of the United States and of Denmark/Greenland.

Canada's internal waters

As indicated earlier, the straight baseline applied by the Canadian Government to define its internal waters has caused controversy. While Canada's sovereignty over the archipelago outside its coast has been recognized by the international community,[144] the controversy is linked to the method used by Canada, which implies a baseline to the outer perimeter of the archipelago and thus making the Northwest Passage Canadian internal waters.

The first official claim to these internal waters was, as indicated earlier, made in 1973; more precisely, by stating that the waters of the Canadian Arctic Archipelago are Canada's internal waters by historical title.[145] A similar statement was made in 1975, where it was further stated that the regime of transit passage did not apply in these waters,[146] which in itself could be seen as an indication of these waters having earlier been defined as territorial. Straight baselines around the archipelago were made official in a statement of 10 September 1985, effective as of 1 January 1986, indicating that "[t]hese baselines define the outer limit of Canada's historical internal waters".[147] As indicated earlier, these baselines are today laid down in the Oceans Act where reference is made to the definition found in UNCLOS as well as to historic title.[148]

Protests against the straight baseline method came from the United States in February 1986, stating as its position that "there is no basis in international law for the Canadian claim".[149] Protest also came from the European Community in July 1986, stating that "[t]he Member states cannot recognize the validity of a historic title as justification for the baseline".[150]

Canadian authorities have, as indicated, referred to two legal bases in support of its internal waters. One is by virtue of historic title, which enables a state to supersede purely geographical conditions when claiming sovereignty and thus prevent the application of rules and regulations that would otherwise be binding. The other legal base is by virtue of the fact that their internal waters are on the landward side of the baseline drawn in 1985.

It has been argued that the historic title argument does not fulfil the requirements either of Canada having exercised exclusive jurisdiction over the actual waters for a sufficiently long period or of the need of acquiescence from foreign states.[151] As for the other argument, namely the actual drawing of straight baselines in 1985, it is argued that these lines are consistent with international maritime law laid down in UNCLOS as the archipelago meets the criteria of being sufficiently closely linked to the land and also that it does not depart to any appreciable extent from the general direction of the coastline. The Canadian argument is that the baseline also met the criteria of customary international law in 1985.

The United States' position is that the Northwest Passage qualifies as an international strait, thus giving foreign ships the right of transit passage.[152] In order for a strait to meet the qualification of being international, there are geographical as well as functional criteria. The geographical criterion is that of a waterway between adjacent land masses that links two bodies of the High Seas or EEZs.[153] Its functional definition, not laid down in UNCLOS but based on customary international law,[154] defines the criterion to be a strait used for international navigation.[155]

Relating these criteria to the Northwest Passage, the crucial factor is the functional criterion. Based on analyzing the ship traffic in this Passage it has been concluded that "it is evident that the Northwest Passage has not had a history as a useful route for international maritime traffic".[156] However, it has been pointed out that the navigation of foreign commercial ships could increase through the Northwest Passage, therefore being sufficient for making it an international strait in the future.[157] It has thus been maintained that it is hard to argue that the ship traffic in the Northwest Passage so far meets the functional criterion of an international strait, but that this may be the case in the future.[158]

It goes without saying that, for ship traffic, it makes a difference whether the actual waters are defined as internal waters or international straits. It seems, however, that the Canadian authorities have a rather pragmatic attitude when it comes to the actual sailing through these waters. Their main concern seems to be national jurisdiction in order to pass appropriate legislation to protect and control the marine environment as well as to introduce other safeguarding measures.

As for the legal status of these waters, protests from other countries have not changed the Canadian position on its straight baseline; rather, Canadian authorities have reasserted their baseline claim, partly by stressing the historic usage of ice and waters in the Archipelago by the Canadian Aboriginal peoples. It has been indicated that implementing the Nunavut Land Claims Agreement[159] could help bolster the Canadian claim, as the Agreement clearly states that "Canadian sovereignty over the waters of the Archipelago is supported by the Inuit use and occupancy".[160] This occupancy can thus be said to represent "an important component of the strategy to assert, affirm and express Canada's Arctic sovereignty", of which Inuit leaders are fully supportive.[161]

Even though not yielding to protests from other countries, Canadian authorities have been open for talks on the matter, and have welcomed cooperation with the United States or other Arctic states on the condition that the Canadian claim is fully respected. It has been stated that "any cooperation with the United States, or with other Arctic nations, shall only be on the basis of full respect for Canada's sovereignty".[162] Talks with the United States resulted in an agreement in 1988.[163] Referring to the uniqueness of ice-covered maritime areas and to the search for increased knowledge of such areas through the use of icebreakers, Canada and the United States agreed to facilitate navigation by their icebreakers in their respective Arctic waters and to develop cooperative procedures for this purpose. The agreement states that all navigation by US icebreakers within waters claimed as internal by Canada will be undertaken with the consent of the Government of Canada.

The agreement does not interfere with the two countries' diverging positions on the legal matter, as it is clearly stated that neither their cooperative endeavor laid down in the agreement nor any practice thereunder affect the position either of the Government of the United States or the Government of Canada on international maritime law in this or other maritime areas or their respective positions regarding third parties. In other words, since 1988 the two countries have maintained their respective legal positions, but at the same time had a practical solution in place.

Even though there has been no formal change in the United States' position, and the dispute between the two countries on this matter goes on, it has also been suggested that the US could come to regard the Canadian claim as more palatable politically and legally.[164] This is related to concerns regarding the necessity of policing the Northwest Passage, and that this is better taken care of through the Canadian legal claim over the Passage than by it being defined as an international strait. This might relate to what has been indicated above, namely that the Canadian government seems to support international shipping through the Northwest Passage as long as Canadian regulations are followed. It goes without saying that it is the pollution and other possible negative implications of shipping that is considered most threatening in the Northwest Passage.

An alternative between the two positions has also come forward: namely, to define the Northwest Passage as neither internal waters nor international strait, but rather as territorial waters, implying that the Passage would be open for innocent passage.[165] Canada's position seems, however, still to be that the country has "an unfettered right to regulate the North West Passage as it would its land territory". [166]

Unsolved delimitation lines

There are at present two unsolved delimitation lines related to Canadian waters. In parts of the Beaufort Sea, Canada claims that the maritime boundary between her own maritime zone and the maritime zone of the United States runs along the 141st meridian. The argument is that the extension of the boundary agreed with United States regarding the land territory is the proper method also to draw the maritime delimitation. The United States challenges this position, arguing that the maritime delimitation must be determined by using the equidistance principle, as it more closely reflects the direction of the respective coastlines.

Canada's other unsolved delimitation line is in the waters between Ellesmere Island and Greenland. When settling the delimitation line between their maritime zones in 1973, Canada and Denmark agreed to delay a final solution regarding a small (1.3 square km) island – Hans Island – in the Lincoln Sea, which is claimed by both countries. In regard to these controversies the Canadian Government has expressed that "...although these disputes are well managed, an early resolution to the issues will allow for more certainty in terms of navigation and exploration". [167]

6.4 The Transpolar Passage

Unlike the Northern Sea Route and the Northwest Passage, the Transpolar Passage is international waters and outside geographical reach of national jurisdiction. This has not been challenged by Russia or Canada with basis in the Sector theory. The Russian and Canadian laws which have been passed are limited to the coastal zones.

Thus, the legal regime governing the Transpolar Passage is the regime of the High Seas as expressed by UNCLOS and other conventions, most prominently the SOLAS convention, the STWC convention and MARPOL. The IMO Guidelines have the same significance for activities in the Transpolar Passage as they have for any other Arctic sea route. But since no national legislation exists, it becomes even more important to transform the IMO Guidelines into binding rules. The importance of good corporate governance and corporate social responsibility also increases in lieu of national legislation.

As indicated earlier, international rules are only binding for state parties, not for private enterprises. It therefore becomes important that flag states with registered ships operating in the Arctic are bound by the relevant treaties and conventions. Without having made an extensive survey, it can still be said that many of the states offering "flags of convenience" do not impose strict regulations in areas such as safety and environmental issues.

6.5 Indigenous peoples' rights

The point of departure in the second theme of this chapter is that "Arctic marine shipping involves several distinct activities, each with different interactions with local residents and thus different implications and likely impacts".[168] Having lived in a remote and rather challenging area for ages, Arctic indigenous peoples have been left more or less undisturbed, and thus been able to maintain many aspects of their traditions and way of life. Climate change and modern technology, resulting in industries able to operate in harsher surroundings, have gradually made this area more accessible, with the implication that traditional local communities now are "facing an unprecedented combination of rapid and stressful changes involving both environmental forces like climate change and socioeconomic pressures associated with globalization."[169] Arctic indigenous peoples are, in other words, now confronted with dramatic changes inflicting on their culture and traditional way of life.

As can be seen from Figure 6.5, indigenous peoples in the Arctic represent a great variety of ethnic groups. In the northern part of the Russian Federation, there are altogether

Source: http://amap.no/acia/Files/PeopleoftheArctic_150.jpg

Figure 6.4 Indigenous peoples in the Arctic.

26 minority groups with the Nenets, the Chukchi, the Evenks and the Events as the largest. The Sami are found in the Russian Federation, but with the largest group living in the northern part of Norway. In Greenland, the major part of the population is Inuit, a group constituting a large part of Aboriginal peoples in Canada together with the Natives and the Métis. The Alaska Natives include various groups such as Inupiat, Yupik and Aleut.[170]

Notwithstanding cultural and ethnographic differences, indigenous peoples in the Arctic share a common history of attempts to assimilate them into their various mother states and a lack of recognition of rights on traditionally occupied territories. Against this background, the indigenous peoples of the Arctic have united in working steadfastly towards self-determination. Their primary objective has been to ensure that development and activity on traditional lands and waters take into consideration indigenous peoples' traditional way of life, as well as to give them their fair share of the economic benefits of this development.

The main focus here is on what rights indigenous peoples have acquired in regard to this objective in their respective Arctic states. These rights, laid down in national laws and land claim agreements or settlements, form the legal and political framework within which other actors, such as the shipping industry, should operate when in the Arctic area.

It has been acknowledged that there is still a lot to be learned regarding the actual interaction between shipping and indigenous peoples in the Arctic.[171] Some of the knowledge that has come forward in the AMSA Report and from other research will be referred to. Even though scarce and incomplete, it indicates what the interaction might be.

As uniting across national borders, and working in the international arena proved to be a good strategy for indigenous peoples to acquire their rights, it seems appropriate to start with an overview of international conventions and declarations securing such rights, as well as with an overview of their regional cooperation, before turning to what rights indigenous peoples have actually acquired in their respective mother states.

6.5.1 International conventions and declarations

When dealing with international development of indigenous peoples' rights, it is important to bear in mind that their demand was not to be given equal opportunities in what they described as a "Western culture", but rather to be given the right to maintain a different culture.[172] This formed the background for most of the indigenous peoples' organizations that emerged in the 1970s, such as the Inuit Circumpolar Council (ICC) of 1977.[173] Together with the Sami Conference dating back to 1953, the ICC played an important role in the United Nations Working Group on Indigenous Populations and eventually in formulating the United Nations Declaration on the Rights of Indigenous Peoples in 2007.

The international human rights movements, initiated in the 1950s,[174] even though dealing with the rights of minorities, did not fully recognize the right of indigenous peoples. The two International Covenants adopted on 16 December 1966 by the United Nations on Civil and Political Rights (ICCPR) and on Economic, Social and Cultural Rights (ICESCR)[175] nevertheless formed the basis for indigenous peoples arguing their rights. Both Covenants state the right of all peoples to freely determine their political status, to freely pursue their economic, social and cultural development, to freely dispose of their natural wealth and resources and not to be deprived of their own means of subsistence. Of special importance to indigenous peoples was that the ICCPR made it clear that minorities within a state should not be denied the right to enjoy their own culture, religion and language.[176] Of great importance was that the United Nations Human Rights Committee, monitoring the ICCPR, in 1995 stated that culture manifests itself in many forms including a particular way of life and that, especially in the case of indigenous peoples, might include the right to traditional activities as well as effective participation in decisions affecting them.[177]

The first convention dealing specifically with indigenous peoples' right to self-determination was the ILO Convention No 169 of 1989.[178] Using the term "peoples",[179] this Convention confirms that there can be several peoples within a state, and goes far in setting standards for nation states in their relations with indigenous peoples within their borders. More precisely, the Convention states that national governments have a duty to implement measures recognizing indigenous peoples' right to ownership and possession over lands they traditionally occupy,[180] and also that their right to the natural resources pertaining to these lands should be specially safeguarded, including participation in their use, management and conservation.[181] Also safeguarded is the indigenous peoples' right to decide their own priorities for the process of development as it affects their lives as well as the lands they occupy or otherwise use.[182]

Twenty states have ratified the ILO Convention 169, thus having an obligation to implement the Convention in their national legislation. Of the five states bordering the Arctic Ocean, only Denmark[183] and Norway[184] have ratified.

After more than 20 years of negotiations between nation states and indigenous peoples,[185] in September 2007 the General Assembly adopted the United Nations Declaration on the Rights of Indigenous Peoples. In this Declaration, rights already recognized in international human rights law were specifically recognized for indigenous peoples, one such right being the right to autonomy and self-government in matters relating to their internal and local affairs.[186] The Declaration contains some specifications in regard to self-government.[187]

The UN Declaration goes far in stating indigenous peoples' right to own, use, develop and control the lands and resources in territories that they possess by reason of traditional ownership or other traditional occupation or use.[188] Also secured is the right to redress, by means of restitution or fair and equitable compensation, such lands, territories and resources that have been confiscated, taken, occupied, used or damaged.[189]

Of particular importance in regard to new activity moving into local communities is indigenous peoples' right to determine and develop priorities and strategies for the development or use of their lands and resources. Before the final approval of a project affecting such lands and resources, the state has to consult and cooperate with the indigenous peoples in order to obtain their consent.[190]

Viewed as a strong statement embodying the most important rights of indigenous peoples,[191] the UN Declaration is nonetheless a declaration, and as such does not put any obligations on the nation states. However, with 143 states having voted for the Declaration, it is described as an important guide in future relations between nation states and their indigenous peoples,[192] and thus as an expression of what can be said to constitute "good governance" in regard to indigenous peoples' issues. Of the five Arctic states, Russia, Denmark and Norway voted in support of the UN Declaration while Canada[193] and the United States[194] gave their support in 2010.

Worth mentioning is also that the three Arctic states – Denmark, Norway and the Russian Federation – together with members of the European Council and other European states, are signatories to the Framework Convention for the Protection of National Minorities of 1995. Aiming at non-discrimination the Convention states that "[t]he Parties undertake to guarantee to persons belonging to national minorities the right of equality before the law and of equal protection of the law. In this respect, any discrimination based on belonging to a national minority shall be prohibited".[195] Also aiming at protecting national minorities' right to maintain their own culture and way of living, the parties to the Convention undertake "to promote the conditions necessary for persons belonging to national minorities to maintain and develop their culture, and to preserve the essential elements of their identity, namely their religion, language, traditions and cultural heritage".[196] The Framework Convention provides for a monitoring system which means that parties to the Convention

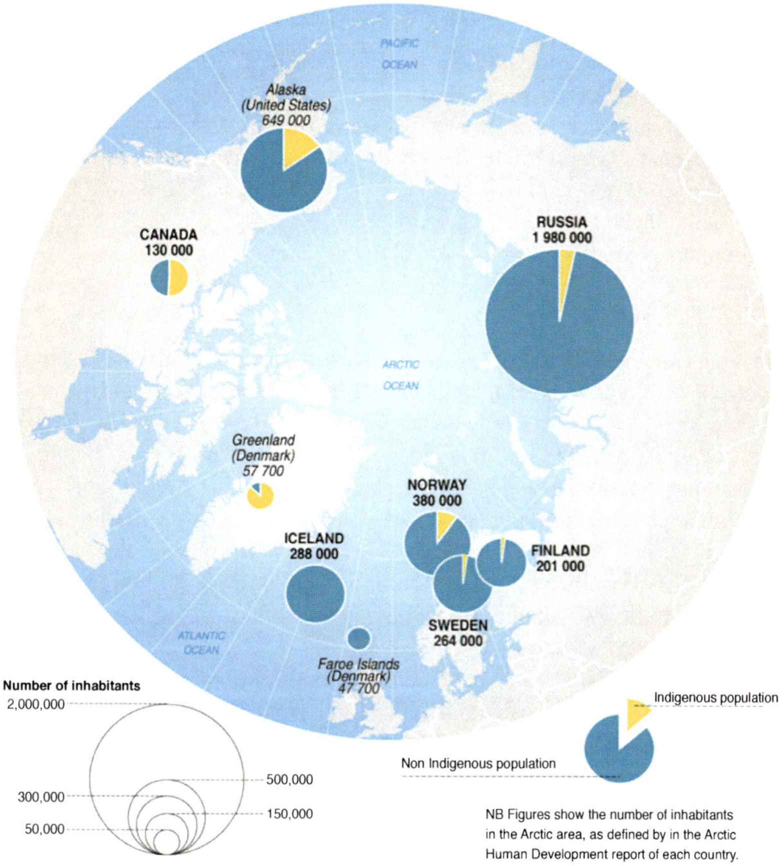

Source: https://qed.princeton.edu/getfile.php?f=Population_Distribution_and_Indigenous_Peoples_in_the_Arctic,_2002.png

Figure 6.5 Population distribution of indigenous peoples in the Arctic.

have to report to an Advisory Committee every five years on how the Convention treaty is implemented in their respective legal systems.

Expressions of international law as it comes forward in the international conventions and declarations presented above have of course no formal bearing on individual shipping companies, which are bound only by national regulation in the coastal waters where they sail or by own flag-state jurisdiction. In lieu of the fact that what has been laid down in these instruments has strong support in the international community and thus can be said to express what could be designated as "good governance" in relation to indigenous peoples, the shipping industry might find it in its interest to consider these instruments when interacting with indigenous communities in the Arctic.

Here it may be interesting to note that the European Bank for Reconstruction and Development (EBRD) in 2010 published the so-called Performance Requirement, aimed at providing practical guidance "on the processes and standards that should be applied and considered when planning and implementing projects that involve indigenous peoples".[197] Even though the actual scope in this note was practical guidance relevant for Russia, the requirements nevertheless should be considered to express good practice in relation to indigenous peoples everywhere.

6.5.2 Regional cooperation

The international instruments mentioned above form an important basis in the continuous engagement of indigenous peoples in the Arctic to further what they consider to be their rights and interests. One example is the Circumpolar Inuit Declaration on Arctic Sovereignty adopted in April 2009. Referring to their rights as an indigenous people embodied in international instruments as well as their rights as citizens in an Arctic state, the Inuit of Greenland, Canada, Alaska and the coastal regions of Chukotka in Russia in this Declaration make it clear that issues of sovereignty and sovereign rights in the Arctic "must be examined and assessed in the context of "our long history of struggle to gain recognition and respect as an Arctic indigenous people having the right to exercise self-determination over our lives, territories, cultures and languages".[198]

The Declaration is considered a permanent means of reminding the international community about indigenous peoples' rights to self-determination and thus having a say in regard to lands and natural resources in the Arctic. This was demonstrated at the United Nations Framework Convention on Climate Change (UNFCCC) in Copenhagen in 2009. In a "call-of-action" to the parties to this Convention,[199] indigenous peoples called attention to the fact that they were living in areas most vulnerable to the impacts and root causes of climate change, which they described as having "profound and disproportionate adverse impacts" on their cultures and traditional livelihoods, and demanding that agreements and principles in regard to climate change need to reflect the "spirit and the minimum standards" contained in the United Nations Declaration on the Rights of Indigenous Peoples. Specific reference was made to the right to "free, prior and informed consent" in regard to specific programs and projects affecting indigenous peoples' lands, territories and natural resources, laid down in this Declaration.

The Inuit Circumpolar Council plays an important role at the regional level, representing Inuits across the Circumpolar North. Thus it constitutes a forum for cooperation on common matters and presents Inuit views on issues related to economic wellbeing as well as the preservation of their traditional culture and way of life. Another forum is the Arctic Council, on which six indigenous peoples' organizations are Permanent Participants with a common support secretariat, Indigenous Peoples' Secretariat (IPS).[200] In addition to the Inuit Circumpolar Council, other Permanent Participants are: the Aleut International Association (AIA) representing Aleut on the Russian and American Aleutian, Pribilof and Commander Islands, the Arctic Athabaskan Council (AAC), representing the interests of United States and Canadian Athabaskan peoples; the Gwich'in Council International (GCI), representing all the Gwich'in people in the Northwest Territories, Yukon and Alaska; the Saami Council, which covers Sami member organizations in Finland, Russia, Norway and Sweden.

The last Permanent Participant, the Russian Association of Indigenous Peoples of the North (RAIPON), represents 41 different indigenous peoples in the Russian Federation, in the North, Far East and Siberia. The organization is the representative of the northern indigenous people and Russian authorities and government. There are also large indigenous peoples' organizations at the national level which are not Permanent Participants, such as Inuit Tapiriit Kanatami representing four Inuit regions in Canada,[201] and the Alaska Federation of Natives.

Worth mentioning also is the Norwegian Barents Secretariat, which hosts the Secretariat for the Working Group for Indigenous Peoples (WGIP), and promotes a variety of projects aimed at connecting indigenous peoples across borders in the Barents Region.

Also worth mentioning is that there has for a long time been close contact between Sami in the three Nordic countries.[202] Initiative has been taken for a Nordic Sami convention, and a draft presented by a group of experts[203] confirms the Sami peoples' right to self-determination. It is envisaged that a Nordic Sami convention will expand the scope of functional autonomy from central governments to the respective Sami parliaments. A working group, appointed by the Norwegian Government to scrutinize the convention draft, concluded that some of the provisions go beyond national as well as international law, thus implying extended duties on the part of Norwegian authorities.[204] This may be part of the reason why a Nordic Sami convention is still pending. Being a Nordic convention, the Sami of the Federation of Russia is not a party, but the group of experts has recommended some form of cooperation within the convention with Russian Sami, but obviously leaving an actual initiative to the political authorities.

From what is outlined above it seems fair to conclude that indigenous peoples in the Arctic have a strong legal and institutional basis on the international and regional level for cooperation in promoting their traditional rights to lands and waters in their respective mother states.

6.5.3 The Northeast Passage: Indigenous peoples

Indigenous peoples living along the Northeast Passage/the Northern Sea Route are found in communities along the coasts of Norway and the Russian Federation. One such people is the Sami, who live in what is usually described as a Sub-Arctic area. It is estimated that approximately 40,000[205] Sami live in Norway, of which more than half live in the region of Finnmark.[206] Sami in Russia, on the Kola Peninsula, total about 2000.[207] The largest group of indigenous peoples along this sea route is, of course, peoples of the Russian North. As indicated earlier, there are altogether 26 minority group[208] living in the Russian Arctic and Sub-Arctic which are affected by the Northern Sea Route.[209]

The Sami and the Finnmark Act

From around 1850, the Norwegian authorities' attitude towards the Sami was characterized by an assimilation policy, one that was gradually eased after the Second World War. While the Norwegian Government initially had argued that Sami use of land and resources did not establish any formal rights, this was later modified, not least due to ILO Convention 169, which Norway was the first to ratify.[210] In 1980, the Norwegian Government, in cooperation with Sami organizations, established a Sami legal commission to study a very broad range of aspects related to the Sami rights to land and resources on territories that they occupy, as well as a Sami representative body.[211] The commission's work, relying heavily on international law relating to the right of minorities, led to the adoption of the Sami Act of 1987.[212] The Act proclaimed the Sami's right to preserve their culture and way of life and contained provisions for a Sami Parliament. Such a Parliament was established in 1989, and functions as an advisory body in Sami-related matters. The following year, Sami rights were included in the Norwegian Constitution.[213]

The question of land rights continued to be of importance after the Sami Act, and was strengthened under the Finnmark Act of 2005.[214] The purpose of this Act is "to facilitate the management of land and natural resources in the county of Finnmark in a balanced and ecologically sustainable manner for the benefit of the residents of the county and particularly as a basis for Sami culture, reindeer husbandry, use of non-cultivated areas, commercial activity and social life".[215] Through this Act, the major part of Finnmark county was transferred to the county's inhabitants, that is both Sami and ethnic Norwegians, to be managed by the Finnmark Estate which took over the ownership of the transferred territories. Further investigations are now being conducted[216] in order to clarify ownership to lands in Finnmark by virtue of historic title by Sami or other inhabitants of Finnmark. Even though it applies to all inhabitants of Finnmark, it is assumed to relate primarily to the Sami population.

The Finnmark Act acknowledges that the Sami people has acquired rights to land and resources in Finnmark, and therefore also a right to be heard in matters related to resource management within their territories, hence giving the Sami Parliament a rather significant role in the Finnmark Estate. By this influence of the Sami Parliament, as well as by relating this to the acquired Sami rights, it was concluded that the Finnmark Estate meets the requirement in the ILO Convention regarding indigenous peoples' rights and that the Finnmark Act thus is in accordance with international law.

Through the Finnmark Act a legal and organizational framework has been established for the managing of natural resources in Finnmark, taking into consideration the particular rights acquired by the Sami as an indigenous people in this area. This is, however, not to say that conflict of interest will not occur in relation to the development of resources in Finnmark.

One example relates to plans to exploitat mineral resources in this region, such as copper and gold. This has raised the question of possible impacts on the traditional reindeer herding. Pointing to the need for new ways of economic development in this region, also for the Sami population, and having mineral exploitation as one of the objectives in its High North policy, the Norwegian Government seems to be of the opinion that mineral exploitation and reindeer herding can live side by side in Finnmark.[217]

Whilst not totally rejecting mineral exploration, the reindeer herders are stressing that living side by side might be possible in some areas, but definitely not in others. Their demand is, therefore, that consulting with the reindeer herders will have to be an integratal part of mineral exploration plans in Finnmark. On their part, the mineral companies have expressed positive experiences in consulting with reindeer herders, but are also expressing some impatience, as well as of pointing to the fact that, in areas where mineral exploration and reindeer herding cannot live side by side, "one will have to move". [218]

Even though Sami rights to uphold their culture and way of life as well as their ownership to land to a large extent is ensured in Norwegian legislation, other activities may come as a challenge, not least when such activities imply economic development. As such development could be of value in the form of employment opportunities for other segments of the Sami population, it implies a need within the Sami themselves to balance a traditional way of life with economic development.

As the Finnmark Act includes rights to land territory only,[219] implications of such rights for navigating along the coast of Finnmark will primarily relate to implications on land, such as the need for the infrastructure that follows in the wake of ship traffic.

Minority groups: Russian Federation

As indicated above, there are 26 minority groups constituting indigenous peoples in Russian Arctic and Sub-Arctic, all affected by the Northern Sea Route. The groups belong to various forms of political-administrative units. The largest minority groups in the Arctic are the Nenets, who live primarily in the autonomous districts (okrug) of Nenets and Yamal-Nenets, and the Chukchi, living in the autonomous district of Chukchi. Other large groups are the Evenks and the Events living in the republic (respublik) of Sakha.

Indigenous peoples in the northern part of Russia have been subject to different forms of governance. While the first years after the Russian revolution were characterized by autonomy for minority groups, a leftover from the Tsarist regime, the late 1920s saw the beginning of a change towards forced integration into Soviet society. Traditional reindeer herding, hunting and fishing were forcefully transformed into collective farms. Local uprisings were put down, a number of national areas were dissolved, and the North was divided between various ministries.[220] The system of collectivism thus transformed the system of the traditional indigenous communities, which suffered a decline in their traditional forms of economic activity as well as a fall in the quality of life compared to other social groups.[221]

The most significant blow to the traditional way of life was done in the 1960s when the government resettled people from small villages into large communities, a relocation that destroyed the former historical and ecologically balanced settlement structure of the indigenous peoples.[222] Another feature that added to the transformation of the social structure was the development of natural resources and forms of agriculture. These brought in newcomers who soon outnumbered the indigenous population.[223]

The democratization of Russia from the late 1980s has seen a development of self-consciousness among indigenous peoples of the North and the organization of regional associations to defend indigenous interests. An important initiative by indigenous peoples themselves was the establishment in 1990 of the above-mentioned Russian Association of Indigenous Minorities of the North, Siberia and the Far East (RAIPON). RAIPON, building on a territorial and territory-ethnical principle, organizes 41 indigenous groups from 34 regional and ethnic organizations, totalling around 250,000 people. Its objective is to guarantee these peoples the protection of traditional way of life as well as their right to self-governance "according to the national and international legal standards".[224] As indicated earlier, the organization is the representative of the northern indigenous people to Russian authorities and government.

What can be said to characterize these peoples' situation in the last decade of the 20th century has been described as "demonstrating a willingness to correct the mistakes of the previous assimilation policy".[225] Russia's Constitution of 1993[226] has been amended so as "to guarantee the rights of the indigenous small peoples according to the universally recognized principles and norms of international law and international treaties and agreements of the Russian Federation".

This includes the right to land and other natural resources viewed as "the basis of the life and activity of the peoples inhabiting the corresponding territory" as well as the protection of their traditional living environment and lifestyle.[227] In 1999, a Federal law was passed which established legal guarantees for the social, economic and cultural development of indigenous peoples in the Russian Federation, and protection of their living environment and their traditional way of life and handicrafts.[228] In 2009, a new Federal law[229] introduced some changes, among them a new paragraph that reinforced the protection of indigenous people through listing the places where they traditionally lived and their traditional economic activities. A Federal law giving minority groups the right to organize in so-called communities of native minorities was passed in 2000.[230] The purpose of this law was to secure their traditional patterns of life, rights and legal interests. Another Federal law related to the traditional territories and the use of natural resources of indigenous peoples was passed in 2001.[231] It seems, however, that this latter law has not been implemented; rather, what came forward at a roundtable discussion in 2011 focusing on the results of the Russian Government's obligations to improve the legislation on the indigenous' rights[232] was that the Ministry of Regional Development is drafting a new federal law on the territories of traditional nature which will reduce the indigenous peoples' rights. More precisely, the new draft law will not give such territories the special status of environmental areas which protects them from industrial projects, or allow for such territories to be created at the regional municipality level. This will make it more difficult for indigenous peoples to create such territories.

It seems that the policy towards the northern regions of the Russian Federation is still not fully settled. Power and legal authority has no doubt been shifted to the regional level in the form of autonomous governments, but this does not necessarily mean increased self-determination for indigenous peoples. In some instances, they may have ended up being a minority within the new governments and thus have little say in the control or management of lands and waters, or that the actual application of what should be their rights has not been fully implemented. It has been maintained that there is "little reason to expect these governments to share power willingly with the indigenous minorities within their borders despite the claims of indigenous peoples to international and national rights to protect their land, economic base and culture".[233]

Relating the transformations in the northern parts of Russia and its impact on the indigenous population to the Northern Sea Route, it has been concluded that this sea route "has produced direct and indirect physical, cultural and social impacts".[234] It was, however, not the actual opening of the sea route that affected indigenous peoples the most, but rather "the whole pattern of economic development and industrialization of the Russian North" that this route made possible.[235]

Studies of the impacts on regional communities by industries that have developed in connection with the NSR have identified that oil, gas and mining complexes have removed land from use by indigenous peoples engaged in herding, hunting and trapping. In addition, rivers have been polluted and gas pipelines have formed barriers to reindeer herds.[236] A more recent concern relates to nuclear-powered lighthouses being situated along reindeer-herding routes, implying that the site could contain high levels of radiation.[237]

To illustrate what seems to constitute the situation for indigenous peoples in the northern part of Russia today, we have chosen to look at the Nenets Autonomous Okrug, characterized by the indigenous peoples depending on reindeer husbandry, fishery and hunting for their livelihood while at the same time oil and gas exploitation has been developing. More precisely, the illustration is based on what has come forward in a study on the development of traditional indigenous land use in the Nenets *okrug*.[238]

The study makes the initial statement that Federal laws devoted to indigenous peoples' rights have a general declarative character and therefore do not specify the duty of indigenous resources extractors – such as oil and gas companies – to preserve these rights. One right has been that indigenous peoples engaged in traditional economic activities were entitled to use the land for free and unconditionally. However, as of 2001 reindeer pastures can be leased to companies by the state as long as traditional land users are compensated. Even though the traditional land user is supposed to play a role in the leasing decision, the study raises the question of how "voluntary" this leasing is. According to the study, "[m]ost of the license agreements have been found to comply poorly with legal requirements to consider indigenous peoples' rights" in the Nenets *okrug*, and that only one of 38 analyzed agreements stipulated the license holders' liability to compensate for losses as demanded in the legislation.

The study also makes it clear that "it is widely understood that unlawful conditions prevail in connection with many oil installations", certainly true of older facilities built to low safety requirements, and that many companies withhold information about environmental damage. The relevant government agencies have neither the practical possibility nor the funding to really control such pollution.

Reference is made to numerous examples of good relations at the local level between companies and reindeer herders, and that indigenous peoples have a large capacity to adapt to environmental changes. However, the conclusion is that, despite legal guarantees, indigenous people seem to have no opportunity to change major politically-approved decisions.

6.5.4 The Northwest Passage: Indigenous peoples

Indigenous peoples living along the Northwest Passage are to a large extent Inuit, who constitute the main residents in Greenland and also a large part of the Aboriginal peoples of Canada. Other indigenous peoples along this Passage are the Alaska Natives.

Self-government of Greenland

As the world's largest non-continental island, Greenland covers an area of 2,166,066 square kilometers, most of which is covered with ice, leaving only the 410,494 square kilometer, ice-free coastal areas habitable for human beings. A relatively warm current keeps the southern part of the waters on the west coast clear of ice all year round. Exceptions are the southernmost coast where ice masses drift down along the east coast, and that water in harbors freeze during winter.

It has been estimated that the first peoples of Greenland are ancestors of an Inuit population coming from Alaska, Siberia and Canada. Today, 89% of the population is either a descendent of the original Inuit inhabitants or defined as "Danish or others".[239] The last 50 years has seen a move of the population from settlements to towns,[240] indicating a change from more traditional ways of life for a large number of Greenlanders. Another demographic feature is that by far the majority of the people in Greenland live on the west coast.[241]

Greenland is in a special position in regard to indigenous peoples' rights, as the island has gradually gained a high degree of self-determination, most recently with a new Act on Greenland Self-Government.[242] Without going too much into the historical background, suffice it to say that when Norway was separated from Denmark in 1814 the issue of sovereignty over Greenland was not included in the settlement. Greenland, starting out as a Danish colony,[243] was in 1954 made an integrated part of Denmark.[244] The Greenland Home Rule Act of 1978[245] had given the Home Rule legislative and executive powers within a number of fields,[246] which in the new Act of 2009, based on a special commission's report,[247] and after a referendum,[248] largely extended the fields[249] over which the Greenland Self-Government authorities may decide to resume legislative and executive responsibility. Resuming such responsibility implies a right to the assets, as well as the duty to finance, the actual fields. The new Act has been described as a means for Greenland "to slowly mature and become the kind of society that Greenlanders themselves aspire to".[250] Entering into force on 1 January 2010, a special Act on mineral resources was enacted for Greenland's land territory and territorial sea as well as for its continental shelf and EEZ,[251] replacing the former Danish Act on mineral resources in Greenland.

Having gained self-government, Greenland still depends on an annual subsidy from the Danish Government which, according to the Act of 2009, amounts to DKK 3.439.6 million.[252] The Act on Self-Government makes it clear that revenues from mineral resource activities shall accrue to Greenland's Self-Government authorities, with the consequence that the annual subsidy from the Danish Government will be reduced. However, if the subsidy is reduced to zero, negotiations shall be initiated between the Self-Government authorities and the Danish Government about the distribution of revenue from mineral resources activities in Greenland.[253] The subsidy's part of the total GDP[254] has, however, gradually diminished, and the ultimate objective of the Self-Government authorities is to gain economic self-sustainability, which in turn implies looking for new sources of income.

Greenland's main industry today is fish and fish products, which constitute 87% of all exports.[255] The main landing for coastal fishery is in the area around Illulisat (Jakobshavn). While local fishermen are fishing near the coast, typically with small and semi-big boats, offshore fisheries is characterized by larger trawlers, mostly owned by foreign companies. At least 25% of the catch has to be landed for processing in Greenland.[256]

Even though it constitutes an important source of income, exploitation of living marine resources will not be a sufficient source of income for Greenland to gain economic self-sustainability. There is thus today an increased focus on the country's potential for exploiting other resources such as petroleum and minerals. In October 2009, the Minister for Industry and Mineral Resources in Greenland[257] made the statement that "[the] development of oil activities is one of the most important areas, if we are to create enough revenues to replace the block subsidy from Denmark". The Bureau of Minerals and Petroleum in Greenland is responsible for the whole chain of tasks linked to the production and transportation of minerals and petroleum. The task is to ensure the legal and political framework for reliable, environmentally sound and clean exploitation of energy and minerals resources. According to the Bureau, Greenland has experienced unprecedented international interest in its mineral resources potential. Some of the largest oil companies of the world have now been given exploration and exploitation licenses in Greenland waters and, according to the Bureau, there are at present 13 such licenses in the sea west of Greenland covering 130,000 km2. In addition, application rounds have been prepared in Baffin Bay off north-west Greenland and in the Greenland Sea off north-east Greenland.[258] Greenland Self-Government's participation in hydrocarbon licenses is handled by the National Oil Company of Greenland, Nunaoil. With the Act of 2009, the Self-Government took over shares formerly owned by the Danish Government, which means that Nunaoil now is fully owned by Greenland Self-Government.

Economic development in Greenland also involves mineral resources other than hydrocarbons. More than 70 licenses have been issued to date, mostly to foreign companies, for exploration and exploitation of various minerals. Such minerals are zinc, lead, iron, molybdenum, gold, platinum and uranium, all natural resources to be found on the island. Even though most of the licenses are for mining in west Greenland, there are also licenses issued for mining in the east and in the north. In some instances, these are mines that had previously been closed down but where increasing international demands, not least from growing economies like China and India, makes it sense for them to re-open. (For more on China and India, see Chapter 2).

It goes without saying that both petroleum activity and mining create a need for shipping, to bring relevant equipment to Greenland for use in the actual exploration phase and, not least, in the exporting of the products. It is perhaps particularly interesting in regard to shipping to note that precisely such expanding economies as China and India, having expressed an interest in shipping in the Arctic, seem to be in demand of some of the minerals found in Greenland.

Also related to shipping is, of course, the third means of extended income: the goal of promoting tourism in Greenland. This, of course, also implies more sailings to and from the island.

In general, it is interesting to note that the Government of Greenland Coalition Agreement for 2009–2013[259] makes clear its intention to "create an optimal environment for the development of business and industry and thereby enable growth of the Greenland economy". One measure of importance here is the establishment of an international free port, and the development of the country's transport system including the expansion of harbors.[260] These are, of course, measures that will improve the infrastructure for ship traffic and thus also mean better communication for indigenous peoples in a country where roads are very poorly developed.

With regard to economic development, the Self-Government of Greenland has declared that exploiting mineral resources is to be based on legislation ensuring its people direct influence and involvement in the decision-making process. It has also made clear that such development has to take into consideration the county's special nature and traditional culture.

Due to the Act on Self-Government, the Inuit of Greenland have the tool to gradually extend its self-determination in regard to the development of its traditional lands and waters. Characterizing the situation is the balance between economic development and the preservation of its traditional culture, including environmental considerations. The fact that being economically independent is an important goal for the Self-Government of Greenland the former may get the upper hand in such a balance. This was demonstrated in the 1980s when the Greenland Home Rule strongly opposed a Canadian plan for year-round icebreaking tankers transporting oil and gas through the Davies Strait, as it threatened the Greenland's subsistence economy of fishing and whaling. When, however, possible oil prospecting outside East Greenland came on the agenda not long after, "[t]he prospect of obtaining new revenues from petroleum if found in sufficient quantities, made the Home Rule Government in principle accept oil transportation by icebreaking tankers" passing through waters that were biologically rich and highly productive.[261] An important difference is, of course, that the latter was a decision by, and thus under the control of, its own government.

Aboriginal peoples of Canada: Land claims and self-government agreements

Aboriginal peoples of Canada[262] are defined in the Canadian Constitution of 1982 as First Nations,[263] Inuit and Métis. These are three distinct peoples with unique histories, languages, cultural practices and spiritual beliefs. The Inuit constitute the largest group in northern Canada, with more than 40,000 of a total in all Canada of approximately 60,000. Having lived in these areas for thousands of years, the Inuit is today found primarily in one of four regions collectively known as Inuit Nunaat (Nunaat meaning homeland in the territory of Nunavut). The other two territories in Canada are the Northwest Territories and Yukon,[264] all bordering the Northwest Passage.

Prior to contact with Europeans, Aboriginal peoples relied on a variety of distinctive ways to organize their political systems and institutions. Many of these institutions were later ignored or legally suppressed by the Federal Government's attempt to impose a uniform set of Euro-Canadian political ideals on Aboriginal societies and to assimilate Aboriginal peoples into the non-Aboriginal society.[265] Many Aboriginal peoples thus see self-government as regaining of pre-existing rights, rooted in their long occupation of territories.

During the past four decades, Aboriginal peoples in northern Canada have made important efforts to gain autonomy over their traditional lands and resources through comprehensive land claims and self-government agreements. Such agreements are based on the assertion of continued Aboriginal rights to land and are constitutionally protected documents,[266] defining a wide range of rights and benefits such as ownership of lands, exclusive rights to wildlife harvesting and participation in land, water and environmental management within the settlement area. Apart from securing jurisdiction over traditional lands and resources, the land claims agreements also play an important economic role as they generally contain economic development provisions.

Today, the majority of Aboriginal peoples in northern Canada are signatories to land claims and self-government agreements.[267] Land claims could be comprehensive, meaning that they always involve land, but also specific, claims which are not necessarily land-related. With regard to land claims, treaties are negotiated between the Aboriginal group, the

Federal Government of Canada and the government of the Province or the Territory. Comprehensive claims settlements also include self-government agreements whereby Aboriginal groups are to govern their internal affairs and assume greater responsibility and control over the decision making that affects their communities. Such agreements could take many forms based on the diverse historical, cultural, political and economic circumstances of the Aboriginal groups, regions and communities involved. Each agreement thus reflects local and regional circumstances, but usually have in common that they provide ownership from 10–30% of the settlement area, cash compensation, economic development opportunities, a share of royalties from development of Crown land and natural resources as well as management participation of land, water and wildlife. Also included are means to support self-government.[268] These agreements have been said to make Aboriginal governments in the territories "key players in local and regional economies".[269]

The Canadian Parliament has enacted many laws on land claims as well as on self-government. In the three Territories, such laws are the Northwest Territories Act of 1985,[270] the Nunavut Act and the Nunavut Land Claims Agreement Act, both of 1993,[271] and the Yukon First Nations Land Claims Settlement Act and the Yukon First Nations Self-Government Act, both of 1994.[272] Other Acts related to Aboriginal matters such as the First Nations Land Management Act of 1999, First Nations Oil and Gas and Moneys Management Act of 2005, Northwest Territories Waters Act of 1992, and the Yukon Environmental and Socio-economic Assessment Act of 2003[273] have also been passed. Responsibility for administering the various Acts, as well as the negotiated agreements and relevant legal decisions lay partly with the Aboriginal Affairs and Northern Development Canada (AANDC), one of the Federal Government departments responsible for meeting Canada's obligations and commitments to First Nations, Inuit and Métis.

It seems, however, that problems occur when it comes to fulfilling obligations and achieving objectives that require cooperation between Aboriginal peoples and the Federation of Canada. Part of the problem is that upon ratification of an agreement, the Canadian Government immediately obtains full title to land and natural resources in the agreement, while Aboriginal peoples' benefits from the agreement still needs some clarification. Other problems seem to have been lack of capacity, inadequate funding, institutional timidity and, not least, disagreement as to the meaning of provisions in the actual agreement.[274] Some efforts have been made to solve the problem. In 2007, the Government launched an initiative to improve the system,[275] and again in 2008[276] with recommendations "on restructuring and improving the regulatory system in the North to facilitate responsible development of resources".[277]

Looking more closely at the Nunavut and the Northwest Territories, Nunavut was established as a territory through the Nunavut Act of 1993, mentioned earlier. It formally separated from the Northwest Territories and was established as a new territory in the Canadian Federation with its own territorial government serving both Inuit and non-Inuit. In 1993, the Nunavut Land Claims Agreement, the largest land claims agreement in Canadian history, was settled. Through this agreement Nunavut, where 85% of the population is Inuit, was given title to approximately 356,000 square kilometers of land, constituting about 18% of Nunavut. The Nunavut Land Claims Agreement gives title to Inuit-owned lands of which about 35,000 square kilometers include mineral rights. According to the agreement, Nunavut Inuit are to receive capital transfer of $1.148 billion, payable over

14 years, as well as a share of federal royalties from oil, gas and mineral development on so-called Crown land.

The agreements established a joint land use and resource board ensuring that the Inuit were part of the decision-making process in regard to natural resources.[278] As indicated above, it seems, however, that this sharing of decision-making has not always been easy to realize. The Government of Nunavut has been described as finding itself "in the middle of a bilateral treaty relationship between the Inuit and Canada where it has major implementation responsibilities, but where it is not recognized as a full partner and where it does not feel adequately resourced to fulfil its treaty obligations or the expectations of the Inuit."[279]

The situation in the Northwest Territories[280] has been different than in Nunavut. In these Territories, where Aboriginal peoples constitute 50% of the population of which 20% is Inuit, only four of about 30 Aboriginal communities have settled land claims agreements.[281] The challenging part of the negotiations has been to find solutions whereby the Aboriginal peoples' desire for self-government can be combined with the public government structure already in place. A special effort in the form of a constitutional package solution did not succeed.[282]

In other words, there are still agreements to be settled as well as settled agreements to be implemented, in order for the Aboriginal peoples of Canada to get a say in the development of their traditional territories. To have such a voice is considered important, not least as communities which have achieved access to and jurisdiction over lands and resources enjoy greater economic benefits than communities where the matter is unsolved.[283]

As for Aboriginal peoples' participation in the economy in northern Canada, it has been described as a mix of traditional activities such as hunting, fishing and trapping, and wage employment, particularly in small businesses, the mineral sector and tourism.[284] Aboriginal peoples are, however, described as gradually bettering their position when it comes to having a share in the resource revenues from development on their traditional lands. One example is that of being the principal partners in the proposed Mackenzie Valley Pipeline project. This 1196-kilometer natural gas pipeline system along the Mackenzie Valley of Canada's Northwest Territories is planned to connect northern onshore gas fields with North American markets. Considering that, in the 1970s, Aboriginal peoples strongly opposed the development of such a pipeline, this is an indication of a larger incentive for economic development. While this shows that there is more optimism regarding what benefits economic development can bring to the Aboriginal communities,[285] awareness also exists among Aboriginal peoples of the adverse effect of economic development on their traditional culture and way of life, as subsistence economy continues to be a vital part of the livelihood of Aboriginal peoples in northern Canada. The Inuit describe their entire culture and identity as being based on free movement on sea and sea-ice for hunting and fishing, and thus as their primary means of transportation "whether thickly frozen or open in the summer". Challenging this free movement is, on the one hand, the thinning of ice and, on the other, shipping. When confronted with thinning ice the Inuit persist in continuing their traditional hunting and fishing further out or at different times. As for shipping, some incidents have been reported[286] as causing concern, both in relation to marine pollution and also as reported by hunters in Arctic Bay in Nunavut, because of the consequences for hunting on ice when ships passing through are "wrecking the area". In open water, shipping is sometimes seen as a threat to hunters in small boats, while another concern relates to marine mammals, as shipping along the Northwest Passage goes through waters also traversed by the beluga whale. More research seems to be needed in order to know exactly where and when shipping will overlap with the hunting of the belugas.

Another example of colliding interests relates to the consequences of establishing a key hub for supply ships in a harbor "teeming of fish", as is the case in Tuktoyaktut on the Beaufort Sea coast. More precisely, the consequence was that the fishermen could not set their nets on this important fishing ground at the time when "ships were going back and forth". Interesting to note in this regard is that, even though the community obviously had a say as to when ships were to pass, the actual need for fuel and supplies to be distributed to various other communities gave them no choice but to let the ships pass at times when it hampered their own fishery.

Skepticism has also been expressed in regard to tourist ships. Acknowledging the possible selling of handicraft, the arrival of such ships in large quantities has been described as having a potential cultural and social disturbance. In the context of hunting, these ships could cause disturbance, as was expressed by hunters in Pangnirtung in Nunavut who complained about cruise ships scaring away mammals and whales, and expressed the belief that "last summer we hardly saw any whales around". On a more modest level, tourist ships were, however, seen in a more positive light, referring particularly to the opportunity of the selling of local handicraft. Ship traffic was also seen in a positive context, acknowledging that shipping offers many benefits for local communities, such as a more effective distribution of goods and materials and also by creating more jobs in construction and in ports activities.

Alaska Natives: Land claims settlements

Shipping along the Northwest Passage also runs along the coast of Alaska, purchased from Russia in 1867 and attaining statehood in 1959. Through this period, people living in this area were ruling themselves independently, trying to create a system of government. In the initial period the Native peoples of Alaska, today making up 22% of the total population of Alaska, was not regarded as citizens. In the treaty with Russia in 1867, it says that "[t]he uncivilized tribes will be subject to such laws and regulations as the United States may, from time to time, adopt in regard to aboriginal tribes of that country".[287] It was not until 1915 that the territorial legislature came up with a complicated procedure for Natives to become citizens of the United States, the condition being that they had "severed all tribal relationship and adopted the habits of civilized life".[288] In other words, the objective was to encourage the Natives of Alaska to assimilate into the White culture.

This line of thought was rejected in the 1930s; rather, the provisions under the Indian Reorganization Act (IRA) was extended to Alaska in 1936, and the Federal Government urged Native communities to adopt constitutions for self-government under this Act. From the 1930s to the 1950s, about 70 so-called IRA villages were set up, and still exist as one type of Native authority recognized by the Federal Government.

In the Alaska Statehood Act of 1959 the new, and 49th, state of the United States was given the right to select 103.5 million acres of land. The Natives grew anxious that their traditional territories for hunting and fishing would be threatened, and this saw the founding of the Alaska Federation of Natives in 1966, defending their rights. Discovery of oil in Prudhoe Bay on the Arctic coast in 1968 created a situation where the Native peoples' way of life met with the dominant society's interests, in this case a pipeline to carry the oil across Alaska which would cross land involved in the dispute with the Natives in Alaska. It was thus urgent to reach an agreement, and in 1971 a settlement, the Alaska Native Claims Settlement Act (ANCSA),[289] was reached whereby Alaskan Natives received 44 million acres of land representing approximately one-ninth of the original claim, and were paid $962 million in cash. By this Act, the Natives, land claims in Alaska were transformed into a system of 220 villages ranging from 25 to 2000 people and 12 Alaska Natives Regional Corporations, described as an extraordinary national experiment in federal relations with Native Americans, but also as "a bewildering institutional array".[290]

The Act thus provides for the conveyance of both title to land as well as a monetary award to be divided between the villages and the regional corporations. As for the distribution of the monetary award, the regional corporations were to receive the $962 million divided among them on the basis of their Native population, and which the corporations in turn had to distribute not less than 50% among the villages or, more precisely, among corporations formed by these villages.[291]

The Alaska Native Claims Settlement implied that while the Natives' earlier use and occupancy of land had made them co-owners of shared lands, they had now become shareholders in corporate-owned land.

At the time, most Native Americans were unaware of the complexities of the new system and the Alaska Federation of Natives' leaders were enthusiastic about what they considered a large monetary settlement. Critics of the settlement saw, however, the possibility of larger multinational corporations taking over valuable lands and resources currently held in Native hands, followed by a challenge to the Native culture.[292] It has also been argued that by this Act "the government's long term goal of assimilating Alaska's Native population into the larger mainstream would finally be achieved", referring to a Native educator expressing that "[w]hat is happening to Native people in Alaska is not a new story; it is a new chapter in an old story".[293]

Within the new Act, the Native governments exercise power and operate programs in relation to health, education and other governmental services, but they do not have jurisdiction over lands conveyed to village and regional corporations. This has been made clear in a case put before the US Supreme Court in 1998.[294]

What can be said to characterize the rights achieved by the Natives of Alaska is that they have a share in the economic development of the lands conveyed in the Settlement Act through their shares in the village and regional corporations, but that their Native governments have no ownership to the actual lands, merely an administrative function with regard to public services.

The Natives of Alaska thus seem to still be working to achieve what they consider to be their legal and political goals, as can be seen from their reference to the present US administration having promised "to engage in discussions about protecting subsistence opportunities for the Natives of Alaska and to break what is described as 'the endless cycle of litigation'".[295] Reference has here been made to the Alaska National Interest Lands Conservation Act of 1980 which provides comprehensive management guidance for all public lands in Alaska with provisions on subsistence specifically dealt with in title VIII, but also including provisions on wildlife, transportation and utility corridors, oil and gas leasing, mining, public access, hunting, trapping and fishing. In regard to title VIII on Subsistence Management and Use,[296] it was stated that in order to fulfil its obligations under the Alaska Native Claims Settlement Act, Congress needed to invoke its constitutional authority to provide for the continuation of the opportunity for subsistence uses of resources on public and other lands in Alaska. According to the Alaska Federation of Natives (AFN), what Congress envisioned when it enacted title VIII of this Act has not been fulfilled; rather, the AFN believes "it is imperative that the Federal Government once again act to safeguard our villages' essential food resources and traditional way of life".[297] It seems in other words that, according to the AFN, the US Government has not fulfilled its commitment in the above-mentioned Act to establish and implement a comprehensive federal program that would protect the traditional way of life for Alaska Natives. Other objectives expressed by the Natives have been to "advance solutions for changing the investment and tax climate, which continues to hinder the development of sustainable Native economy and to sustain the traditional way of life".[298]

With regard to shipping, it might be of interest to note that nine environmental and Alaska Natives groups recently challenged an air permit given to a Shell drill ship, maintaining that the permit failed to ensure that all air pollution controls were in place and that all standards were met for what they described as a "major new source of pollution in the Arctic". [299] Shell's plans were to use the *Noble Discoverer* to drill three exploratory wells in the Chukchi Sea off Alaska's Northwest coast during the open water season. The nine groups have, however, asked the court to send the granted permit back to the Environmental Protection Agency for reconsideration. But also in Alaska shipping in the form of cruise ships has been seen to represent an economic alternative, one example being a Native village corporation transforming the former premises of a declining fishing and logging industry into a cruise ship port, with hundreds of thousands of visitors since 2004, and with the creation of more than 130 seasonal and permanent jobs for the community.

6.5.5 The Transpolar Passage

When dealing with coastal communities, it goes without saying that transpolar shipping is less relevant compared to shipping along the Northeast Passage/the Northern Sea Route or the Northwest Passage. That is, however, not to say that transpolar shipping is totally without effect on local communities. This is due not least to the currents in the Arctic Ocean which could enable oil spills or other forms of pollution from transpolar shipping to reach coastal shores.

One relevant example is the experience of an Aleut community in Russia.[300] In 2003, a 20-tonne container filled with hazardous chemicals, which had been lost at sea, hit their shore. The container cracked, leaking approximately 15 tonnes of chemicals. There was no disaster response plan and thus no preventive measures in place to meet the consequences of this pollution. This incident is an example of the fact that transpolar shipping, although passing far out at sea, also could constitute a potential for environmental impacts on local activities. What was demonstrated as well in this incident was the importance of having in place a legal framework in order to clarify responsibility and secure compensation related to the damage caused by the pollution.

Another impact on local communities is that, even though the ship traffic is transpolar, it has a need for infrastructure on land, such as port and harbor facilities. This in turn impacts on local communities; it causes disturbances and challenges the traditional way of life but, as has been demonstrated, also contributes to economic development welcomed by local communities.

Notes

1 UNCLOS, article 5 (1)
2 UNCLOS, article 7
3 UNCLOS, article 10 (4)
4 UNCLOS, article 7 (5)
5 UNCLOS, article 10 (6)
6 UNCLOS, section 3
7 UNCLOS, article 21 (1)
8 UNCLOS, article 21(2)
9 UNCLOS, article 22
10 UNCLOS, part V
11 UNCLOS, article. 56 (1) a
12 UNCLOS, article 87 (1) a
13 The interpretation of article 234 is thoroughly discussed in Brubaker (2007), p. 67.
14 UNCLOS, article 236
15 Molenaar and Corell (2009)
16 UNCLOS, article 77
17 UNCLOS, article 78
18 UNCLOS, article 60 (6).
19 www.imo.org
20 Jensen (2008) pp. 107–114.
21 Meaning any vessel covered by the International Convention for the Safety of Life at Sea (SOLAS), thus excluding ships less than 50 gross tonnage and naval vessels.
22 The initial intention was also to cover the Antarctic
23 IMO Guidelines (2002), p. 1
24 As of 2007 no states had implemented the Guidelines as national legislation
25 Guidelines, introductory remarks
26 Van der Zwaag et al. (2008)
27 Jensen (2008), pp. 107–114
28 Jensen (2007)
29 Jensen (2007)
30 Jensen (2007)
31 Reference is made here to requirement of relevant experience in the Canadian Arctic Shipping Pollution Prevention Act
32 Jensen (2007)
33 Fleischer (2007)
34 WWF.org (2010), Limiting a Little, Save a Lot
35 The Ilulissat Declaration, Arctic Ocean Conference, Iluissat, Greenland, 27–29 May 2008
36 The conference was hosted by Per Stig Møller, Danish Minister of Foreign Affairs, and Hans Enoksen, Prime Minister of Greenland. Other attendees were: Sergey Lavrov, Russian Minister for Foreign Affairs; Jonas Gahr Støre, Norwegian Minister for Foreign Affairs; Gary

Lunn, Canadian Minister for Natural Resources; and John Negroponte, American Deputy Secretary of State.

37 The Illulissat Declaration. Arctic Ocean Conference, Illulissat, Greenland 27– 9 May 2008. http://www.oceanlaw.org/downloads/arctic/Ilulissat_Declaration.pdf

38 I bid

39 Covened in Tromsø, 19–20 January 2009

40 LOV-2003-06-27-57

41 LOV-1976-12-17-91

42 FOR-1977-06-03-06

43 Foreign Minister Knut Frydenlund in the Storting.

44 For more on the fishery protection zone outside Svalbard see: Fløistad, Brit. "The controversy over the applicable regime outside Svalbard's territorial sea". Oslo 2007

45 LOV-2007-02-16-9

46 LOV-1981-03-13-6

47 FOR-1997-09-19-1061

48 LOV-2001-06-15-79

49 LOV-1996-11-29-72

50 Federal Act of the internal sea waters, territorial sea and contiguous zone of the Russian Federation, 17 July 1998, article 14

51 Federal Act of the internal sea waters, territorial sea and contiguous zone of the Russian Federation, 17 July 1998 http://www.un.org/depts/los/LEGISLATIONANDTREATIES/PDFFILES/RUS_1998_Act_TS.pdf.

52 Federal Act of the Exclusive Economic Zone of the Russian Federation of 2 December 1998, http://www.un.org/depts/los/LEGISLATIONANDTREATIES/PDFFILES/RUS_1998_Act_EZ.pdf

53 Federal Act of 17 July , article 4

54 Federal Act of 17 July, article 11

55 Federal Act of 17 July, article 15

56 Federal Act of 17 July ,article 12

57 Federal Act of 17 July, article 23

58 Federal Act of 2 December, chapter V

59 Federal Act of 2 December, article 32

60 Federal Act of 2 December, article 33

61 Franckx (2009) p. 337

62 Franckx (2009) p. 338

63 Franckx (2009) p. 337

64 Franckx (2009) p. 339

65 http://www.docstoc.com/docs/21229289/According-the-Standing-on-Federal-rate-service_-ratified-by-

66 Frankx (2009) p. 340

67 Castel (1976), p. 160

68 Osherenko and Young (1989), p. 38

69 Skagestad (1975), p. 116

70 Smith (1966), p. 226.

71 Castel (1976), p. 167

72 Lakthine (1928), pp. 15, 35, 47

73 Elferink (1994), p. 225

74 Among others: V.L. Lakthine, S.A. Bergvinov, E.A. Korovin, S.S. Sigrist

75 Lakthine (1928), p. 37

76 Lakthine (1928), p. 46
77 Timtchenko (1997), p. 30
78 Butler (1978)
79 Among others: Kozhevnikov, Zudro, Baraboyla, Ignatenko
80 Among them: Uustal
81 Franckx (1993), p. 88
82 Yakovlev, (1997), p 20. The italics are added by the author.
83 P.D.Baraboyla et al (1966), p. 273
84 Franckx (1993), p. 169
85 Timtchenko (1997), p. 30
86 Vartanov, Roginko, Kolossov (1999), pp.70–71
87 Timtchenko (1997), p. 33
88 Franckx (1993), p. 170
89 Luton (1986), p. 399
90 Timtchenko (1997), p. 31
91 Brubaker (2005), pp. 54–67
92 Timtchenko (1997), p. 31
93 Olenikoff (19), p. 2
94 Theutenberg (1984), p. 36
95 Castel (1976), pp. 178–179
96 For a more thorough discussion, see Østreng (2010)
97 Smith (1966), p. 249
98 *Regulations for Navigation* (1991)
99 Kolodkin and Kolosov (1990), p. 160
100 Kolodkin and Kolosov (1990), p. 164 (italics added by the author)
101 Kolodkin and Kolosov (1990), p. 164
102 Frolov, Borodachev, Alexandrov (2007), pp. 21–22
103 Kolodkin and Kolosov (1990), p. 166
104 Kolodkin and Kolosov (1990), p. 165
105 Brubaker (2000), pp. 231–232
106 Dahle (2010), p. 22
107 Støre (2008), p. 50
108 Støre (2008), p. 49
109 Doe Jones Newswire (2008)
110 Ilulissat (2008)
111 Østreng (2012)
112 Baker (2008), p. 253
113 *Fundamentals of Russian Arctic Policy* (2009), p. 1
114 Østreng and Prydz (2007)
115 Evensen (1978)
116 Tresselt (1988), p. 81
117 Vartanov, Roginko, Kolossov (1999), pp. 70–71
118 LOV-1996-05-22-411
119 Anordning nr. 1005 of 15 October 2004
120 BEK nr. 864 av 08/08/2006
121 Lov nr. 476 of 30 June 1993
122 Anordning nr. 1035 af 22 October 2004 om ikrafttræden for Grønland om beskyttelse af havmiljøet
123 Grønlands Kommando operationscenter: (GLK)

124 Oral statement from employee in the Ministry of Environment
125 Lov nr, 929 of 24/09/2009
126 Amending its Territorial Sea and Fishing Zones Act of 17 April 1970
127 Pharand (2007), p. 11
128 Oceans Act, 18 December 1996 , http://laws.justice.gc.ca/eng/O-2.4/page-1.html
129 Fishing Zones of Canada (Zone 1, 2 and 3), (Zone 4 and 5), (Zone 6)
130 Fisheries Act of 1985, http://laws.justice.gc.ca/eng/F-14/page-1.html
131 Canada Shipping Act of 2001, http://laws.justice.gc.ca/eng/O-2.4/page-1.html
132 Canada Shipping Act, section 8
133 Marine Transportation Security Regulations [SOR/2004-144]
134 Arctic Waters Pollution Prevention Act (R.S, 1985, ca. A-12) http://laws.justice.gc.ca/eng/A-12/page-1.html
135 § 2. "Arctic waters" means the internal waters of Canada and the waters of the territorial sea of Canada and the exclusive economic zone of Canada, within the area enclosed by the 60th parallel of north latitude, the 141st meridian of west longitude and the outer limit of the exclusive economic zone; however, where the international boundary between Canada and Greenland is less than 200 nautical miles from the baselines of the territorial sea of Canada, the international boundary shall be substituted for that outer limit
136 11. (1) Subject to subsection (2), the Governor in Council may, by order, prescribe as a shipping safety control zone any area of the Arctic waters that is specified in the order, and may, as the Governor in Council deems necessary, amend any such area
137 Shipping Safety Control Zones Order (C.R.C., c. 356) http://laws-lois.justice.gc.ca/eng/regulations/C.R.C.,_c._356/
138 Order Amending the Shipping Safety Control Zones Order, 9 December 2009
139 Amended by Bill C-3, An Act to amend the Arctic Waters Pollution Prevention Act, which received Royal Assent on 11 June 2009
140 Northern Canada Vessel Traffic Services Zone Regulations (SOR/2010-127) http://www.gazette.gc.ca/rp-pr/p2/2010/2010-06-23/html/sor-dors127-eng.html
141 The waters of Ungava Bay, Hudson Bay and Kugmallit Bay that are not in a shipping safety control zone; the waters of James Bay; the waters of the Koksoak River from Ungava Bay to Kuujjuaq; the waters of Feuilles Bay from Ungava Bay to Tasiujaq; the waters of Chesterfield Inlet that are not within a shipping safety control zone, and the waters of Baker Lake; and the waters of the Moose River from James Bay to Moosonee
142 Regulatory Impact Analysis Statement, *Canada Gazette*, 27 February 2010
143 Marine Liability Act 2001, http://laws.justice.gc.ca/eng/M-0.7/page-1.html
144 Based partly on the cession of Northern territory by the United Kingdom at the end of the 1880s, and partly on Canadian activity on the Arctic islands after the cession
145 Pharand (2007), p. 11
146 Expressed by the Bureau of Legal Affairs (Pharand (2007), p. 11
147 Adopted in an Order-in-Council, Pharand (2007), p. 11
148 Oceans Act, section 5
149 State Department File No. P86 0019-8641
150 Via the British High Commission Note no. 90/86
151 Pharand (2007)
152 UNCLOS, article 38 (1), (2)
153 UNCLOS, article 37. This section applies to straits which are used for international navigation between one part of the High Seas or an exclusive economic zone and another part of the High Seas or an exclusive economic zone
154 Established in the Corfu Channel Case (ICJ, 1949), p. 26

155 Discussing various definitions of "use of international navigation", Pharand concludes that for a strait to meet the functional criteria it must have "a history of useful route for international maritime traffic", Pharand (2007), p. 351
156 Pharand (2007), p. 42
157 Pharand (2007), p. 44. Referring to the rather high threshold for being defined as an international strait, Pharand states the opinion that special factors call for a lower threshold in these waters
158 Coté and Dufresne (2008)
159 Dealt with later in this chapter
160 Fenge (2007–2008)
161 Fenge (2007–2008)
162 Pharand (2007), p. 12
163 Agreement between the Government of Canada and the Government of the United States of America on Arctic cooperation, 11 January 1988
164 Coté and Dufrese (2008)
165 Coté and Dufrese (2008)
166 Address by Minister Cannon to Third Annual Arctic Shipping North America Conference, 15 November 2010
167 Ibid
168 AMSA (2008)
169 Arctic Human Development Report, November 2004
170 Also Eyak, Tlingit, Haida, Tsimshian, and a number of Northern Athabaskan cultures
171 AMSA (2008), p. 18
172 Eide (no date)
173 The Original name was Inuit Circumpolar Conference, but was changed to Inuit Circumpolar Council at the Assembly in 2006 as there had been some perennial confusion over an organizational name that sounded more like a past meeting
174 The starting point was the Universal Declaration on Human Rights of 10 December 1948 (United Nations) and the Convention for the Protection of Human Rights and Fundamental Freedoms of 4 November 1950 (European Council)
175 The ICCPR entered into force 23 March 1976 and the ICESCR 3 January1976. The former is signed by 72 states and the latter by 69 states. All the five states bordering the Arctic Ocean have signed and ratified both Convenants, the only exception being that United States has not ratified the ICESCR
176 ICCPR, Article 27: "In those states in which ethnic, religious or linguistic minorities exist, persons belonging to such minorities shall not be denied the right, in community with the other members of their group, to enjoy their own culture, to profess and practice their own religion, or to use their own language"
177 CCPR/C/"!/Rev.1/Add.5, General Comment No 23
178 Indigenous and Tribal Peoples Convention, 1989 (No. 169) Concerning Indigenous and Tribal Peoples in Independent Countries. The Convention replaced the ILO Convention No 107 of 26 June 1957: Concerning the Protection and Integration of Indigenous Peoples and Other Tribal and Semi-tribal Populations in Independent Countries, which had as its goal the assimilation of indigenous peoples rather than to acknowledge their right to their own tradition and culture. The Convention entered into force . September 1991
179 According to article 1 (3) of the Convention, "[t]he use of the term peoples in this Convention shall not be construed as having any implications as regards the rights which may attach to the term under international law"
180 ILO Convention, article 14

181 ILO Convention, article 15
182 ILO Convention, article 6
183 Ratified 1 April 1996
184 Ratified 24 April 1990
185 This process has been described in detail by Eide (2006), pp. 155–212
186 UN Declaration, article 4
187 UN Declaration, article 46 (1) "Nothing in this Declaration may be interpreted as implying for any State, people, group or person any right to engage in any activity or to perform any act contrary to the Charter of the United Nations or construed as authorizing or encouraging any action which would dismember or impair, totally or in part, the territorial integrity or political unity of sovereign and independent States"
188 UN Declaration, article 27
189 UN Declaration, article 28
190 UN Declaration, article 32
191 Expressed by the Chair of UN Permanent Forum on Indigenous Issues
192 Second International Decade of Indigenous Peoples 2005–2014. Resolution 59/147 by the General Assembly, Partnership Action and Dignity, 24 Februrary 2005. (The First International Decade ... from 1995–2004. Resolution 48/163 of 21 December 1993)
193 Canada's Statement of Support for the United Nations Declaration on the Rights of Indigenous Peoples, 12 November 2010
194 Announcement of President Obama on 16 December 2010
195 The Framework Convention, article 4 (1)
196 The Framework Convention, article 5 (1)
197 Indigenous Peoples Guidance Note, London 2010: http://www.ebrd.com/downloads/research/guides/indp.pdf
198 "A Circumpolar Inuit Declaration on Sovereignty in the Arctic", adopted by the Inuit Circumpolar Council, April 2009
199 The Anchorage Declaration, 24 April 2009. The Inuit Circumpolar Council, 20–24. April 2009, http://www.indigenoussummit.com/servlet/content/declaration.html
200 IPS does not speak for the Permanent Participants. Instead, it creates opportunities for the Indigenous Peoples' Organizations to speak for themselves, and helps provide them with necessary information and materials
201 Nunatsiavut (Labrador), Nunavik (northern Quebec), Nunavut, and the Inuvialuit Settlement Region in the Northwest Territories
202 Finland established a Sami parliament in 1973 and Sweden in 1993
203 Expert group led by former Chief Justice Carsten Smith. "Nordisk samekonvensjon" http://www.regjeringen.no/Upload/AID/temadokumenter/sami/sami_samekonvensjon_norsk.pdf
204 "En gjennomgåelse av utkast til nordisk samekonvensjon" http://www.regjeringen.no/Upload/AID/temadokumenter/sami/sami_arbeidsgrrapport_gjennomgaelse_nordisk_samekonvensjon.pdf
205 38,248 in 2009 (Statistisk Sentralbyrå)
206 Sami in Norway live all over the country, but as uncertain criteria make it difficult to determine who is Sami, the official statistics only include Sami in the three most northern counties, and based on so-called SNT (Sametingets tilskuddsordninger for næringsutvikling: the Sami Parliament's funding for resource development)
207 Eide (no date)
208 Nenets, Evenks, Khanty, Evens, Chukchi, Nanaitsy, Koryaki, Manci, Dolgany, Nivhi, Selkups, Ulchi, Itelmens, Udegeitsy, Saami, Eskimos, Chuvantsy, Nganasany, Yukagris, Kety, Orochi, Tofalary, Aleuty, Negidalts, Entsy and Oroki (Sokolova and Yakolev (1998)

209 Sokolova (1998)
210 Eide (no date)
211 "Samerettsutvalget", headed by the former Chief Justice Carsten Smith
212 LOV-1987-06-12- 56: Lov om Sametinget og andre samiske rettsforhold (Sameloven)
213 Article 110a
214 LOV-2005-06-17-85 Lov om rettsforhold og forvaltning av grunn og naturressurser i Finnmark fylke (finnmarksloven)
215 Section 1
216 "Finnmarkskommisjonen", established 14 March 2008
217 "Gruvedrift nødvendig for samisk kultur". NRK Sameradio, 13 April 2010
218 Ibid
219 Finnmark Act, article 2. The Act shall apply to real property and watercourses with natural resources in the county of Finnmark. On the shoreline, the Act shall apply as far out to sea as private right of ownership extends
220 Diatchkova (2001)
221 Dallmann (no date)
222 RAIPON http://www.raipon.org/
223 Diatchkova (2001)
224 RAIPON, http://www.raipon.org/
225 Ibid
226 The Constitution of the Russian Federation of 25 December 1993 (amended 9 January 1996, 10 February 1996 and 9 une 2001)
227 Third Report submitted by the Russian Federation pursuant to art. 25 para 2 of the Framework Convention for the Protection of National Minorities, 9 April 2010
228 Federal Law of 30 April1999, No. 82-FZ "On Guarantees of the Rights of Indigenous Peoples of Russia"
229 Federal Law No. 40-FZ of 5 April 2009
230 Federal Law of 20 July 2000, No. 104-FZ "On General Principles of Organization of Communities of Indigenous and Minority Peoples of the North, Siberia and Far East of the Russian Federation"
231 Federal Law on the Traditional Territories and the Use of Natural Resources by the Indigenous Small Numbered Peoples of the North, Siberia and the Far East of Russia, 7 March 2001
232 13 September 2011, Russian Public Chamber hosted the roundtable "The legislation development on the rights of Indigenous Peoples of the North, Siberia and the Far East: Issues and horizons". The meeting was organized by the Russian Association of the Indigenous Peoples of the North (Raipon) and the Russian Indigenous Peoples' Alliance (Rica) in cooperation with the Russian Public Chamber and International work group for indigenous affairs (IWGIA).
233 Osherenko (1997)
234 Osherenko (1997, p. 55
235 Osherenko (1997)
236 Osherenko (1997), p. 58
237 AMSA (2008)
238 The project "Monitoring the development of traditional land use areas in the Nenets Autonomous Okrug" is a cooperation between the Association of Nenets People and the Norwegian Polar Institute during the Polar Year 2007–2008
239 All figures are from "Greenland in Figures 2009" and Statistics Greenland
240 47,093 of a total population of 56,194, that is, about 85%, live in towns

241 Less than 4000 live on the east coast, 2998 in Tasiilaq (Ammassalik) and 485 in Illoqqortoormiit (Scoresbysund)
242 Lov 473 of 12 June 2009
243 Greenland was made a colony in 1721. Had Greenland remained a non-self-governing territory under Chapter XI of the United Nations Charter it would have gained the right to opt for independence, but the UN took note of Greenland's integration into the Kingdom of Denmark in 1954
244 This process has caused some discussion as to its validity under international law
245 Lov 577 of 29 November 1978
246 Such as internal administration, direct and indirect taxation, fishing and hunting, social welfare, education, culture and protection of the environment
247 The Greenland–Danish Self-Government Commission. Report on Self-Government in Greenland, April 2008
248 75.5% voted for, 23.6% voted against
249 These areas of responsibility are listed in a Schedule forming an appendix to Lov 577 of 29 November 1978, and cover almost all fields, the exception being that the Danish Government will retain a significant influence in foreign- and security-policy issues, as well as in constitutional matters and monetary policy issues
250 From Remote Island to Self-Government – Greenland's journey towards independence. *Arctic Portal*, December 2009
251 Lov nr. 7 of 7 December 2009
252 The subsidy will adjusted annually in accordance with the increase in the general price and wage index
253 Section 8 (1)
254 In 2007, Greenland's GPD was DKK11,063 million (Statistics of Greenland)
255 The main species, Greenland halibut, a stock shared with Canada in the Davies Strait and with Iceland and the Faeroe Islands, constitutes more than half of the catches in Greenland's waters
256 The most important fishery is prawns, primarily taking place in the waters outside the south-western part of Greenland around Sisimiut (Holsteinsborg), but also on the east coast, characterized by unusually large, and therefore valuable, shrimps. FAO. Fishery country profile. Greenland. http://www.fao.org/fi/oldsite/FCP/en/GRL/profile.htm
257 Ove Karl Berthelsen
258 Bureau of Minerals and Petroleum. Annual Report 2009
259 The Coalition Agreement was entered into on 10 June 2009 http://uk.nanoq.gl/Emner/Government/~/media/9E0C6F8EBEFE42758CB5D9FC29FC2358.ashx. The current Cabinet is based on political cooperation between Inuit Ataqatigiit, Demokraatit, and Kattusseqatigiit Partiiat The two additional parties in Greenland are Siumut and Inuit Ataqaligiit
260 There are today ports in 16 towns and harbors in 60 settlements in Greenland
261 Østreng (1999)
262 In all of Canada, Aboriginal peoples number 1,172,790, 60% of which is First Nations peoples, 33% is Métis and 4% is Inuit. Statistics of Canada (2006)
263 First Nations are primarily Indians
264 In addition to these three Territories, Canada consists of ten Provinces
265 Hurley (1996)
266 The Constitution Act of Canada of 1982 section 35 states that the existing aboriginal and treaty rights of the Aboriginal peoples of Canada are recognized and affirmed, further stating that "treaty rights" includes rights that now exist by way of land claims agreements or may be so acquired

267 Twenty-one such agreements are being implemented covering more than one-third of the country in Yukon, Northwest Territories, Nunavut and Labrador, northern Quebec and northwest British Columbia (Fenge, 2008)
268 Fenge (2008)
269 Simone (2008)
270 Northwest Territories Act (R.S.C., 1985, c. N-27), last amended 2005. http://laws-lois.justice.gc.ca/eng/acts/N-27/index.html
271 Nunavut Act (S.C. 1993, c. 28), last amended 2003. http://laws-lois.justice.gc.ca/eng/acts/N-28.6/index.html
Nunavut Land Claims Agreement Act (S.C. 1993, c. 29) last amended 2004. http://laws-lois.justice.gc.ca/eng/acts/N-28.7/index.html
272 Yukon First Nations Land Claims Settlement Act (S.C. 1994, c. 34), last amended 2011. http://laws-lois.justice.gc.ca/eng/acts/Y-2.3/index.html
Yukon First Nations Self-Government Act (S.C. 1994, c. 35), last amended 2006. http://laws-lois.justice.gc.ca/eng/acts/Y-2.6/index.html
273 For a list of Acts, see http://www.aadnc-aandc.gc.ca/eng/1100100032317
274 Fenge, Terry, "Implementing comprehensive land claims agreements". (*Policy Options,* July–August 2008)
275 "Northern Regulatory Improvements Initiative", November 2007
276 "Road to improvement"
277 Simone (2008)
278 In February 1999, residents of Nunavut held their first election for members of their Legislative Assembly
279 Dewar (2009). http://www.tunngavik.com/documents/publications/NLCA%20&%20Nunavut%20Dewar%202009.pdf
280 Northwest Territories Act 1985, http://laws.justice.gc.ca/eng/N-27/page-1.html
281 Inuvial (1984), Gwich'in (1992), Sahtu Dene and Métis (1994) and Ticho (2005)
282 Presented in "Partners in a New Beginning" in October 1996 by a Constitutional Working Group
283 Simone (2008)
284 Simone (2008)
285 Simone (2008)
286 AMSA (2008)
287 Governing Alaska. Native Citizenship and Land Issues. Native Citizenship. Information courtesy of the Alaska Humanities Forum. No date
http://www.gov.state.ak.us/ASCC/pdf/Governing%20Alaska%20-%20Native%20Citizenship.pdf
288 Ibid
289 http://www.lbblawyers.com/ancsatoc.htm#top
290 Governing Alaska. Native Citizenship and Land Issues. A "New Deal" for Alaska Natives. Information courtesy of the Alaska Humanities Forum. No date
291 Jones (1981) http://www.alaskool.org/projects/ANCSA/reports/rsjones1981/ANCSA_History71.htm
292 Chance (no date), http://arcticcircle.uconn.edu/SEEJ/Landclaims/ancsa1.html
293 Chance (no date)
294 Alaska v. Native Village of Venetie Tribal Government - 522 U.S. 520 (1998) http://supreme.justia.com/us/522/520/case.html
295 Alaska Federation of Natives, http://www.nativefederation.org/
296 http://www.wilderness.net/NWPS/documents/publiclaws/PDF/96-487.pdf

297 Alaska Federation of Natives. 2011 Federal Priorities

298 Alaska Federation of Natives. http://www.nativefederation.org/

299 *The Washington Post*, 2 February 2010 http://www.washingtonpost.com/business/industries/alaska-native-environmental-groups-challenge-arctic-air-permit-granted-to-shell-drill-ship/2012/02/22/gIQA6sDkTR_story.html

300 The village of Nikoloskoye on the Commander Island situated east of Kamchatka

7 The Northeast, Northwest and Transpolar Passages in Comparison*

The purpose of this chapter is to make a comparison of the prospects of commercial shipping operations along the Northeast, Northwest and Transpolar Passages as expressed in the issues discussed in previous chapters: geopolitics; resources and modes of transportation; environmental challenges; infrastructure; ocean law; jurisdictional matters and indigenous peoples. The comparison will be conducted on the level of *sub-issue areas*, which basically are identical to the main section headlines of the individual chapters. These sub-areas will be assessed and compared for each of the three Passages, with the aim of deciding which of the Passages offers the best shipping condition within the respective field. This approach will provide an image of what the state-of-the-art of regional shipping is for the Arctic Ocean as such, for the individual Passages and for each of the sub-issue areas.

Sea routes in southern waters connecting with the passages – through *destination Arctic* and *transit shipping* – are part of the assessment. Apart from defining the state-of-the-art of the present, the analysis will wherever possible be conducted to cover three time spans: *Short* (2012–2020), *medium* (2020–2030) and *long term* (2030–).

The chapter ends with a simplified commercial comparison of the three routes for two ship types on two trades compared to using the Suez Canal. The setting for the analysis is a future in which ice is no longer the major hindrance for shipping. Since it is far beyond the scope of this study to make a detailed empirical analysis (empirical data are not available in sufficient quantity and quality), the comparisons rest to a large, if not exclusive, extent on the assumptions being made, but they will give an indication of the order of magnitude in cost and time savings using these waters.

*This chapter is edited by Willy Østreng, Ocean Futures. The authors are as follows: Introduction and Sections 7.1 and 7.2 are written by Willy Østreng; Section 7.3 is written by Arnfinn Jørgensen-Dahl, Ocean Futures; Sections 7.4 and 7.5 are written by Karl Magnus Eger, Ocean Futures. Section 7.5.1 is written by Morten Mejlænder-Larsen, DnV; Section 7.6.1 is written by Lars Lothe, Ocean Futures; Section 7.6..2 is written by Brit Fløistad, Ocean Futures; and Section 7.7 is written by Tor Wergeland, Ocean Futures (affiliated).

The issues discussed are condensed versions of those assessed in a more comprehensive manner in previous chapters. This calls for a note of amplification or even caution. Condensation implies that details get lost in order to highlight and convey the core message of a text. The practice of condensation is to simplify complexity without losing out in validity. Whether we have succeeded in this is for the readers to decide. In any event, going back and forth in the text is a rewarding way of study to get a firmer grip on the details underpinning and substantiating the core messages.

7.1 The operational relevance of definition

In this book, the concepts of *Arctic*, *Arctic waters* and *Arctic Ocean* are defined to fit the operational needs and constraints of the shipping industry in ice-infested waters. Definitions based on subjective interests of a political nature rather than on natural characteristics give little or no guidance to the unique challenges of shipping operations under polar conditions. To enhance the relevance and effectiveness of the ensuing discussion we apply a version of the IMO definition, considering waters with "sea ice concentrations of 1/10 coverage or greater.. and which pose a structural risk to ships" to be Arctic in character (including also the Barents and White Seas). In this definition, the NEP, NWP and TPP – which can be delineated by fairly exact historical coordinates – are all safe within the operational confines of what the shipping industry will encounter in Arctic waters. At the same time, connecting transportation corridors extending from the Arctic to southern markets – the NMC, NPC, FC and DC – are singled out on the basis of their own operational characteristics. By this definition, partly blue water corridors are made different from ice water passages.

7.2 Geopolitics and power constellations

Geopolitics concerns the relationship between *geographical space* and *international relations* – between physical space and national interests. As has been demonstrated, the Arctic attracts political interest from an increasing number of states, European as well as Asian, big as well as small, polar as well as tropic. The Arctic space is gradually assuming a new role in the geopolitics of the world. Six geopolitical features of the region stand out:

- its spatial location between three continents;
- its abundance of strategically-important industrial and mineral resources;
- its dwindling sea ice regime, opening up new spatial areas of economic exploitation and prospecting;
- its sea lanes – inside and outside of the region;
- its unique environmental fragility and eco-systemic interconnections with eco-systems in southern latitudes;
- its regulatory affinity to existing global ocean conventions, in particular to UNCLOS.

The interaction of these features is what makes the Arctic space significant in international relations. They all relate to geographical spaces of which four will be discussed in this section: the location of the Arctic Ocean; the water spaces of transportation within the Passages; the operational spaces of military and civil activities; and connecting corridors in southern waters. The other geopolitical features of the Arctic will be summarized in subsequent sections.

7.2.1 The location of the Arctic Ocean

There is an obvious and at times considerable distance advantage involved in using the Arctic Ocean between ports in the Pacific and those in the Atlantic as compared to the Suez and Panama Canals. Most of the North American west coast, the Russian east coast, Japan, North Korea and China are all closer to the European Union/European Economic Area through the Arctic than by way of the Mediterranean. Thus viewed, the Arctic Ocean is indeed an "industrial Mediterranean" – a shortcut between the most industrialized regions of the world. At first glance, this geographical feature applies differently to the three Passages.

The Northeast Passage

Most north Pacific ports in Asia are closer to London on the Atlantic via the NEP than through the Suez. The distance between Yokohama in Japan and Hamburg in Germany is only 6600 nm by way of the NEP, as against 11,400 nm through the Suez Canal. This implies a 42% reduction in freight distance. Between the town of Tromsø in northern Norway and Vancouver on the Canadian west coast 3350 nm can be saved using the NEP instead of the Panama Canal. These savings in distance have resulted in significant time-saving benefits. As has been demonstrated time and again, Russian vessels have in summertime been able to save between 10 and 15 days at sea by using this Passage between Japanese and North American ports instead of going through the Mediterranean. The accelerating sea ice melt will extend the number of months in which such time savings can be achieved.

The Northwest Passage

On the condition of geography only, most of the freight routes of the NWP are comparable in distance to those of the NEP. From Baffin Island in the east to Bank Island in the west – the most troublesome stretch of the Passage – the distance is about 1300 nm. In addition come the waters of Baffin Bay, the Beaufort Sea and the six other routes that break off at Banks Island and take a more southerly course through the archipelago before it reaches the Beaufort Sea. Thus, in terms of distance only, the same saving benefits could apply to the NWP as to the NEP.

The Transpolar Passage

Voyages between Atlantic and Pacific ports across the North Pole will provide an additional saving of 700 nm as compared to the other Passages. This equals a saving in freight distance of one third of the whole stretch of the NSR or half of the coast of Norway (without fiords and bays), which is the biggest chunk of the Northern Maritime Corridor. In this calculation, the TPP is 1500 nm long measured from the Fram Corridor across the Pole to the Bering Strait. A distance reduction of such a magnitude could under certain specific conditions be converted into further savings in transport time. However, this calculation does not take into account the corridor features of the Passages and the types of ships employed in Arctic navigation.

The Passages in comparison

Due to the presence of sea ice, none of the Passages can offer ships a single navigation lane to follow. In practice, ships are forced to use the channel that offers the best ice and navigational conditions at any one time and place. Thus, each one of the Passages is more like a broad transportation corridor, containing several alternative navigational channels and fairly huge expanses of water. A voyage along the TPP can therefore be longer than those conducted along the coastal corridors and vice versa. As long as sea ice is a dominant feature of the Arctic Ocean it is at the outset hard to decide which of the Passages offers the shortest sailing distance. Navigational zigzag around ice floes to find suitable waters of operation makes the distance of voyages unpredictable and uncertain. To this come differences in the icebreaking capabilities of ships. Icebreakers are more capable of cutting through thick ice floes and taking a straight course than cargo ships. Ice conditions also vary between the Passages. First-year ice is more frequent in the coastal areas than in the Central Arctic Ocean. This makes it "easier" to break ice in the NEP and NWP than in the TPP.

Multiple factors are involved in deciding shipping distances in Arctic waters. Therefore, the focus should be on the individual voyages and ships rather than on the length of the Passages when it comes to calculating the distances between ports in southern waters through the Arctic. This is clearly illustrated by the fact that the difference between the shortest and the longest navigational channels of the NSR is 700 nm which is the same length by which the TPP is geographically shorter than the other Passages. As of the present, sea ice conditions, ship types and geography are the main interaction factors deciding the ultimate length of any voyage. If ice conditions and ship types are not part of the calculation it is not possible to decide the distance to be covered. However, in the long term improving ice conditions will gradually alter the premise of this conclusion and enhance the role of geography, which will play out fully only if the Arctic Ocean gets ice free. Indications are that voyages between the Bering Strait and Kirkenes in North Norway undertaken under ice free conditions and north of the archipelagos could take less than 10 days.

As long as sea ice conditions prevail in the Arctic Ocean, none of the Passages have a definite distance advantage over the others. In the relations between the three Passages, the TPP can only provide a distance advantage if the ocean is free of ice. Geopolitically speaking, the shipping industry is bound to focus on a complexity of interacting factors to calculate the length of any Arctic voyage.

7.2.2 The water spaces of transportation within the Passages

All IPCC climate models predict that the Arctic Ocean will continue to freeze over in winter. What climate change will affect is that (a) the volume of sea ice will be thinner and contain smaller fractions of multiyear ice, and (b) the coastal areas will gradually experience longer periods of open waters. All indications are that the blue water portion of the Arctic Ocean will increase over time and provide better sailing conditions for cargo shipping.

Therefore, coastal areas will be in the center of industry attention in the short and medium terms, and the focus will be on destination Arctic and intra-Arctic sailings. The TPP will be of less interest due to its permanent ice cover. No commercial ship has ever completed a voyage across the Central Arctic Ocean, and prospects are that years will pass before that happens. It is the ice-free spaces along the NEP and NWP that carry the most immediate potential for increasing shipping activities in the short and medium terms, not least because this is where the most easily accessible resources are located and where developmental projects will be started. This overall trend applies differently to the Passages.

The Northeast Passage

In recent years, stretches of open water extend in summer westward from the Taymyr peninsula to the Barents Sea and eastward from the same peninsula to the Bering Strait. Satellite images of the tip of the Taymyr peninsula show that fast ice is present and reluctant to give way to global warming, requiring voyages between the eastern and western part still to go through a limited stretch of sea ice in summer. Whereas multiple ports along the whole Passage have been the recipients of intra-Arctic shipping ever since the October revolution, destination-Arctic shipping is to a large extent located to the western part. The Kara Sea Route has been operated on an annual basis ever since 1978, taking nickel from Dudinka to Murmansk, and since the turn of the century oil and gas have been added to the list of cargoes being transported westward along the same route. It is believed that these shipments will be stepped up in the years ahead due to the rich resource base of western Siberia and northwest Russia and the accessibility of ships to these waters on a year-round basis. The ice forming during winter in this part of the route is mostly first year, which is relatively easy to overcome for ice-strengthened oil and gas tankers. For the period up to 2020 and beyond Russia will, according to her policy plans of 2009, use her Arctic Zone as a strategic resource base for the socio-economic development of the country as such. A main goal is to expand the resource basis of the Zone to satisfy Russia's needs for oil and gas and other types of strategic minerals, and to transform the Zone into a leading national strategic resource base in the period 2016–2020. In this overall scheme of development, the NSR will be used as a unified national transport link and will still be open for international navigation. The likelihood is that the eastern part of the NSR, which is the least developed, will reap some benefits from this scheme. Up until the present, this part of the route has mostly attracted intra-Arctic shipping lasting from only three to four months every year.

In the Barents Sea, two shipping routes have been established between the archipelago of Svalbard and cities on the Russian and Norwegian mainlands. Cargo ships from Murmansk supply the Russian community in Spitsbergen and bring coal back to the city of Murmansk, whereas Norwegian ships do the same between the capital of Svalbard, Longyearbyen and the city of Tromsø in North Norway. These routes have been in operation throughout summer ever since the early 1900s, and in recent years the Norwegian route has been operated on a commercially-viable basis. In addition, a rapidly increasing number of cruise ships calls on different locations in the archipelago during summer, and the Barents Sea has for decades been an area important for fisheries.

The Northwest Passage

The Canadian part of the NWP consists of seven different routes running through a massive archipelago with narrow and often ice-clogged straits. Sea ice conditions within the archipelago vary dramatically from one year to another, from season to season and within seasons, so any surface operation is faced with unpredictable conditions. In the summers of 2007 and 2008, most of the archipelago was for the first time in centuries ice free, promising to open the Passage to commercial shipping. There is mounting evidence that the sea ice reductions recorded so far will continue, but the uncertainty is great as to the rate at which the ice will diminish. According to the Canadian Ice Service, it is quite likely that the latter half of this century will still experience occasional summers with ice conditions as severe as those witnessed in the 1980s. In such events, multiyear ice in low concentrations will present a major hazard to shipping. The AMSA report concurs, stating that ice conditions make it nearly impossible to schedule sailings with any degree of certainty of reaching the desired port on schedule. Although the area is supposed to be rich in resources – not least in oil and gas – Canadian authorities seem doubtful whether the route will be opened up to high volumes of international commercial shipping, at least in the medium term. The bulk of shipping will continue to be mostly intra-Arctic, restricted to a few summer months. For all practical purposes, the Canadian NWP will be "closed" for shipping for at least eight months each year in that Canada withdraws its operational resources (icebreakers, etc.) from the area at the end of summer.

The Transpolar Passage

The waters of the TPP are ice covered on a year-round basis. This, however, does not deter the EU Commission from stating that transit sailings will considerably shorten trips from Europe to the Pacific, save energy, reduce emissions, promote trade and diminish the pressure on main trans-continental navigation channels. To secure the possibility of transits in all time spans, the Commission makes it clear that no country or groups of countries have sovereignty over the North Pole or the Arctic Ocean around it. This is to say that the freedom of navigation and the freedoms of the High Seas shall rule the waters of the TPP. To achieve this, the EU will work with the Arctic states to develop a "cooperative Arctic governance system". According to the Commission, transit sailings will also act as a countermeasure to avoid transportation routes through troubled southern waters – the Suez and Panama Canals, East China and Indian Seas.

The US also denounces the drafting of an Arctic Treaty of broad scope. Such a treaty is, according to the White House, inappropriate and unnecessary due to "geopolitical circumstances". On the question of the freedom of the High Seas, the EU and the USA have taken a joint position. This position will be challenged if the Russian government officially declares the NSR to extend to the North Pole as suggested by her polar scientists and some Russian international lawyers.

The Passages in comparison

The water spaces available to shipping operations along the two coastal Passages differ. Despite climate change, the NWP will continue to be controlled by ice conditions in the most narrow of straits even in summer. It could also take years before the Canadian Arctic matches the volume of resources extracted when compared with Alaska and the Russian North. As sea ice conditions are widely expected to improve more rapidly in the NEP than in the Canadian Archipelago, the NWP will most likely not open up for high volumes of international transit shipping or, for that matter, be a serious competitor to the NEP in destination-Arctic shipping, in neither the short or medium terms. Although expectations are that Arctic destination shipping will increase also for the NWP, the bulk of shipping activities in this area will stay moderate and mostly be intra-Arctic in character (as is the case today). The waters running along the Alaskan coast are the only part of the NWP that could host a realistic potential for a notable increase in destination shipping in the medium and long terms. Transit sailings might be undertaken through all three Passages. In terms of distance, the one could be as good as the others (see above), and in terms of manoeuvrable sea ice the best shot in general is the NEP.

The NEP is by far the most attractive of the three Passages when it comes to offer spaces of manoeuvrable ice, both in terms of transit and destination-Arctic shipping. Geopolitically, this implies that Russia, controlling the waters of the NSR, has a key role in regional shipping in all time perspectives. If the sea ice disappears – seasonally or permanently – the power position of the Russian government will be weakened when it comes to multinational transit sailings. In such a scenario, the TPP will assume a more prominent position of utility.

7.2.3 Operational spaces of military and civil activities

During the Cold War, the Arctic was transformed from being a military vacuum in the post-World War I period to a military flank in the 1950–1970 periods to a military front in the 1980s. This was due to, among other things, the spatial location of the Arctic Ocean between three continents providing the shortest attack route between belligerent superpowers and blocks. The coastal areas of the Eurasian and North American continents got militarized for surveillance, monitoring and response purposes. The NSR was also used for some transfers of naval ships between the Northern and Pacific Fleets and, according to witness reports, the deep water portions of the NWP was occasionally visited by small foreign submarines, probably both US and Soviet. Due to the hegemonic features of the East–West conflict, civil and military issue areas were tightly interlinked and civil affairs were subordinated to military requirements. In short, military strategic interests got the upper hand and dominated the utilization pattern of the region for decades. Arctic waters were put in the context of military strategic rivalry and civil exploitation happened only on the premises of the military establishments. It lasted until Mikhail Gorbachev made an explicit distinction between military and civil security in his speech in Murmansk in 1987 and the military grip on civil activities in Arctic affairs loosened. At the same time, the arms race had created changes in military technology that affected the deployment pattern of naval vessels to Arctic waters. This made for a gradual "de-securitization" of the NSR and NWP, and for a strong militarization of the TPP.

The Northeast Passage

The Arctic Ocean is suited for submarines, not for surface warships. Surface warships are constructed with thin hulls because of speed requirements, whereas cargo vessels are ice-strengthened to cope with sea ice at the expense of speed. This is, according to Russian naval authorities, why the NSR was never used for operations and battle training by Soviet battleships and why transference of Russian warships between the Pacific and Atlantic fleets had nearly stopped in the early 1990s. What is more, the coastal areas are too shallow to accommodate the operational needs of the big strategic submarines (150 m long). As stated by US security experts, any submarine longer than 105 m is probably incapable of meeting manoeuvrability requirements under ice in shallow waters. As the strategic submarines grew bigger and acquired polar operational capabilities, and the Northern and Pacific fleets became self-sufficient with surface battle ships, the military establishment gradually lost its dominant influence on NSR affairs. The NSR became an object of increasing civil/economic interest and political attention, which could explain why the route was opened to international shipping in July 1991, and is still – 20 years later – open to foreign ships. In the sphere of military security, defence and protection the Arctic Zone is regarded as vital to the interests of the Federation in preview of various military and political situations. The policy is to create an integrated security system to protect the territory, population and sites critical for Russian security within the Arctic Zone.

The only area of the NEP where clashes of military and civil interests could still occur is in the Barents Sea, where submarines and surface naval vessels in a future scenario could have their thoroughfare restricted and narrowed in by oil, gas and other civil installations

The Northwest Passage

The waters of the Canadian archipelago are on average somewhat deeper than the offshore areas of Siberia, but several of the channels of the NWP are both too narrow and shallow to allow for the access and effective operation of strategic submarines. No Admirality in its full senses would gamble with the survivability of its valuable strategic submarines in such waters as long as better alternatives are immediately at hand in neighboring sea areas – in the Central Arctic Ocean. Submarine intrusions into the channels of the archipelago do not serve any important operational military purpose. Canadian as well as US cities can be attacked and hit by intercontinental submarine missiles launched from waters far away in the Arctic Ocean and from waters just north of the Archipelago. US submarines that might have traversed these straits are first and foremost small and, second, their main mission has most likely been to be spotted to assert US right to operate in these waters without Canadian consent. As seen from the recent US point of view, the freedom of navigation in these waters (including the NSR) is a "top national priority" which is vital to support the United States' ability to exercise these rights throughout the world, including through strategic straits. Here, US Arctic policy is defined in the context of her global military strategic interests. The *Presidential Directive* of 2009 states that US military interest in the Arctic is to secure a maritime presence to be able to project sea power throughout the region, and to do so independently by US actions alone or in conjunction with other states.

Canada has announced that she will build up her military and surveillance capabilities in the Archipelago to improve, among other things, her ability to counteract submarine intrusions into Canadian internal waters.

The Transpolar Passage

The Central Arctic Ocean is an ideal environment for strategic submarine operations. This area offers depths of thousands of meters and has a constantly-moving ice sheet producing more than enough ambient noise to provide a certain degree of noise protection for strategic submarines. The United States announced in 2009 that strategic deterrence is high on the national agenda for the region. Surprisingly enough, Russia did not. During the later part of the Cold War it was Soviet strategic submarines that had the need to use the water columns of the Central Arctic Ocean for strategic operations, not US submarines or, for that matter, French or British ones. Although this need is not reiterated in the Arctic policy document of 2009, missiles launched from Russian submarines in the Arctic Ocean in later years indicate that strategic deterrence will continue to be a fundamental Russian interest in the Central Arctic Ocean. In 2007 and 2008, 14 longer Russian submarine patrols were reported in the Arctic Ocean, and in 2009 the surfacing of a US attack submarine was reported in the vicinity of the North Pole. The increased Russian submarine activities in the Central Arctic Ocean come after nearly two decades of little Russian naval activity in these waters.

The Passages in comparison

In terms of available space, to a large extent nature sees to it that the civil and military sectors can fulfil their objectives and activities in separate parts of the Arctic Ocean, avoiding interference in each other's activities. The shelf areas seem to be "reserved" by nature for the civil sector – shipping and resource mining – whereas the Central Arctic Basin is the prime operational space for strategic submarines – a strategic sanctuary of submarines, in particular Russian submarines. This natural separation scheme is in no way absolute, but it seems to suffice the needs of the two sectors to conduct most of their activities without continuous and undue interference from the other. Thus, geographical features provide for a separation of activities lowering the level of potential inter-sector conflicts. The only possible clash to be anticipated is if, in the future, the TPP is used for regular freighting purposes, the shipping industry will share ocean space with strategic submarines.

As seen from the point of view of the EU, the main causes for conflicts in the Arctic as of the present relate to climate change and the quest and competition for regional resources, in particular hydrocarbons. The Commission takes climate change as a "threat multiplier", altering the geo-strategic dynamics of the Arctic with potential consequences for international stability and European security interests. The threats are civil in character and in the separation scheme discussed above, civil interest comes to expression mainly in the coastal areas along the NEP and NWP. A new element of conflict has, according to this perspective, been introduced to the region.

To an astonishing extent, natural parameters provide for cohabitation between military and civil actors in the Arctic Ocean, and for both sectors to fulfil their obligations on their own terms without necessarily interfering too heavily in the activities of the others. In the short, medium and long term, the NSR and NWP could open up for economic activities without risking conflicts with the military sector. This also applies to the TPP in the short and medium term, but not necessarily in the long. Perceptions of conflicts and threats are now identified by the EU to be located in the coastal areas related to global climate change and the quest for resources. These threats may be real, but it is in the national interest of all Arctic states to keep the level of conflict low and controlled to get access to resources that are crucially and strategically important to the survivability of their respective societies.

7.2.4 Connecting corridors in southern waters

Four transportation corridors connect the Arctic Passages to world markets: The Northern Maritime Corridor, the "Northern Pacific Corridor", the "Davis Corridor" and the "Fram Corridor". The NPC connects with all three Passages in the East, as does the NMC in the West. In the short and medium terms, the Fram Corridor serves only as an alternative to the NMC due to sea ice conditions. For this reason we will concentrate our attention on the NMC and NPC only.

The Northern Maritime Corridor

The Northern Maritime Corridor stretches from the White Sea in the north, with partners in Murmansk, Nenets and Archangel regions, to multiple ports in the North Sea. The corridor involves 22 regions in eight countries and was approved as an inter-regional project by the EU in 2002. The NMC connects northwest Russia to the European continent and the European continent to the Pacific Ocean. The NMC is the highway of the EU to and from the Arctic – a visible sign of her economic and political interest in Arctic affairs. Non-Arctic states have been active in establishing a connecting "bridge" between the Arctic Ocean and themselves.

The Northern Pacific Corridor

The Northern Pacific Corridor has not yet been established or officially baptized. It got its name for the purpose of this study. As of today, it is a High Sea area pending the establishment of a cooperative corridor among the littoral states to the North Pacific. If and when established, it most likely will extend in two directions from the Bering Strait: southward to somewhere in South East Asia and eastward to the North American west coast.

The waters of the North Pacific have been used for transportation purposes for decades and will continue to be so also in the future. As such, there is no pressing need of a formally established corridor. However, as of now, these waters lack the political and logistical framework of an inter-state corridor. Such a corridor will add some extra navigational value to these waters, contribute to a reduction of regional political tensions and provide more safety of navigation.

There are signs in regional politics that these waters may be converted into an inter-state corridor through cooperation among North Pacific countries, in particular Japan, South Korea and China. If this materializes, non-Arctic states will be instrumental also in the North Pacific to link the Arctic to their own markets.

The Corridors in comparison

The NMC is by far the safest and most developed of the four corridors. It was established by the EU in collaboration with Russia and Norway. Regional tensions and lack of pressing needs explain why the countries of the North Pacific have not yet established a similar corridor in the Pacific Ocean. Indications are that this may change in due time, possibly with China as a prime mover.

If and when that happens, non-Arctic states in the North Atlantic and North Pacific clearly signal their interest to take part in Arctic resources developments and to pool resources with the Arctic five.

7.3 Shipping, resources, economic trends and alternative means of transport

7.3.1 The resources and the economies of the Arctic

Arctic shipping is both a contributor to the economies of the Arctic regions, and itself dependent on and a beneficiary of the vitality and health of these economies. To varying degrees, the Arctic economies rely heavily on the abundance and accessibility of mineral and biological resources. The presence of vast resources is by and large a well-established fact but access to them has proven to be very demanding indeed. Mother Nature has not been kind to those who are set on venturing into the Arctic to exploit its resources for whatever motives.

But to assess the prospects of Arctic shipping demands more than a focus on the Arctic itself. The Arctic regions and their economic activities and life are but parts of the states to which they each belong. The economies of these states in turn are parts of and (in the case of some) vital to the overall world economy, the ups and downs and health of which inevitably reflect back on these economies and their Arctic parts.

7.3.2 Developments in the world economy

Economic activity and power has always been unevenly distributed. The North Atlantic area – Western Europe and North America – has been the center of world economic power and influence for the last three to four centuries. This state of affairs is changing. During the last three decades, the economies of countries such as China, India and Brazil have grown at rates impressive enough to indicate that in the not too distant future these countries will become a new center of gravity in a global economic system.

The emergence of these and other developing countries is changing international trade patterns as they lay claim to an increasingly larger share of the world's Gross Domestic Product (GDP) and natural resources. After having had double digit GDP growth rates until 2008, China's rate slowed down under the influence of the global economic recession but still managed to grow by 8.7% in 2009 and by 10% in 2010. India's growth rate slowed to 5.6% in 2009 but grew by 7.7% in 2010.

The still robust growth rates in countries like China and India could not, however, prevent the world financial crisis that began in the US to spread to almost all other parts of the world. It caused the advanced economies to shrink by 3.2% in 2009 with the result that the world economy suffered its first recession since the 1930s when it declined by 0.8%. The collapse of demand for goods and services in markets as important as those of the advanced economies inevitably influenced economic activities everywhere, including the Arctic parts of the world.

The shipping sector was among those hit hard. After having enjoyed a decade of increasing demand for its services during which massive over-contracting of new tonnage took place, the economic conditions of 2008–2009 served notice that harder times lay ahead. In a sense, shipping is double hit with both a reduction in demand and a supply side growing too fast. With a record high order book by the end of 2009, the world fleet will grow by 6–8% per year the next 2–3 years, the order cancellations taken into consideration, and the prospects for demand growth is way below 6–8% on average, so there are tough years ahead.

The Northeast Passage

Resources

Russia is the world's leading natural gas producer and in 2008 had the largest natural gas reserves at 23.4% of total world reserves. Almost 90% of the natural gas comes from one region in northern West Siberia. The production in these fields is declining, however, and the search for new supplies is focusing on the Yamal Peninsula in particular, and on the development of the Shtokman gas and condensate field in the Barents Sea. Higher exploration and recovery costs could slow the pace of the search for Arctic gas in the time ahead. The international market for gas, especially in the US, no longer seems as promising as it did not so very long ago because of the appearance of alternative sources of supplies, very important among which are large quantities of shale gas in the US.

Russia will nevertheless continue to export large quantities of gas well into the foreseeable future from fields already in production. With 23.4%, 6.3% and 19% of the world's proved reserves of gas, oil and coal respectively, Russian resources are indispensible. Most of these resources are to be found in the Russian Arctic. About 91% of her natural gas production and about 80% of her natural gas reserves are in the Arctic. It is also estimated that 90% of Russia's offshore reserves of hydrocarbons are in the Arctic. According to the United States Geological Survey (USGS) 22% of the world's undiscovered oil and gas are to be found in the Arctic, and most of it is thought to be in offshore fields in Russian-controlled Arctic waters.

To what extent already discovered and undiscovered resources will soon be exploited will depend on a number of factors, including their location and accessibility, the cost of exploration and recovery in the Arctic compared to other parts of the world, and the general and particular global supply and demand situation.

The areas in the Arctic least difficult and presumably less financially demanding from an exploration and recovery point of view could be in the Barents Sea. This stretch of sea is about 1,760,000 km^2 and covers a continental shelf most of which lies under less than 500 meters of water. Large parts are not ice infested and other parts much less so than other offshore areas of the Arctic. According to the USGS, the Russian and Norwegian Barents Sea shelf areas could contain about 11 billion barrels of undiscovered oil, 380 billion cubic feet of undiscovered natural gas, and 2 billion barrels of undiscovered natural gas liquids.

Most of this, approximately 68% of the oil, 85% of the natural gas, and 65% of the natural gas liquids, are estimated to be in the Russian-controlled East Barents Basins Province. The resources of the Barents Sea shelf are assumed to be recoverable regardless of the presence of sea ice or depth of water.

Russia is also a leading producer of many important minerals. In 2006, 25 mines were operating in Arctic Russia. The Kola Peninsula contains a remarkable abundance of various minerals. In 2007, the mineral raw material sector produced about 30% of Russia's GDP and about 70% of the country's budget revenues. The oil and gas sector accounted for about 20% of the GDP. Norilsk Nickel, the world's leading nickel producer, has mining and processing operations on the Kola Peninsula and on the Taymyr Peninsula in Western Siberia. A good deal of the Russian minerals industry is struggling with outdated production technology that makes it hard to compete with counterparts in the industrially-advanced countries. Oil and gas will remain the most important sector of the Russian Arctic minerals industry but at a slower rate of growth in the immediate years ahead because of depleting proven reserves, lack of sufficient investment, and challenging international market conditions.

Tracts of the vast forest resources are also found in the territories regarded as Arctic. Yet the Russian forest industry is lagging behind its counterparts in Europe, North America, and elsewhere. Old-fashioned methods of harvesting, outdated equipment and a lack of a trained workforce to handle modern equipment have been prominent features of the Russian forest industry. In 2007, a New Forest Code was enacted to modernize the industry away from an emphasis on export of round wood towards an industry relying predominantly on value-added timber products.

Shipping and alternative modes of transportation

When one turns to the transportation picture of Arctic Russia, one is at once confronted with a more complex picture than in the North American Arctic. The region stretches some 7000 km from the border with Norway in the west to the Bering Strait in the east, and the demographic picture is much different from its counterpart regions in North America. Arctic Russia has a much bigger population. Fairly substantial volumes of goods and merchandise have to be transported into the region every year to sustain human life and settlements. A new census was conducted in Russia in 2010 and the Russian Arctic or parts thereof are losing population at an alarming rate. Since the year 2000, the population of the Russian part of the Barents region has decreased by 462,000 or by almost 11 percent. The Murmansk Oblast recorded a decline of 10.4%, the Komi Republic of 9.3%, the Arkhangelsk Oblast a decline of 9.2%. At the beginning of 2009, the total population of Barents Russia was almost 3.8 million and a year later 24,000 less. One should note, however, that the Barents region includes territories that are not regarded as Arctic. In the population of 3.8 million there are substantial non-Arctic populations. Yet the biggest decline was recorded in the parts of the Barents region that are usually regarded as Arctic. The decline has been steady and continuous since the start of the century, and if it continues will influence not only the economy of the Russian Arctic but also shipping prospects.

Apart from transportation by sea, large amounts of goods are carried by railways, roads, inland water ways, and pipelines. The distribution of freight on the various modes of transport has remained fairly steady since the middle of the 1990s until 2008. How much of the freight volumes originates in the Arctic and how much is resupply of the Arctic is difficult to determine. Oil and gas figure very prominently in Russian economy and export accounts. Of the 1.3 billion tons of freight carried by the Russian railways in 2008, some 200 million tons (18%) were oil. The only significant amounts of oil exported by railway go to China. Much the largest amounts are exported by pipelines to Europe. Most of the oil transported by sea goes from the port of Primorsk in the Baltic Sea and the port of Novorossiysk in the Black Sea. The shipment of Russian oil and gas and other commodities from the Black Sea area has made the straits of Bosporus and Dardanelles busier than the Suez and Panama Canals.

About 36,000 tons/d of crude oil and products were transported from Murmansk by sea in 2007. During the last decade, the amount of oil exported by sea from terminals in the Russian Arctic, including the Barents area, has increased substantially. In 2002, some 4 million tons were sent by sea, in 2008 some 11.5 million tons. In a 5–10 years' perspective it is thought that exports of oil by ship from Russian Arctic terminals could reach 100 million tons a year.

The Northwest Passage

Resources

Like other Arctic parts, the economies of the North American Arctic are resource based.

Arctic Canada – Yukon, the Northwest Territories (NWT) and Nunavut – all rely heavily on mining. Gold, uranium, and diamonds are the most important products. Though valuable, the quantities mined are relatively small and by and large transported by means other than ships, for example, by aircraft. Some mines have been closed in the Northwest Territories and Nunavut. The development of the iron ore mine at Mary River on Baffin Island is the most promising mining project in the Canadian Arctic. The mine is not supposed to be operational before 2020 and is then expected to produce about 12 million tons of ore a year.

The Mackenzie Delta in the NWT could contain huge amounts of natural gas, equivalent to about 10–11% of Canada's proven natural gas reserves. Gas from this source is meant to be transported by the much-heralded but yet not built Mackenzie Gas Pipeline to Alberta, where it will be fed into the existing natural gas distribution system. After years of hearings, the federally-appointed Joint Review Panel delivered its results in December 2009. The final decision to build the pipeline is still pending. The costs of constructing the pipeline and the uncertain market outlook for gas in the years ahead are two of the factors likely to weigh most heavily in the decision.

The Alaskan economy is completely dominated by the petroleum industry. Some 80% of state revenues come from the oil and gas industry. Oil production in Alaska is centred around the Prudhoe Bay field on Alaska's North Slope. In 2007, Alaska contributed about 20% of US oil production but only 3% of natural gas production. The principal mode of transport of the oil is by overland pipeline from Prudhoe Bay to Valdez from where it is shipped by tankers to the lower 48 states.

Alaska has non-fuel mineral production, the value of which in 2007 was about $3.3 billion or 4.4% of the US total. The most important mineral both in terms of value and volume is zinc. The Red Dog Mine in north-western Alaska is the world's largest zinc mine and accounts for more than 70% of the value of Alaska's mineral production. The mine has its own loading facilities and the shipping is seasonal because of ice conditions. The output is stockpiled and shipped out between July and October. What is not sent out by pipelines of Alaska's exports, as in the case of oil and gas, is sent south through the Northern Pacific Corridor and not through the NWP.

Shipping and alternative modes of transportation

As in other parts of the world there are in the Arctic various ways of transporting goods and cargoes. In the Canadian Arctic, very small but extremely valuable volumes of precious minerals are transported by air, and other types of cargo are sent by ship through the NWP. In 2007, about 35,000 tons of cargo were carried by water transport and loaded and unloaded in the Northwest Territories. In Nunavut, some 159,000 tons were carried by water transport the same year. Small populations partly account for the small volumes. The Northwest Territories has a population of about 42,000, Nunavut about 28,000 and Yukon about 31,000. Only a very modest increase in shipping activity is envisaged by Transport Canada in the years ahead until 2020.

Much the same situation prevails in Alaska. As in the case of minerals, goods are transported partly by road and partly by sea but, as far as oil and gas are concerned, by pipeline. To the extent goods are transported into Alaska from outside, the volumes are limited mostly because of the relatively small population, estimated in 2008 to be about 690,000. Little or nothing is transported by sea through the NWP.

The Passages in comparison

Which mode of transport will become the leading one in the Euro-Arctic area is a rather open question. As far as oil and gas are concerned the choice is between shipping and pipelines. Gas is the dominant resource in the Russian Arctic, especially offshore. How it will be transported also depends very much on the location of the principal markets. If these markets are mainly in Europe and Asia pipelines are likely to be the favored mode of transport. Oil and gas for markets outside Europe and Asia will be sent by ship.

7.3.3 Some reflections on the future of resources and shipping routes

The world economic recession has hit all countries. It is not just a blip that will quickly fade away in a year or two. Among the Arctic countries, the US is experiencing rates of unemployment of 10% and even more when under-employment is taken into account. The situation in many parts of Europe and the EU is even worse. Russia has been hit hardest because of a much reduced demand for Russian oil and gas on which she is very dependent. At the same time, proven reserves of gas in particular have increased substantially elsewhere, especially in the biggest consumer of all – the United States. This in turn has augmented the rather bleak outlook for Russian gas exports in particular caused by the economic recession. An immediate result has been the temporary shelving of gas investment projects in Arctic Russia. The Arctic parts of North America have not suffered the same consequences, in the case of Alaska because it is not dependent on export of oil and gas to other countries, and Arctic Canada simply because it does not as yet produce significant amounts of oil and gas.

The Northeast Passage

Of all Arctic sea routes, the NEP and the NSR are much the most promising from a shipping point of view but are also the more vulnerable to changing global and national economic conditions than alternative shipping routes in the Arctic and in other parts of the world. The hope that the exploitation of Arctic gas resources in particular would prove a boon to Arctic shipping is less strong today than in the recent past. Planned projects like Shtokman have been put on hold because of world economic conditions. New sources of supply of gas are developed in those parts of the world that have the highest consumption. Should new Arctic natural gas provinces be developed, the question still remains about which mode of transport will be chosen, maritime transport or pipeline. The pipeline network is continuously being expanded to such an extent that at the moment it could have considerable idle capacity.

Intra- and destination-Arctic shipping along the NSR is likely to continue in the near future at about the same level as now and in the recent past, especially taking into account the depopulation of the parts of the Russian Arctic we now are witnessing. Whether this will continue is very much a function of how economic conditions in the rest of Russia will fare in the years ahead. To supply the Arctic is also not only a question of shipping but also of rail, road and river transport.

The use of the NSR as a transit route between the Atlantic and the Pacific and vice versa appears now more promising than it has been in a long time. The *Beluga* shipments set an example that other shippers and ship owners might like to try to repeat and perhaps emulate. Cargoes have to be transported and should the NSR prove competitive so much the better. Beluga Shipping of Germany, Sovcomflot of Russia and Tschudi Shipping started pilot shipments along the NSR in the summer of 2010.

The Northwest Passage

The maritime traffic in the North American parts of the Arctic is likely to continue at the same modest level it has in the past. The increase in the amount of Arctic cargo foreseen by Transport Canada over the next ten years will not change this picture. The only development on the horizon substantial enough to do so would be the commissioning of the Mary River iron project that could each year generate perhaps 60 shiploads of iron ore, each of some 200,000 tons. But according to current plans this will not happen before 2020. Alaska does not generate any cargo needing to use the NWP.

The Transpolar Passage

The study carried out by Det Norske Veritas on the feasibility of using the TPP as shipping routes offers little hope for shippers and ship owners alike in the near future. Along these routes there simply exist no Arctic resources and no cargo. In a 15–30 years' perspective, ice conditions might have eased sufficiently to make these routes viable alternatives. But being more-or-less direct trans-ocean routes, they do not have a cargo base of natural resources that need to be shipped in or out of the Arctic. Their main utility will rest on their ability to be useful and profitable transit routes between the Atlantic and Pacific oceans and vice versa.

The Passages in comparison

Of the three alternative sea routes in the Arctic, the NEP and the NSR stand out as a first choice from a shipping point of view. Trying to use the NWP and TPP would entail facing natural obstacles of a much more severe kind for a considerably longer period of the year than those found along the NSR. The shipping infrastructure (harbor facilities, icebreakers, etc.) is almost nonexistent along the NWP and totally absent along the TPP in addition to which both routes have little access to cargoes of any significant kind. This, of course, is a function of the lack of resources (as in the case of the TPP) or a lack of exploitation of existing resources (as in the case of the NWP).

There are three factors that point to the NSR as the sea route that will be favored in the foreseeable future. Great stretches of the route go along or through parts of the Arctic where great amounts of resources are found or likely to be found both onshore and offshore. The extraction of resources produces cargoes. The second factor that favors the NEP is the already-established and continuously developing infrastructure along the route. And the third factor is the less challenging ice conditions and the longer sailing season during summer.

7.3.4 Cruise shipping

The Arctic will remain a remote and wild area of breathtaking beauty with a unique flora and fauna. If climate change leads to less ice, more of the Arctic will be accessible to cruise ships. The cruise industry is a concept industry which is always on the lookout for new themes and destinations. It is to be expected that if the ice is retreating some cruise operators will follow the edge of the ice. In a longer-term perspective, Arctic cruises would seem to be a growth industry with a great potential.

In the shorter perspective, specialist operators will continue to offer cruises in all three passages. With powerful icebreakers one can reach the North Pole or circumnavigate the entire Arctic Ocean. With expedition ships and ice-classed passenger/cruise vessels, trips are offered into the NWP and in the NSR trips go to Novaya Zemlya and Franz Josef's Land in the west and to the Wrangle Island in the Chukchi Sea in the east. The vessels are fairly small – 50 to 400 passengers – very small compared to the largest cruise vessels of more than 5000 passengers. The larger vessels concentrate on the blue waters around Svalbard and Greenland, which is the target for most Arctic cruises, while the smaller vessels go on expeditions in ice-covered waters. Arctic cruises are a small niche product, indeed, compared to the mass markets of the Caribbean or the Mediterranean.

The typical cruise ship for Arctic operations carries zodiacs and hovercrafts to enable exploration of shallow waters. Inevitably such cruise activities will one way or the other interfere with the cultures of the indigenous peoples of the Arctic, disturbing hunting and fishing grounds and even be a danger to people in small crafts such as canoes or kayaks. In this respect, the cruise operators have a social responsibility to perform.

The cruise vessels used in Arctic operations are very much self-contained units that are not dependent on land infrastructure. This is fine when everything goes smoothly, but is a major problem the day an accident occurs. The coastal states face a big challenge when it comes to providing adequate search and rescue service if cruise activities expand in the Arctic.

The Passages in comparison

Specialist operators currently offer cruise experiences in all three Passages with vessels and icebreakers accommodating typically 50–150 passengers. For the NSR the busiest areas are in both the western and eastern extremes of the route, but trips are also made along the entire NSR. For the NWP, only occasional trips are made for the entire Passage, but regular sailings are made in the eastern part of the Passage. There are regular sailings to the North Pole with a nuclear icebreaker. Larger vessels with up to 400 passengers operate on a regular basis, mostly in the corridors just outside of the passages, mainly around Greenland and Svalbard, which stand for the main Arctic cruise traffic.

With the prospect of less ice-covered waters in the Arctic in the future, it is expected that cruise operators will follow the edge of the ice. The ships are self-contained and most of them carry zodiacs and even hovercrafts, enabling the cruise passengers to explore shallow waters. The chances are high that cruise traffic will interfere with the traditional hunting life of the indigenous peoples, mainly along the NEP and the NWP, so the cruise industry must show social responsibility to avoid direct conflict.

7.4 Environmental challenges and other operating conditions

The purpose of this section is, firstly, to compare the shipping activities along the three Arctic transport corridors and how they could impact the ecosystem and, secondly, to compare the physical characteristics and the operating conditions of the routes from a navigational point of view.

7.4.1 Arctic shipping and impacts on marine ecosystems

Large Marine Ecosystems (LMEs) are national and regional focus areas of a global effort to reduce the degradation of linked watersheds, marine resources, and coastal environments from pollution, loss of habitat and over fishing. LMEs have been developed to identify areas of the oceans for conservation purposes only, which means that organisms across the LMEs must be understood as they are functioning together. Arctic waters are divided into 17 LMEs.

The Northeast Passage

The following LMEs are defined on the NEP: (1) the *western part* including the Barents Sea and the Kara Sea; and (2) the marginal seas of *eastern part* of the NEP/NSR including the Laptev, East Siberian and Chukchi Seas.

There is a relatively large volume of shipping in the Barents Sea while the shipping activity is considerable less in the Kara Sea. Moreover, the Barents Sea LME is considered as a high productivity ecosystem, with biological activity determined mainly by seasonal changes in the temperature and light regimes, advection and ice cover. Furthermore, vulnerable areas in the Barents and Kara Seas have been identified in relation to oil and gas activities, based on where there are aggregations of animals that could potentially be impacted by oil spills or disturbances from activities. However, significant impacts on the ecosystems by shipping activities, like disturbance (noise, potential collisions with animals, etc.) or emissions from ships has not been documented on the western NEP/NSR.

The number of known species decreases from the western part of the NEP/NSR to the eastern part of the NSR. This is partly a result of harsher environmental conditions eastward on the NSR, but also because the fauna of the eastern part are some of the least studied in the world. From a commercial shipping point of view, fish resources on the NSR play an important role for the local communities. Nevertheless, on a global scale these resources are insignificant. The fish fauna are so sparse and difficult to access that no commercial fishing takes place in the open parts of the seas, except in the western Kara Sea and occasionally in the western Chukchi Sea. In addition, no offshore fishery takes place in the Laptev Sea. The commercial fisheries on the NSR are restricted to the lower parts of the large rivers and estuaries. Moreover, studies focusing on species that might be affected by shipping activities and where changes in the populations could occur as a result of such impacts have been performed since the beginning of INSROP in 1993. Even if vulnerable areas are identified, there are no indications of permanent damage to ecosystems directly caused by shipping activities.

The Northwest Passage

The following LMEs are located in the NWP: Baffin Bay/Davis Strait, the Canadian Archipelago and corresponding seas and the Beaufort Sea. Furthermore, a number of geographical restrictive areas are located here, used as shipping lanes for various destination-Arctic and intra-Arctic purposes. It has been identified that certain parts of the NWP, like the Lancaster Sound, are used by various species of whales, marine birds and seals for seasonal migration. It is where shipping routes travel through these geographically restrictive areas, or chokepoints, that marine mammals are most vulnerable and where it is the greatest potential for conflict between ships and migrating marine mammals. The migration routes used by bowheads and belugas from their wintering areas in the southern extent of the ice and into the Arctic are also broadly the routes used for intra-Arctic and destination-Arctic shipping. On the other hand, there is a relatively low level of shipping occurring on the NWP. Currently, there has not been documented any significant damage on the NWP ecosystems as a result of shipping activities.

The Passages in comparison

Effective reduction of ship emissions can be achieved through the application of feasible and best available technologies and, most importantly, through effective implementation of relevant IMO regulations.

Increased shipping could pose potential impacts on the ecosystems. For instance, the migration corridor used by marine mammals and birds corresponds broadly with the main shipping routes into and out of the Arctic. However, studies performed during INSROP identified that effects of increased shipping on the NSR could be both adverse and positive for commercial fisheries. Accidental discharges such as oil spills belong to the first category. On the other hand, NSR could serve as a means for transportation of fish products to markets outside the area, and also ensure a supply of fishing gear, equipment, etc. which can facilitate exploitation of fish resources that currently are considered less attractive.

Vulnerable LMEs are identified on the two Arctic transportation corridors. Nevertheless, no significant negative impacts on the ecosystems caused by shipping activities have been documented.

7.4.2 Oil spill response on Arctic transportation passages

Oil spill response in Arctic waters is particularly challenging due to lack of infrastructure and because the natural conditions severely constrain effective response. Moreover, there are limited port reception facilities to dispose of oily sludge along the main shipping routes in Arctic.

The Northeast Passage

The Russian system of oil spill contingency response consists of two subsystems: state and private. The federal response system was established in the 1980s and the activity was then organized into operative zones and responsibility areas shared between Murmansk and Sakhalin. These areas include the seaways from the Russian part of the Barents Sea, through the NSR and parts of NPC to the La Pérouse Strait. In addition, Russia has oil spill contingency agreements with Norway for the Barents Sea and with the United States for the Bering Sea. Furthermore, the Russian federal response system and the available equipment used for oil spill elimination have not changed much since the 1980s, despite the increased marine traffic and oil and gas activities. Private systems include oil spill response contingency performed by oil companies and operators of oil terminals on the NEP (i.e., Varandey, Murmansk, Kandalaksha and Arkhangelsk). Nevertheless, the response systems performed by a majority of oil companies and operators in the Arctic are not sufficiently developed.

The Northwest Passage

An operative oil spill contingency response along the NWP is currently absent. This reflects the lack of adequate infrastructure. Moreover, it also makes the area extremely vulnerable should any oil spill accident occur. The Canadian legislation shows how strong the concerns are. The Arctic Water Pollution Prevention Act (AWPPA) 1970 and other regulations under this act, in particular the Canada Shipping Act 1985, have introduced regimes for oil spill response. In addition, a zone/date system is used in order to regulate vessels entering various parts of the NWP. Furthermore, Canada and the United States have agreed to establish a joint marine contingency plan for the Beaufort Sea. The plan sets out regional contacts and procedures in case of oil spills. This agreement has periodically been revised, most recently in 2003. However, if any accidental oil spill should occur at this stage, there is no operative contingency to perform quick and timely response.

The Passages in comparison

The Russian federal response system covers a relatively large area, including the NSR, parts of the NPC and parts the NMC. Nonetheless, the lack of preparedness, adequate equipment and facilities for oil spill elimination makes the areas highly vulnerable should any accidental oil spill occur. Canada has a set of strict regulations and requirements for ships operating along the NWP in order to prevent oil spills. However, like the Russian oil spill contingency on the NEP, the Canadians also lack the capability to perform an effective oil spill response along the NWP.

At this time there is not an adequate preparedness for oil spills, either on the NEP or the NWP. The lack of infrastructure, response and necessary combating equipment makes the ecosystems vulnerable should any accident occur.

7.4.3 Natural conditions and shipping on the Arctic transport passages

The three potential transport corridors share some of the same natural challenges. However, a set of detailed natural characteristics makes each corridor unique, as well as more or less favorable in terms of shipping operations.

The Northeast Passage

Of all the alternative routes for navigation on the NEP, the selection is strongly affected by natural conditions. One of the primary factors when selecting a favorable shipping lane on the NSR is the distribution of sea ice. The ice conditions vary greatly between seasons, between years, and between the marginal seas.

A major percentage of the recent NEP commercial shipping activities do not involve the NSR. Under normal conditions, winter navigation on the eastern NSR is not beneficial, and commercial operations are restricted to the summer season. Currently, the only year-round transport on the NSR is the Kara Sea Route. All other transport is carried out only during summer. The summer season has traditionally been defined as June–October, but favorable ice conditions and technological improvements have gradually allowed an extended summer season.

Navigation during the winter season (November–May) is generally much more difficult than in the summer season, due to the thicker ice cover. An important feature of winter navigation is the *fast ice* – stable, immovable ice which is fastened to the coastline. Fast ice is very difficult to pass through, and normally it is preferable to avoid it by using northerly routes. Nonetheless, if off-shore winds prevail, one will often during winter find open leads at the edge of the fast ice – so called *polynyas* – which are very suitable for navigation.

While the ports of the Barents Sea are almost completely ice free even in the winter, no parts of the NSR are completely ice free even during the most favorable summer month (September). Each end of the NSR – the south-western Kara Sea and the south-western Chukchi Sea – has the lightest ice conditions, with the East Siberian Sea having clearly the most difficult ice conditions. This corresponds with navigational experience, where the East Siberian Sea has been seen as the most difficult sea to navigate, and also being the main bottleneck in terms of navigation. Moreover, the difficult conditions in the East Siberian Sea has to do with the *ice massifs*, consisting of multiyear ice, and which often extend almost to shore even during summer due to currents and winds.

Variations in water depth and narrow straits are other key factors that decide which navigation course to set on the NSR. Since most of the NSR consists of continental shelves, a major part of the corridor and especially the straits are shallow. Shoal depths less than 20 m are by no means rare and increase the risk that deep-draft vessels could run aground. However, there are great variations between the marginal seas of the NEP. Although the western and eastern Barents Sea has depths of 100–300 m, the Kara Gate (the main shipping strait between the Barents and Kara seas) is a mere 20–30 m deep. Furthermore, the Kara Sea has an average depth of 90 m, but some areas are as shallow as only a few meters. Of key importance to shipping on the NSR is that four straits require mandatory icebreaker escort, organized by the MOH: the Vilkitskii Strait, the Dmitri Laptev Strait, the Shokalskiy Strait and the Sannikov Strait. The Vilkitskii Strait is a key NSR Strait between the Kara and Laptev Seas. Although it has adequate water depth for deep-draft vessels, it is mainly ice covered, except for a short period in the summer season. Thus, Shokalskiy Strait is a second possible route north of Vilkitskii Strait. The Laptev Sea has an average depth of 600 m, but as one draws near to the New Siberian Islands the sea becomes shallower than 20 m. The route alternatives between the Laptev and the East Siberian are either through the Dmitri Laptev Strait or the Sannikov Strait. Both straits have minimum depths of 12–13m, a limiting factor in terms of maximum ship draft on the NSR. Long Strait links the East Siberian and the Chukchi seas. In this part of the NSR, draft should not be a problem for any ship that keeps well offshore from the mainland.

The Northwest Passage

The NWP is difficult to navigate due to the pattern of ice drift from the central Arctic Ocean. As Arctic warming continues, it is likely that much of first-year ice in the Canadian Archipelago will break up much earlier than is currently the case. This will allow old hazardous ice from the permanent ice pack to enter the NWP in greater quantities.

There are potentially seven routes through the NWP, of which three are considered as being practical for regular navigation: the M'Clure Strait, the Prince of Wales Strait and the Peel Sound. The presence of sea ice and depth restrictions constitutes the critical factor in terms of NWP navigation. Ice conditions on the NWP are very complex. Observations of minimum sea-ice extent from year to year in the eastern and western regions of the NWP vary significantly. Thus, inter-annual variability is not conducive in order to plan a reliable marine transportation system.

Navigation through the M'Clure Strait is the shortest and deepest NWP route alternative with an average depth of 400 m. However, it is also the most difficult passage due to the severe ice conditions. Old, hard and dangerous ice is present most of the time which can seriously delay the vessels' progress and potentially damage even ice-strengthened ships. In addition, even during the most favorable shipping seasons (10–15 days), there will be a risk of encountering multiyear ice.

The route that passes through the Prince of Wales Strait is an easier alternative deep-draft route which avoids severe ice in the M'Clure Strait. However, a limiting factor is the depth variations which are less than 20 m in some parts of the strait. Normally this strait is open during September; however, the presence of multiyear ice is still seen as a threat to ships. Moreover, as the melting continues, there is likely to be an increase in the presence of old ice as the decrease in land-fast ice in the western part of the Archipelago would allow more old ice from the Arctic Ocean to pass into channels between the islands.

The route that passes through the Peel Sound is the longest NWP transit and the most frequently used. However, navigation of this route is restricted to ships having a maximum 10 m draft. Currently, this route is passable from mid-August to mid-September. However, the presence of multiyear ice in this sound is seen as a significant challenge and is likely to continue to be so.

The Transpolar Passage

When considering the natural characteristics and shipping on the TPP, the central Arctic Ocean has no draft limitations, narrow straits or complex archipelago to put constraints on navigation. Nonetheless, the sea ice conditions are the key factor of limitation in terms of any potential commercial shipping routes. No commercial cargo ship has yet crossed the central Arctic Ocean. There are huge uncertainties as well as variations between different climate models trying to predict the development of ice conditions. In terms of any regular shipping on these routes, however, DNVs Arctic Container Project (ARCON) simulations indicate that the ice will be too heavy and the calculated costs too high for any regular transport. Models indicate that the ice conditions will continue to be heavy during winter and spring seasons, even in 2050, and the route is not expected to be completely ice free in summer. Currently, the TPP is not an alternative.

The Passages in comparison

As demonstrated, both the NEP and the NWP face large water depth variations and complex archipelagos which put constraints on the technical construction and capacity of the vessels operating on these corridors. Moreover, the sea ice conditions vary significantly both between and within each of the three passages, presenting different challenges to navigation on the NEP, the NWP and the TPP.

The availability of icebreakers is an important feature. Russia has a considerable larger and stronger icebreaker fleet to support operations on the NSR than Canada has for the NWP. Russia has five nuclear icebreakers available for NSR support with, in addition, several river and diesel icebreakers operating on a year-round basis. Canada operates five diesel-powered icebreakers only for seasonal (May–November) NWP operations.

From a navigational point of view, the NWP will be the last area where the multiyear ice will disappear and shipping through this Passage will remain risky even in the summer season. It is not very likely that the TPP is going to be used as a regular transport corridor. The ice models indicate that the ice conditions will be too heavy for any commercial shipping. Currently, the NEP/NSR seems to offer the best operating conditions.

7.5 Shipping and Arctic infrastructure

The purpose of this section is to make a comparison of the available infrastructure on the NEP, the NWP and the TPP. Arctic infrastructure is defined broadly to address the major aspects of Arctic shipping, including: technical requirements; crew training and competence requirements; search and rescue capabilities; available Arctic ports; the costs and bureaucratic procedures of Arctic shipping; and marine insurance.

7.5.1 Technical requirements for ships operating in Arctic waters

Excepting the areas along the coastline, where national requirements apply, the same international maritime regulatory regime applies in the Arctic as for other oceans. The most important international convention concerning the safety of merchant ships is the SOLAS convention, which specifies minimum standards for construction, equipment and operation of ships. Flag States are responsible for certifying that ships under their flag comply with the requirements. All Arctic states have ratified the SOLAS convention.

The IMO Guidelines for Ships Operating in Arctic Ice-covered Waters

The SOLAS convention does not specify any additional construction regulations and equipment required for operations in Arctic waters. Recognizing the need for additional recommendatory provisions related to ships operating in Arctic ice-covered waters, IMO initiated the work resulting in the voluntary *Guidelines for Ships Operating in Arctic Ice-covered Waters* (the IMO Guidelines), issued in 2002.

Baltic and Polar ice classes

All vessels are subject to various rules and standards depending on the type of vessel and operation. Compliance with these standards has to be certified and documented, normally by issuance of a certificate.

Ships intended for operation in ice-covered waters will normally be built with an additional ice class notation. The different ice classes include requirements for the strengthening of the hull, rudder and the propulsion system. Specification of an ice class notation is voluntary and the actual ice class decided by the owner will depend on type and area of operation. In addition to an ice class, the vessel can be assigned an Icebreaker notation including additional strengthening for vessels for which icebreaking is the main purpose.

Ice classes are divided into two different regimes. The Baltic ice class rules for ships operating in the Baltic Sea and other areas with lighter first-year ice conditions including areas like St. Lawrence and the southern part of Sakhalin. The Polar Ice Class rules for all areas with heavier and multiyear ice. The Baltic Ice Class rules were developed by the Finnish-Swedish Maritime Authorities, while the Polar Ice Class rules were developed by the Classification Societies (see Table 5.1).

The IMO Guidelines include no direct technical requirements for hull and propulsion, but refer to the additional requirements in the Polar Ice Classes developed and issued by the main members of the *International Association for Classification Societies*, IACS.The additional requirements are divided into different steps depending on actual ice conditions in the area of operation. The Polar Classes are divided into seven levels, where PC7 is the least capable, limited to vessels operating in summer/autumn in thin first-year ice (with old ice inclusions), whereas PC1 ships are capable of operating year-round in all Arctic ice-covered waters. It is voluntary to include an ice class, but national and port authorities can require an ice class before accepting entrance into some national waters, a port or request for icebreaker assistance.

The Northern Sea Route

Russia regulates shipping along the NSR on the basis of UNCLOS, Article 234 and domestic legislation. Regulations adopted in 1990 and 1996 allow navigation on the NSR on a non-discriminatory basis for ships of all nationalities based on *Guide for Navigation through the NSR 1996* (NSR Guidelines). The guidelines include some technical requirements for the vessels. Basically, ships navigating the NSR must satisfy the applicable *Rules of the Russian Register of Shipping*, but equivalent Baltic and Polar Ice Classes are accepted. A transit fee to cover the expenses for the icebreaker support and administration will be charged, but the sum to pay for an NSR transit is currently under evaluation by the NSR administration. The interest in using the NSR for transit sailing has increased significantly during the last years – from two vessels in 2009 and six vessels in 2010 to 34 vessels in the 2011 summer season. Furthermore, according to the Russian regional minister of Property Relations, Yuri Chuykov, the volume of goods shipped through the NSR is expected to almost double in 2012 (from 111,000 tons in 2011 to 1.5 million tons in 2012).

The Northwest Passage

Canada divides the Arctic area into 16 shipping safety control zones based on the ice conditions in each zone. Canadian maritime jurisdiction is divided into non-Arctic waters and Arctic waters. Shipping in the NWP is primarily governed by the *Arctic Waters Pollution Prevention Act 1970* (AWPPA) and other regulations under this Act, in particular the *Arctic Shipping Pollution Prevention Regulations* 1978 (ASPPR). The ASPPR provides detailed requirements for ship design, construction and operation into the various zones. Canada's ability to support any shipping activity on the NWP is limited because of lack of icebreaking services during the winter season as the Canadian icebreakers are not in operation during the winter. This means that shipping activities on the NWP have to be performed without icebreaker assistance during the winter season.

In 1996, the *Arctic Ice Regime Shipping System* (AIRSS) was introduced as a regulatory standard providing a formula to calculate an Ice Numeral based on known ice conditions in the zone and the Ice Classification of the vessel. The results of the formula will advise the master whether or not to go.

Canada recognizes the need for harmonization and a process to adopt the IACS/IMO polar class rules into their own regulations has been initiated. At the same time, Canada has encouraged other nations to do the same. This means that the CAC system will be outdated in the future.

The Passages in comparison

Generally, one can say that the mandatory technical requirements when operating the transpolar passage are similar to any other international waters. The specification of an ice class is voluntary and up to the owner/operator to decide. The additional requirements for the NSR are more focused on the operational and practical aspects supporting a safe and uninterrupted voyage while the special requirements along the NWP have the environmental aspects to prevent pollution as the main objective.

> *There are today no additional requirements to ice class, equipment or training of crew for vessels entering Arctic waters, except from national requirements for internal waters. Several nations are behind an initiative and see the need for a set of mandatory requirements to ensure future safe shipping in the Arctic.*

7.5.2 Crew training and competence requirements in Arctic waters

The IMO Guidelines are applicable for seafarers operating in Arctic ice-covered waters, including the NSR, NWP and TPP. In addition, the Russians and the Canadians have set down regulatory frameworks applicable for seafarers operating on the NSR and the NWP respectively.

The IMO Guidelines for *Seafarers Operating in Arctic Ice-covered Waters*

The IMO Guidelines require that the ice navigator shall provide documentary evidence of having satisfactorily completed an approved training program in ice navigation. Yet there exists no international recognized model course for ice navigators or qualification scheme for individuals who are likely to operate in ice-covered waters. Moreover, from an operational safety point of view, it is unlikely that any ship would cross the central Arctic Ocean without an ice navigator and a crew trained for such conditions. However, ships crossing the central Arctic Ocean operate in convoys, which means that there will be another ship to support in the event of an emergency incident. Furthermore, the convoys must rely on self-support as the capability of any external rescue is limited. In addition, there are great challenges when it comes to emergency measures, e.g., salvage capability, due to the long distances to the mainland and operative ports offering a rescue service. If an accident should take place in these waters, it would take time for any external rescue unit to arrive.

The Northern Sea Route

The NSR Guidelines require an inspection of every crew member and ice navigator before they can enter the NSR. These control routines are provided by the Marine Operations Headquarters (MOH), one on each side of the NSR. Their areas of responsibility are divided in the Laptev Sea at longitude 125°E. Ships planning to sail from west to east will be referred to the MOH in Dikson, and those planning to sail from east to west must contact the MOH in Pevek. The only requirement imposed on the crew is the size, which must be large enough to guarantee a three-shift watch. Moreover, the NSR Guidelines focus, in particular, on the skills of the ice navigator. A qualified ice navigator must have experience of steering vessels in ice conditions on the NSR for at least 15 days. If this requirement is not met, the MOH could assign an ice navigator to assist the vessel through the NSR. Currently, there are no requirements regarding qualified training courses for an ice navigator or crew operating on the NSR.

The Northwest Passage

The NWP is a three-state route divided between the US, Canada and Denmark. Canada has set down a regulatory framework for seafarers operating on the NWP: the *Arctic Shipping Pollution Prevention Regulations* 1978 (ASPPR) is the Canadian legal framework in terms of training and competence provisions for seafarers operating on the NWP. In order to navigate within any of the 16 shipping zones of the NWP, the navigator must have served on a ship for at least 50 days, of which 30 days must have been served in Arctic waters. However, the ASPPR does not require that ice navigators must complete an approved training course when operating on the NWP. Currently, the most common training practice of ice navigators is accomplished by "on-the-job experience", which means that less experienced navigators learn from experienced Canadian Coast Guard captains and local ice advisors (i.e., person with local knowledge and experience of the area). What happens in most cases is that a local ice advisor is carried on board the ship in order to provide assistance. In addition, the ASPPR requires no formalized training course for the crew.

The Passages in comparison

As demonstrated, the regulations set down for seafarers operating in Arctic waters differ substantially between the three legal regimes.

Firstly, Canadian authorities do not require any inspection of seafarers before they can enter the NWP. Unlike the Canadian NWP regulations, the NSR Guidelines require a mandatory control inspection of every crew member and ice navigator before they can enter the NSR. In this respect, the NSR regulations are stricter than the NWP regulations.

Secondly, there are different qualification requirements for an ice navigator operating on the NSR versus the NWP. Unlike the Canadian NWP regulations, which recommend that an ice navigator must have at least 30 days of experience of ice manoeuvring operations in Arctic conditions on a general basis, the NSR Guidelines require that the ice navigator shall have at least 15 days of experience, exclusively from the NSR. Moreover, an ice navigator with extensive experience from Arctic waters in general is not necessarily qualified for navigation on the NWP, since the transport corridor represents unique geographical areas in combination with various ice characteristics.

Thirdly, there are certain similarities between the three regimes. Neither the Canadian NWP regulations, the Russian NSR Guidelines nor the IMO Guidelines require any certified training courses for crews or ice navigators in order to operate in Arctic waters. From a safety point of view, cold weather course training would be an essential qualification for any crew member in order to do a satisfactory job on board any ship operating in Arctic waters.

To date, the international and national regulations for the seafarers who operate along the NSR, NWP and TPP are not harmonized. In addition, the current requirements for seafarers operating on the Passages are far from adequate for any regular marine traffic.

7.5.3 Search and rescue in Arctic waters

IMO has divided Arctic waters into search and rescue coordination and responsibility areas. Canada is responsible for providing meteorological information, traffic services and coordination of the eastern part of the Bering Strait, the northern slope of Alaska and the NWP. Russia is responsible for the NEP, from the Barents Sea in the west to the Bering Sea in the east. The first international agreement made exclusively for the Arctic region was signed at the ministerial meeting in Nuuk, 12 May 2011. The SAR Agreement is the first legally binding agreement established under the auspices of the Arctic Council and covers search and rescue of aeronautical and maritime vessels and passengers.

The Northeast Passage

Radio communications and satellites are generally adequate for the NSR. Three satellite systems support NSR operations today: the INMARSAT system, the OCEAN system and the IRIDIUM commercial system. In addition, *Vessel Traffic Management and Information Systems* (VTMIS) are used to distribute vessel traffic and marine transport data to improve the safety of ship operations. The VTMIS have been used on the NSR for decades. Furthermore, the Russians now want to integrate new communication services to the system, such as emergency services, that is currently not part of the system. The Russians recently presented plans for development of communications services until 2020, but it is uncertain whether they will receive financial funding. It has also been recognized that the Russian satellites are not always capable of providing Arctic navigation with reliable data due to technical problems.

The compound geography of the NSR in combination with marine traffic requires the installation of various navigational aids at important waypoints (i.e., radio beacons, lighted marks, radar beacons, radar reflectors, etc.). However, the coast of the NSR is relatively low and flat, which makes it difficult to use radar. Thus, it has been necessary to equip the NSR with high-accuracy positioning systems. The *Global Positioning System* is used by ships on the NSR to determine the precise position for civilian users. However, the GPS constellation is not configured for optimal positioning in high latitudes. For higher accuracy, a *Differential Global Positioning System* (DGPS) has been introduced. The plan is to cover all NSR areas with DGPS.

The Russian icebreaker fleet is the world's strongest, in terms of both icebreaking capability and number of ships. So far, Russia has built ten nuclear-powered icebreakers, but only five remain operative and capable of year-round operations on the NSR. In addition, two nuclear-powered icebreakers are operational on rivers (see Table 5.2). Generally, the NSR icebreaker support operations are divided, west and east, between two organizations, the Atomflot headquartered in Murmansk, and the Far East Shipping Company (FESCO), headquartered in Vladivostok. The nuclear icebreaker fleet services the western section of the NSR extending from Murmansk to River Lena as well as river ports on major Siberian rivers. In addition, Russia holds several diesel-powered icebreakers operating on rivers and supports ships entering various ports on the NEP/NSR.

As a coastal state, Norway has obligations under international law with regard to search and rescue services. A number of agreements on search and rescue have been concluded. For example, in 1995, Norway entered into an agreement with Russia on search and rescue cooperation in the Barents Sea being relevant to shipping on the NEP. As a result of this agreement, Norway and Russia carry out joint annual sea rescue exercises

The Northwest Passage

The Canadian Coast Guard operates two seasonal *Marine Communication and Traffic Services* (MCTS) centers, one in Iqaluit and the other in Inuvik (see Figure 5.10) The MCTS program provides marine safety communications and manages the movement of vessel traffic. Currently, MCTS are seasonal, opening at the end of May and closing at the end of November. However, the system's capability to regulate movements of commercial traffic is rather limited.

A binding regulatory vessel traffic zone referred to as the *Northern Canada Vessel Traffic Services Zone Regulations* (NORDREG) has been adapted for the NWP. Coming into force on July 1 2010, the Canadian Government, pursuant to the Shipping Act of 2001, passed the Northern Canada Vessel Traffic Services Zone Regulations, covering Canada's coastal northern waters. More precisely, NORDREG covers the *Shipping Safety Control Zones* referred to above, but it also consists of additional waters.

For most of NWP, the Canadian Military Forces provide SAR aircraft from Trenton, while the CCG relies primarily on its helicopters and icebreakers. Regardless, it is a long way from Trenton to any area of the NWP, which means it would take a long time to provide any rescue operation by air transport (see Figure 5.4). There are very limited modern aids to navigation along the NWP. There is, for instance, no Differential Global Positioning System available. However, there are some marine navigation aids on the NWP, maintained by the CCG, ranging from radar reflectors to buoys, lights, ranges and beacons. However, most of these are for the benefit of community access, not for transits by vessels between the eastern and western NWP.

Every summer, six CCG icebreakers (two heavy, three medium and one light) are deployed to the NWP at the end of June (see Table 5.3). The anticipated longer active shipping season on the NWP raises a number of service level issues for the Government. One of them concerns the development of icebreaker services. Furthermore, the CCG is faced with an aging fleet and most of the ships date to the late 1970s/early 1980s. However, the current Government has committed 720 million Canadian dollars for a new Polar icebreaker, expected to be launched by 2015.

The Transpolar Passage

When considering possibilities for the TPP, there is currently no active vessel traffic service or other traffic management system in place. Shore-based radio communication services are also limited. Future navigation through the TPP will require significant improvements of marine communication technology. Moreover, once the high latitudinal is reached, voice and data transmission become problematic. Satellite systems such as the IRIDIUM constellations provide worldwide coverage, including the Arctic. The feasibility of communicating ice charts and satellite images to ships in the Arctic via the IRIDIUM system has been demonstrated but communications are limited. In addition, the GPS constellation is not configured for optimal positioning in high latitudes, resulting in a potential degradation of position accuracy. This can make marine transportation on the TPP challenging in the future.

The Passages in comparison

Today, there are different levels of developed search and rescue services between the three transport corridors.

Before any vessel can enter the NSR, the Marine Operations Headquarters require a control inspection in west or east. Today, Canada operates a binding reporting system for vessels entering the NWP.

Russia has a considerably larger and stronger icebreaker fleet to support operations on the NSR than Canada has for the NWP. Russia has five nuclear icebreakers available for NSR support and, in addition, several river and diesel icebreakers operating on a year-round basis. Canada operates six icebreakers (two heavy, three medium and one light) for seasonal (May–November) NWP operations. Both fleets are aging and need improvements.

Overall, there are limited aids to navigation and traffic service systems in order to support navigation for all Passages. Available search and rescue facilities diverge between the NEP, the NWP and the TPP. Global Positioning Systems are adequate for the coastal routes and marine routes of the NSR, but on the NWP it is very limited. The GPS coverage is not optimal in high latitudes which means that position accuracy for the TPP is insufficient.

7.5.4 Arctic ports and terminals

The accessibility of Arctic ports and availability of port facilities differ between the NEP and the NWP. Moreover, there are large differences in the extent to which the ports are operative. These include: poor technical standards; lack of sufficient port facilities; inadequate water depth; limited possibilities for icebreaker assistance; and lack of sufficient connecting infrastructure.

The Northeast Passage

There are some 50 Russian ports on the NEP/NSR open to foreign vessels. However, more than half of the ports are out of operation and only a very few can meet the technical standards required by the international shipping industry. Generally, most NEP ports are open on a year-round basis, but need icebreaker assistance in order to be accessible. Currently, the seaport of Murmansk is the only non-freezing port operative on a year-round basis. Unlike most other NEP/NSR allowing maximum ship draft on average only 9–10 m, Murmansk has water depth adequate for most types of ships (see Section 5.4).

The majority of operative ports and terminals are primarily located in the Barents region. Transport of oil from Russia is from ports in the White Sea, Murmansk and the Pechora Sea (Kolguev, Varandey), and Ob and Yenisei estuaries in the Kara Sea. Very few ports on the NSR are reported to be in a satisfactory technical state. The most adequate ports are found in the Kara Sea on the western part of the NSR (including the Ob Gulf and the Yenisei Gulf). Operative ports on the eastern part of the NSR (including the Laptev Sea, the East Siberian Sea, the Chukchi Sea and the Bering Sea) are currently very limited (see Appendix 5). The most significant port and terminal development (construction and investment) on the western NEP over the last 10–15 years is directly connected to the increase of petroleum exports. Intra-Arctic shipping and destination-Arctic shipping to and from Russia follow a route in the open sea to ports and terminals in Murmansk and the White Sea.

The present network of NSR ports covers the basic needs of NSR settlements only. Yet the only NSR port of major economic importance, Dudinka, seems to be maintained at a relatively good standard by the Norilsk Nickel Company. But capacity and equipment for most of the ports has been rapidly deteriorating during the 1990s and the early 21st century, operating below full capacity and with a great need for improvements.

The Northwest Passage

There are no deepwater ports located on the NWP. Moreover, there are very few ports custom-made for the international shipping industry. Main obstacles include natural factors such as draft restrictions, narrow straits and complex ice conditions, and limited infrastructure. The port of Churchill is currently the only port that can be characterized as a

"Canadian Arctic Seaport". This port, however, is not located on the NWP and its NWP relevance is limited as a link in the trans-shipment of resupply to communities in Nunavut and the Kivalliq region. Although the current port infrastructure on the NWP is insufficient, there are plans for improvements in the years ahead. Recently, it has been announced plans to build two deepwater ports, in Iqaluit and Nanisivik. These ports are intended to serve activities such as cruise tourism and the Canadian Coast Guard. In addition, these ports will play an important role in terms of increased intra-Arctic and destination-Arctic shipping in the Nunavut region.

The Passages in comparison

Although there is potential for improvement in several ports on the NSR, the NWP and NSR ports are mainly covering the same functions, namely as a link in the shipment of resupply to local settlements. Still, most plans regarding port improvements and new buildings along both transport corridors in the years ahead are related to an expected increase of destination- and intra-Arctic transport activities. Most significant port improvements are likely to take place on the NEP/NSR.

Currently, there are no deepwater ports located on the NWP. The most adequate ports are found on the NEP. NSR ports need significant improvements in order to meet the standards required by the international shipping industry.

7.5.5 The costs and bureaucratic procedures of Arctic shipping

The Northern Sea Route

The very first step in arranging for an NSR passage is to obtain official permission. The ship owner must submit an application to the Administration of the NSR in Moscow at least four months in advance. For an additional fee, this time frame could be shortened by up to one month. The application must contain a detailed list of information regarding the vessel: technical specifications, registration, the ownership, crew, etc. Within ten days after receiving the application, the NSR Administration will decide. After preliminary approval has been obtained, a representative of the Marine Operations Headquarters must inspect the ship for ice navigation worthiness. The ship owner is expected to bear all costs associated with getting the ship to an agreed port for this inspection. If the ship is found worthy, the MOH will then schedule the date, set up the specific route and provide the necessary icebreaker and pilotage support. Furthermore, if the ship is not in accordance with the requirements, she can still be guided through the NSR for an additional fee.

The fee system is meant to finance several aspects of public services to make the fairways open to traffic and to protect the environment. The main cost elements would include icebreaker support, hydro-meteorological services, satellite communications and pilotage. The philosophy of the Russian government has always been that the tariffs should be set so that the actual traffic finances the cost elements listed above. Moreover, it is difficult to talk about an average fee, as the fee will depend on a number of different factors, such as: type of ship/cargo, size of ship (larger ships have lower fees per ton), ice class of ship, ice conditions (determines how much icebreaker assistance is needed), competences of captain and crew (inexperienced crew will require more pilotage/ helmsmen), and the actual route.

In the mid 1980s, some 6–7 million tons of cargo were carried along the NSR and the icebreaker fees were around US$3 per ton. In the mid 1990s, the traffic had been reduced to around 2.2–2.4 mill tons, so the fee was increased to US$7.50 per ton in 1995, and the traffic continued to decline to less than 2 million tons in 2000. The resulting loss of revenue meant that the state had to subsidize the icebreaker operations quite substantially. In 2003, the Ministry of Economical Development and Trade issued a decree called "About Changing Rates for Ice-breaking Fleet Services on the NSR", which led to a substantial increase in the icebreaker fee to an average of US$23 per ton. However, in 2009, the fee was set to US$40 per ton of container cargo. This level of fees seems quite unrealistically high, and it is probably fair to conclude that fees are negotiable. However, in 2009, the Beluga ships paid only €60,000 in icebreaker fees, which clearly shows that icebreaker fees are negotiable.

The Northwest Passage

There is no fee system for the NWP. The Canadian government monitors the fairway, but there are examples of ships having navigated the NWP without the knowledge of the authorities. The NWP is a demanding route to navigate, so without a formal system for transit passage, a transit here can only be classified as hazardous. If a ship ends up in a distressed situation, the Joint Rescue Coordination Centers will no doubt respond to any SAR incidents. It is highly unclear, however, what the bill for such an operation will amount to. It is unlikely that the Canadian government would introduce any icebreaker fees in the future. Currently, it is very costly to ship cargo on the NWP, and the Canadian government wants to encourage the economic viability for any future transportation.

The Passages in comparison

The current situation for the NSR is that icebreaker fees are negotiable. Until the NSR is established as a regular transit route, the actual fees one will have to pay will be open to negotiation. If the Russian government wants to attract transit traffic through the NSR, the fee levels must be commercially reasonable.

The NEP is the only one of the passages where fees have formally been introduced. The main fee is icebreaker fee for the passage through the NSR. The system of fees has changed several times, where the principle has been that the total traffic should cover the total costs. In a period of declining traffic, this has led to an increase in fees per ton, and figures from 2003 indicate fee levels that will not attract transit traffic. Currently, the situation is non-transparent, and one could conclude that fees are subject to negotiation, which seems supported by the reported fees paid by Beluga in 2009. If the Russian government wants to attract transit traffic through the NSR, the fee levels must be commercially reasonable.

The NWP has currently no fee system and the Canadian government will most likely not introduce any fees as this would discourage the economic viability for any transportation on the passage.

7.5.6 Arctic marine insurance

Marine insurance allows ship owners and carriers generally to take on the risk of trading on Arctic routes. Of particular importance for Arctic shipping is Protection and Indemnity (P&I) insurance, offered through so-called P&I Clubs. P&I covers a wide range of liabilities, including personal injury to crew, passengers and others on board, cargo loss and damage, oil pollution, wreck removal and dock damage. In addition, underwriters normally charge a surcharge with respect to Hull and Machinery (H&M) and Cargo Insurance.

The Northeast Passage

One of the first attempts to identify the insurance risk aspects of the NSR was the *International Sea Route Programme* (INSROP, 1993–1999), followed by the *Arctic Operational Platform* (ARCOP, 2003–2006). INSROP was important in terms of recognizing required risk coverage related to H&M, cargo insurance, P&I and a number of miscellaneous risks. It became clear that the Russian as well as the global marine insurance market was willing and able to underwrite NSR navigational and related risks, provided that appropriate and adequate information structures of the risks involved could be presented to underwriters. However, the research carried out by INSROP identified that the claims on the NSR would be higher than those experienced in southern waters. In particular, wreck removal, pollution salvage and towage, cargo and crew claims were of particular concern. Estimates showed that the total insurance rate would be expensive for the ship owners and almost twice the comparable insurance rate prevailing in shipping via the Suez Canal.

Overall, there was a lack of reliable statistical data and the insurers suggested that initial premiums would be conservative until a better database is provided. In that respect, Russia has not been willing to share much of its official data, for instance on the rate of damages. However, some statistics are available, based on historical data of accidents. Most hull damages, for instance, have occurred in the eastern NSR, principally the East Siberian and the Chukchi Seas, due to its more difficult ice conditions. Of note is also that approximately 10% of ice damage to ships occurs when navigating independently; 60% of damage occurs when following an icebreaker; and the remaining 30% results from towing, grounding and other operations in ice. Furthermore, it is still unclear to what extent the statistics presented by the Russians are valid and significant. They do not state, for example, how many of the operations have taken place during the winter and/or summer season. The underwriters will need more reliable and extensive statistics in order to make available appropriate risk coverage of the NSR. However, the marine insurance industry is generally unwilling to provide information on risk coverage based on hypothetical or theoretical questions. As a result, the provision of insurance tends to be on a case-by-case basis.

The Northwest Passage

The combination of draft restrictions, narrow straits, severe ice conditions and lack of infrastructure are significant risks that will affect the insurance premium in terms of navigation on the NWP. Draft limited to 10 m excludes many of the standard types of ship unless they are very lightly loaded. The deep draft routes via the Prince of Wales Strait (or M'Clure Strait) would be suitable in this sense, but have the most extreme ice conditions in the archipelago. Furthermore, it is likely that navigation on the NWP will continue to be difficult due to the pattern of ice drift from the Central Arctic Ocean sweeping southward through the Canadian archipelago. Previous observations also show a very high year-to-year variability of sea ice coverage in the NWP, an important factor of uncertainty when considering marine insurance. Moreover, the severe ice conditions would be a potential damage to hulls, which means that additional H&M insurance would be required in order to cover the risk. In addition, the P&I cost would most likely be very high for any ship owner that enters the NWP. When considering the availability of icebreaker assistance it seems also quite limited (from June to November). In light of insurance, this will most likely increase the premium. In additions, there is uncertainty related to the official statistics of incidents on the NWP. Furthermore, there are examples of ships having navigated the NWP without the knowledge of the Canadian authorities. Thus, it is difficult for Canadian authorities to provide a complete picture of the NWP's historical accident statistics. There is limited knowledge and research on the marine insurance aspects of the NWP. However, as demonstrated, there are several risk aspects which imply that the insurance premiums will remain high.

The Transpolar Passage

Currently, there has not been any commercial voyage using the central Arctic Ocean as a shortcut. However, if one considers the possibility of any future TPPs, the insurance rate is likely to be higher than for the coastal routes. Higher risks for ice damage to ships and potential damage to cargoes in extreme cold temperatures, and nonexistent maritime infrastructure in the Central Arctic Ocean will most likely be factors in determining future TPP insurance rates.

The Passages in comparison

With the exception of the NSR experience and the research that was provided through INSROP and ARCOP, there is only limited knowledge about the marine insurance aspects of Arctic shipping. Lack of empirical data and significant statistics makes it difficult for marine insurance underwriters to compose an overall risk assessment related to the risks of the various Arctic shipping routes. Even if lots of the risks associated with shipping were well known and understood by insurers, there are still risks related to Arctic navigation that need to be identified. Underwriters normally base their underwriting premiums on statistics and the frequency of accidents is a key element in the evaluation of risk when navigating in Arctic waters. Knowledge of accident rates can help determine insurance rates.

> *The international insurance market is willing and able to underwrite risks in terms of shipping on the three Arctic transport corridors, but to a very different price. On the basis of the risk factors identified in this study, it is likely that the insurance premium will be higher for the NWP and TPP than for the NEP.*

7.6 Ocean law, coastal waters, jurisdictional controversies and indigenous peoples' rights

7.6.1 International ocean law, recommendations and governance

The Arctic Ocean is subject to international ocean law, the central pillar of which is UNCLOS. This convention deals with all aspects of ocean law.

Each coastal state has internal waters where the state jurisdiction equals that of the land territory. For practical purposes, this is also the case within territorial waters extending 12 nm from the baseline, but here UNCLOS provides for free passage of ships. A coastal state could also proclaim a contiguous zone for an additional 12 nm, for the purposes of controlling immigration, customs, etc.

In addition, coastal states could proclaim an Exclusive Economic Zone (EEZ) extending up to 200 nm from the baseline and with exclusive rights to resources on the continental shelf. Article 234 of UNCLOS gives the states certain rights to regulate activities within the EEZ in order to prevent, reduce and control pollution from vessels navigating in what are considered particularly vulnerable waters. This is achieved by giving coastal states a broader right to control shipping in their EEZs than is otherwise inherent in this regime. Over the last decades, focus has also been steadily growing on corporate governance and corporate social responsibility. This influences corporate behavior, especially in relation to indigenous peoples' rights, environmental issues, and human safety. It is a relevant point that what could be termed "self-restraint in self-interest" has been coming more and more to the fore, at least for major businesses. Further, the growing focus on this issue makes commercial actors much more competent when it comes to compliance with regulations applicable to their activity.

The Arctic has certain special features: It is partly ice covered, it is environmentally vulnerable and is subject to rapid climate changes. It is therefore not surprising that the question of a special Arctic treaty has been raised. The European Parliament and WWF have both suggested variants of this. But all the Arctic states, as well as the European Commission, consider it more feasible to build on the existing international regime and it is very unlikely that a special treaty for the Arctic will be adopted any time soon, or if at all.

The Northeast Passage

The NEP runs between the northern coastline of Norway and Svalbard, but predominately along the coast of the Russian Federation. Coastal zones and national legislation of these two Arctic states thus constitute the framework for sailing along this sea route, as do controversies regarding national jurisdiction in these waters.

Norway's coastal regime covers the EEZ and continental shelf. The Shipping Safety Act of 2007, the Pollution Act of 1981 and the Maritime Act of 1994 together form a national regime, with administrative bodies empowered for their implementation and enforcement. Special rules for Svalbard apply to Svalbard itself and its territorial waters.

Russia's coastal regime also covers the EEZ and continental shelf proclaimed within generally-accepted ocean law. More controversially, Russia has a very expansive policy and legal doctrine regarding its coastal jurisdiction. As discussed in Chapter 6, a full implementation of the Sector theory would extend some sort of national jurisdiction far beyond the EEZ, in fact all the way to the North Pole, limited to lands and islands, but possibly even to sea ice. Russia has never passed laws that apply beyond the EEZ and/or continental shelf but, on the other hand, has never left the Sector theory as a possible position.

There is a longstanding, but not very intense, dispute between Russia and the US related to the ice-covered straits of Russia's Arctic Archipelago. The Russian position is that these waters are internal waters. The US is the only country challenging this explicitly, and contends that the straits are international and that transit passage must be allowed.

Within the EEZ, Russian legislation is extensive when it comes to regulating sailing in the Northern Sea Route, the basis being the *1990 Regulations for Navigating on the Seaways of the Northern Sea Route*. Russia is expansive regarding the geographical scope of national jurisdiction and also expansive when it comes to the content, although emphasizing that it acts with the limits of UNCLOS.

The Northwest Passage

When sailing along the various routes of the NWP, ships pass through national waters of the three Arctic states, Greenland/Denmark, Canada and Alaska/USA. The national legislation of the US is limited in line with her policy of stressing free passage, which is seen as pivotal for national security reasons.

The most important legislation pertaining to the EEZ outside Greenland is the *Danish Act on Protection of the Marine Environment of 1993*.

Much as has been said about Russia for NEP, Canada has also been expansive regarding national legislation applicable to the NWP. Along this passage there are jurisdictional controversies relating to the internal waters of Canada, resembling the controversy over the NEP. The jurisdictional controversy relating to the NWP is whether Canada's straight baseline and, thus, what constitutes Canadian internal waters are in conformance with international ocean law. The waters of the entire Canadian Arctic Archipelago are considered internal waters by Canada. This has been challenged by the US, which expresses the opinion that these waters fall under the regime of international straits. Newly-proposed rules on navigation which, *inter alia,* call for mandatory reporting requirements before entering give Canadian authorities wide authority in implementation and enforcement, could stir further controversy.

The Transpolar Passage

Unlike the NEP and the NWP, the TPP is international waters and outside geographical reach of national jurisdiction. This has not officially been challenged by Russia or Canada with basis in the Sector theory.

Since no national legislation exist, the IMO Guidelines have become central, and perhaps for the TPP the discussion of a special Arctic treaty is of most relevance. As mentioned above, the Arctic states' preference is to build on existing international ocean law and to work within the IMO as well as regionally. The importance of good corporate governance and corporate social responsibility also increases in lieu of national legislation.

The Passages in comparison

All the Passages are subject to the rules set out first and foremost in UNCLOS and the other international conventions mentioned above, as well as the recommended IMO Guidelines and norms for good corporate governance and corporate social responsibility. The most fundamental difference is that sailing through the NEP and the NWP takes place within the Arctic states' coastal zones and internal waters, while sailing through the TPP takes place in international waters. The lack of national legislation, except that of flag states, increases the impact and necessity of good corporate governance and adherence to the IMO Guidelines, comprehensive laws for navigation, protection of the marine environment, etc. This is in no way meant to say that the NEP and the NWP are similar. There are significant differences in the legal and administrative cultures and traditions between the Arctic states. Russia is by far the dominant country as far as the NEP is concerned, as Canada is for the NWP.

There are two jurisdictional disputes relevant to shipping in the Arctic. In the NWP there is a dispute between Canada and the US regarding the outer limit of Canada's internal waters. As defined by Canada, its internal waters comprise the entire waters of the Canadian Arctic Archipelago. The US opinion is that these waters falls under the regime of international straits. In the NSR, there is a dispute between the US and Russia related to whether ice-covered straits in the Russian Arctic Archipelago are Russian internal waters or international straits.

The entire Arctic Ocean is subject to applicable international ocean law, most prominently UNCLOS, but also SOLAS, STWC and MARPOL. In addition, norms for good corporate governance and the IMO Guidelines for Arctic shipping will influence how shipping is carried out. The NEP and NWP are in addition governed by national legislation, most importantly passed by Russia (NEP) and Canada (NWP). In the NWP there is a jurisdictional controversy between Canada and the US concerning the status of the waters in the Canadian Arctic Archipelago – internal waters (Canada) or international strait (the US) – and in the NEP there is a similar dispute between the US and Russia pertaining to Russian ice-covered straits – internal waters (Russia) or international straits (the US).

7.6.2 Indigenous peoples' rights

Facing an unprecedented combination of rapid and stressful changes involving both environmental forces like climate change and socioeconomic pressures associated with globalization," indigenous peoples in the Arctic have felt a need to safeguard their culture and traditional way of life. As can be seen from Figure 6.4, indigenous peoples in the Arctic represent a great variety of ethnic groups. In the Russian Federation there are all together 26 minority groups, with the Nenets, the Chukchi, the Evenks and the Events as the largest. The Sami are found in the Russian Federation, but with the largest group living in the northern part of Norway. In Greenland, the major part of the population is Inuit, a group constituting a large part of the Aboriginal peoples in Canada together with the Natives and the Métis. In Alaska, the natives include various groups such as the Inupiat, Yupik and Aleut.

Notwithstanding cultural and ethnographic differences, indigenous peoples in the Arctic share a common history of attempts at assimilation into their various mother states and a lack of recognition of rights on traditionally-occupied territories. Against this background, these peoples have united in working steadfastly towards self-determination. Their primary objective has been to ensure that development and activity on traditional lands and waters take into consideration indigenous peoples' traditional way of life, as well as their fair share of the economic benefits of development in these territories.

Indigenous peoples in the Arctic have played an important role in establishing two international documents dealing with their right to self-determination. These are the *ILO Convention no 169 of 1989*, which in the Arctic has been ratified only by Norway and Denmark, stating that national governments have a duty to implement measures recognizing indigenous peoples' right to ownership and possession over lands they traditionally occupy, and the *United Nations Declaration on the Rights of Indigenous Peoples of 2007*, which goes far in stating indigenous peoples' right to own, use, develop and control the lands, territories and resources that they possess by reason of traditional ownership or other traditional occupation or use.

It goes without saying that these international documents have no bearing on individual shipping companies, obliged only by state jurisdiction in the waters they sail or by own flag states regulations. As the international standing and support of these documents is widespread and strong, they can be said to express "good governance" in regard to indigenous peoples and thus to have a bearing as such.

Indigenous peoples' second strategy has been to work with their respective national governments to have the rights acquired by indigenous peoples in the various Arctic states laid down in national legislation or in land claims agreements. These would then form the legal and political framework within which other activities and actors, such as the shipping industry, will have to operate.

The Northeast Passage

Along the NEP, indigenous peoples are found in local communities along the coasts of Norway and of the Russian Federation. As indicated, the majority of the Sami live along the coast of Norway, but most of the indigenous peoples along this sea route are of course what are often defined as minority groups of the Russian North, living in the Russian Arctic and Sub-Arctic.

As for the Sami in Norway, the Norwegian Government had initially argued that Sami use of land and resources did not establish any formal rights. This was, however, later modified, not least due to the *ILO Convention 169*, which Norway was the first to ratify. Today, the *Finnmark Act of 2005* forms a legal and organizational framework for the managment of natural resources in Finnmark, taking into consideration the particular rights acquired by the Sami as an indigenous people in this area. The Sami Parliament has thus been given a rather significant role in the Finnmark Estate, the owner of approximately 95% of Finnmark's territories. Further investigations are now being conducted in order to clarify ownership to lands in Finnmark by virtue of historic title by Sami or other inhabitants of Finnmark. Even though applying to all inhabitants, this is assumed to relate primarily to the Sami population.

Sami are found also in the Russian Federation, constituting a minor indigenous group. There are, however, as indicated initially, an additional 25 minority groups in the Russian North. Indigenous peoples in this part of Russia have been subject to different forms of governance. While the first years after the Russian revolution was characterized by autonomy, the late 1920s saw the beginning of a forced integration into the Soviet society, destroying the former historical and ecologically-balanced settlements. Adding to this transformation was that the development of natural resources stimulated newcomers who soon outnumbered the indigenous population. The democratization of Russia from the late 1980s saw a development of self-consciousness among indigenous peoples of the North and the situation in the last decade of the 20th century has been described as "demonstrating a willingness to correct the mistakes of the previous assimilation policy". Russia's Constitution of 1993 has been amended so as "to guarantee the rights of the indigenous small peoples".

Other laws and regulations regarding indigenous rights and territories were passed in the late 1990s/beginning 2000s, but as hardly any has been translated, it is not possible to get the full picture of self-determination for indigenous peoples in Russia. According to indigenous peoples themselves, the support by the Russian Government at the end of the 1980s gave way to opposition towards indigenous self-government. It was maintained that "[i]n many regions strong opposition by the authorities is manifested in complete disregard for indigenous peoples and violation of their lawful rights and interests", and also that Government representatives have been "trying to continue policies of paternalism toward and control over indigenous peoples".

It may seem that the policy towards the northern regions of Russia is still not fully settled. Power and legal authority have no doubt been shifted to the regional level in the form of autonomous governments, but this does not necessarily mean increased self-determination for the indigenous peoples. In some instances, they have ended up being a minority within the new governments and thus might have little say in control of management of lands and waters. It could be that the conclusion of a study of the Nenets Autonomous Okrug – namely that, despite legal guarantees, the indigenous people seem to have no opportunity to change major politically-approved decisions – goes for other minority groups in the Russian Federation as well.

The Northwest Passage

Sailing in the NWP means passing through waters of the three Arctic states Greenland, Canada and the USA, particularly characterized by the many islands and inlets constituting the Archipelago of the Canadian waters. In these waters, indigenous peoples thus to a large extent depend on the sea and sea ice for their traditional hunting and fishing as well as for their supply of fuel and other materials.

Of these peoples, the Inuit, constituting the major population in Greenland, is in a rather unique position as a new Act on Greenland Self-Government largely extends the fields over which the Greenlanders themselves can decide to resume legislative and administrative power. As Greenland still depends on a major economic subsidy from the Danish Government, the ultimate objective of the Self-Government authorities is to gain economic self-sustainability. As the main industry of today, fishery, does not constitute a sufficient source of income, the Self-Government has to find new sources of income. One such source is marine mineral resources. Licenses have been issued for exploration and exploitation in the waters outside south-western Greenland. The areas considered most promising is outside East Greenland where conditions are much more challenging, but some prospecting licenses have been issued in these waters as well. Another source of income is mining, and there are today more than 70 licenses, mostly issued to foreign companies, for exploration and exploitation of various minerals in Greenland, and another 13 licenses for prospecting.

It goes without saying that these are industries that create a need for shipping both to bring relevant equipment to Greenland and to export the products. It is perhaps particularly interesting in regard to shipping to note that precisely such expanding economies as China and India have expressed an interest for minerals found in Greenland. Also related to shipping is, of course, the third means of extended income, namely the goal of furthering tourism in Greenland which will imply more sailings to and from the island.

Having also a focus on Greenland's traditional culture and way of life, it seems that the challenge for the Self-Government will be to balance the goal of economic self-sustainability with the preservation of its culture and traditions. Of importance is that the decision is in the hands of the Inuit themselves.

The Inuit in Canada constitute the largest aboriginal group in the country's North, and have lived in these areas for thousands of years. They have described their entire culture and identity as being based on free movement on sea and sea ice for hunting and fishing. Challenging this free movement is defined as the thinning of ice and shipping. Concern in regard to ship traffic relates to marine pollution, but also to hunting on ice when ships passing through are "wrecking the area". Shipping in open water is seen as a threat to hunters in small boats. Skepticism has also been expressed in regard to tourist ships, especially when they arrive in large quantities. Ship traffic is, however, also seen in a positive context, acknowledging that shipping offers many benefits for local communities such as a more effective distribution of goods and material and by creating more jobs in construction and port activities.

When it comes to the Inuit's self-determination and thus a say on lands and waters they traditionally occupy, the majority of aboriginal peoples in northern Canada are signatories to Land Claims and Self-Government Agreements. Each agreement reflects local and regional circumstances, but usually have in common that they provide ownership of 10–30% of the settlement area, cash compensation, economic development opportunities, a share of royalties from development of Crown land and natural resources as well as management participation of land, water and wildlife. Included also are means to support self-government. These agreements have been said to make aboriginal governments in the territories "key players in local and regional economies". There are, however, problems when it comes to fulfilling obligations laid down in these agreements, not least when it comes to achieving objectives that require cooperation between aboriginal peoples and the Federation of Canada. It seems what is needed is a formal land claims agreement implementation policy on the part of the Canadian Government.

Aboriginal peoples are, however, described as gradually bettering their position when it comes to having a share in the resource revenues from development on their traditional lands. Challenges pointed out in respect of economic development are: limited infrastructure, high transportation costs and remoteness from primary markets which are all aspects related to shipping.

It was not until 1915 that legislation was passed making native peoples of Alaska citizens of the United States, the condition being that they had "severed all tribal relationships and adopted the habits of civilized life". This line of assimilation was later rejected, and the natives were urged to adopt constitutions for self-government under the so-called *Indian Reorganization Act* (IRA). About 70 so-called IRA villages were set up, and still exist as one type of native authority recognized by the Federal Government. Urged on by discovery of oil in Prudhoe Bay, an *Alaska Native Claims Settlement Act* was reached in 1971, whereby Alaska natives received 44 million acres of land, representing approximately one-ninth of the original claim, and were paid $962 million in cash, transforming native land claims in Alaska into a system of more than 200 villages and 12 Alaska Natives Regional Corporations. The Settlement implied that, while the natives' earlier use and occupancy of land had made them co-owners of shared lands, they had now become shareholders in corporate-owned land.

The Passages in comparison

A common denominator for indigenous peoples living along the NEP as well as the NWP is that climate change, in making the Arctic more accessible, implies changes impacting on their traditional way of life. Also sharing a common history of attempts at assimilation into their various mother states and a lack of recognition of rights on traditionally-occupied territories, indigenous peoples of the Arctic have united in working steadfastly towards self-determination. Their primary objective has been to ensure that development and activity on traditional lands and waters take into consideration indigenous peoples' traditional way of life, as well as their fair share of the economic benefits of this development. Today, such rights have been acquired in all the Arctic states in national legislation, and in land claims agreements and settlements. There are thus important similarities between indigenous peoples along the two passages.

What differs is in what form, to what extent and at what pace these rights have been acquired. In general, one might say that what dominates the Northeast Passage are various minority groups in the Russian Federation with rights guaranteed in Russian legislation, but where it is hard to get the full picture of the extent of such rights. As for the Northwest Passage, it is dominated by the Aboriginal peoples of Canada having entered into rather far-reaching land claims agreements, but where the implementation of these agreements is not without problems.

As for shipping, ship traffic could on the one hand impact on indigenous peoples' traditional way of life when going through areas used for hunting and fishing or causing cultural disturbances. On the other hand, shipping is no doubt an integral part of the economic development in the Arctic, welcomed also by indigenous peoples.

Indigenous peoples in the Arctic have worked steadfastly towards self-determination, the objective being to preserve their own culture and way of life as well as having a share in the resource development on territories traditionally occupied. In such development, ship traffic could on the one hand impact on indigenous peoples' traditional way of life, when going through areas used for hunting and fishing or causing cultural disturbances. On the other hand, shipping is no doubt an integral part of the economic development in the Arctic, welcomed also by indigenous peoples.

7.7 Commercial comparisons of Arctic transit routes

The purpose of this section is to look at the commercial aspects of using the Arctic routes. We do that by comparing the use of the NEP, the NWP and the TPP to a similar trip using the Suez Canal. There are many ways one could make comparisons, but the main three usual approaches are:

- calculating the total transportation costs for using each route to obtain a $/ton cost estimate for each route;
- calculating the total cost of setting up a regular service based on an assumed yearly quantity to be shipped;
- focusing on cost differences among the route alternatives.

The first approach focuses on cost savings for the end user of the transport. The second takes into consideration that saving time makes it possible to service a given amount of cargo with fewer vessels (trips) and thus takes into consideration the capital costs of investing in vessels. The third approach is more common in a first commercial feasibility study, i.e. the order of magnitude of cost savings are explored to get a feeling for cost differences and to be able to do simple sensitivity tests.

Since we are considering a hypothetical future where ice conditions have greatly changed in the Arctic, the starting point for a comparison will be to assume Arctic transit *without* icebreaker support. By then calculating the cost advantage of the northerly routes vs. a southern route through the Suez Canal, the implicit willingness to pay for icebreaker support will also be given.

We will therefore use the third approach and look at the main cost components where there will be differences. The comparison would be relevant for a ship owner with a given ship and a choice of which route to sail.

7.7.1 Benchmark Route 1: General cargo ship from Yokohama to Hamburg via Suez

We choose a ship with the same characteristics as the Beluga Fraternity.[1] The main data for the ship are given in Table 7.1. The route-specific data are given in Table 7.2.

The Suez Canal toll has been calculated on the basis of the calculator provided by the Suez Canal authorities,[2] using current exchange rates for SDR[3]/US$.

The insurance figures are based on figures from Drewry (2007),[4] where yearly figures have been converted to $/day figures. The comparison of insurance costs is, however, a tricky one. Currently, the insurance costs for ships passing the Gulf of Aden towards Suez have soared since 2008 due to the risk of piracy. It is claimed that the insurance has increased tenfold for this coastal area between September 2008 and March 2009.[5] If that situation persists, it will increase the advantage of Arctic passages. It is claimed this is one important motivating factor for China's increased interest in the Arctic.

Since we deal with a future scenario, we will disregard this piracy/insurance problem, but keep in mind that this could be a commercial argument in favor of Arctic passages.

7.7.2 Benchmark Route 2: Container ship from Shanghai to Hamburg via Suez

As a benchmark ship we have chosen a Chinese container vessel with a capacity over 4000 TEU,[8] called aptly CSCL Hamburg. Main ship data are given in Table 7.3. The route-specific data are given in Table 7.4.

7.7.3 Northeast Passage comparisons

General cargo ship Yokohama–Hamburg via the Northern Sea Route

There are three parameters that will change if the general cargo ship decides to go via the NSR:

1. The distance (which will affect total bunker consumption).
2. The speed (expected speed reduction during NSR passage)
3. The insurance costs.

For the moment, we will disregard icebreaker costs, as we assume that in the future the NSR can be navigated without icebreaker support. We could include the additional cost of an ice navigator, but this is a minor cost element in the big picture.

The distance will depend on which route is taken through the NSR. Since the ship has a draft of 8 m, we will assume it will go a fairly short route of 2200 nm. The total distance will then be 7200 nm, or a reduction of almost 40%, as given in Table 7.5. We will as-

Gross tonnage GRT	9611
Net tonnage NRT	4260
Deadweight ton DWT	12,672
Suez Canal net tonnage SCNT6[2]	12,915
Draft in meters	8
Service speed in knots	14
Gram fuel per kwh	190
Power in kw	5400
Ton fuel per day at service speed	24,624

Source: www.sea-web.com

Table 7.1 General cargo ship characteristics.

sume that the average speed is reduced somewhat during the NSR passage to 12 knots on average. This will reduce the fuel consumption on this leg. We will further assume, without any particular justification other than the assumption, that although one could pass without icebreaker support, there might still be drift ice on this leg, so hull insurance will increase. We just assume it is tripled compared to the benchmark route. The results of these assumptions are summarized in Table 7.6.

It should be clear from table 7.6 that the main savings from using the NSR are the reduction in fuel consumption in addition to cutting sailing time from 34 to 23 days. Fuel consumption is reduced by some 20%. How much this is worth in US$ will of course depend on the oil price.

Currently the price of low sulphur heavy fuel in Hamburg is $465 per ton, while diesel oil is $695[9]. With the much stricter regulations for sulphur contents in bunker oil coming into effect in 2020[10] it could be that in the future more ships will be using diesel fuel, which currently is 50% more expensive than heavy fuel. At the current price of $465, the savings

Distance in nautical miles (nm)	11,430
Journey days at service speed	34
Fuel consumption in tons	838
Suez canal toll in US$	51,168
Hull and machinery insurance $/day	250
P&I[7] insurance, $/day	200

Table 7.2 Route-specific data, Yokohama-Hamburg via Suez.

Gross tonnage GRT	39,941
Net tonnage NRT	24,458
Deadweight ton DWT	50,790
TEU	4253
Suez Canal net tonnage SCNT	57,387
Draft in meters	12.6
Service speed	23
Gram fuel per kwh	190
Power in kw	36,515
Ton fuel per day at service speed	166.50

Source: www.sea-web.com

Table 7.3 Container vessel data.

Distance in nautical miles (nm)	10,857
Journey days at service speed	20
Fuel consumption in tons	3275
Suez canal toll in US$	135,145
Hull and machinery insurance $/day	600
P&I insurance, $/day	330

The sources are the same as for table 7.2

Table 7.4 Route-specific data Shanghai–Hamburg via Suez.

Yokohama to the Bering Strait	2700
Bering Strait to Novaja Zemlja	2500
Novaja Zemlja to Hamburg	2000
Total	7200

Table 7.5 Distances in nm Yokohama–Hamburg via the NSR.

in our example is $160,300 or more than 15 times the increased insurance costs in order of magnitude. In addition comes the savings of the Suez Canal toll of $51,168, so the total savings amounts to about $200,800. Figure 7.1 shows the development of bunker prices since the turn of the millennium. It will be pure speculation how future bunker prices will develop, but the general sentiment is that they are more likely to be higher than today than lower.

The reduction in bunker consumption will also reduce emissions of CO_2. More use of Arctic passages would, *ceteris paribus*, contribute to more sustainable transport.

A saving of around $200,800 for a ship of almost 13,000 dwt implies that the willingness to pay for icebreaker assistance is limited. A fee of $16 or more per ton will cancel out the cost saving effect. In Section 5.5.1 it was indicated that the icebreaker fee for carrying mechanical engineering products (which seems relevant for this ship type) was $74 in 2003. This is clearly unrealistic from a commercial point of view, as it would imply a cost almost twice that of the Suez Canal toll.

Container ship Shanghai–Hamburg via the Northern Sea Route

We assume the same sailing distance in the NSR of 2500 nm. In addition comes the increased distance from Shanghai to the Bering strait compared to Yokohama–Bering Strait of 814 nm. The Arctic route is thus 8234 nm, or a reduction of 24%.

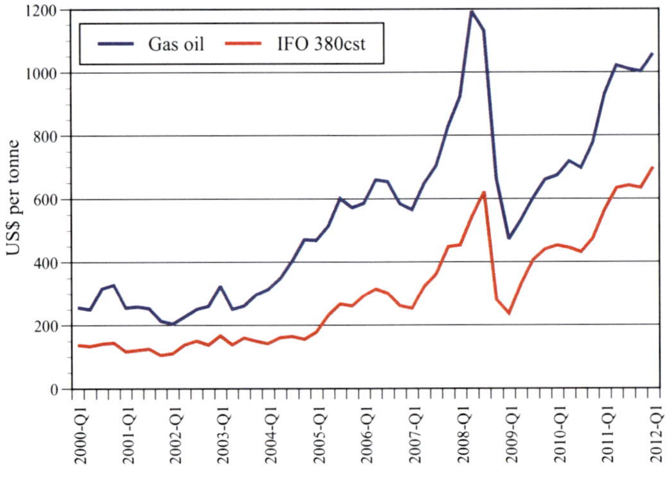

Source: Clarkson SIN - gas oil: Genoa, IFO: Rotterdam

Figure 7.1 Bunker price developments

There is no way a container ship can go through the NSR in 23 knots if there is any ice there at all, so we assume that the average speed through the NSR is 14 knots. This will substantially reduce the bunker consumption and we have used the Admiralty formula as explained in a previous footnote to calculate the consumption. The results are summarized in Table 7.7.

If we again use the April 2010 price of low sulfur-heavy fuel oil of $465 per ton, the fuel cost savings are $606,000. The total cost savings for the container ship sums to $720,750, or $169 per container. The saving in sailing time is only two days, however.

The NSR Administration assumes that a loaded container weighs 24 tons. If that were the case, our container ship would only be able to carry around 2100 TEU or half its TEU capacity. If we then, more realistically, assume an average weight for a loaded container of 11 tons, our ship fully loaded would be willing to pay around $15 per ton for eventual icebreaker assistance. The NSR Administration stipulated a fee for container cargo of $31 per ton in 2003.

This comparison might not be very realistic. Because of the combination of high bunker prices and overcapacity in the container trades, hardly any container ship moves at service speed at the time of writing. The main operators have all introduced super slow-steaming schemes to save on fuel consumption. By reducing the speed from 23–24 knots down to 16 knots, the round trip from the Far East to Europe increases from 63 days to 84 days (port time included). In order to maintain frequency, the operators must use more ships on a specific service. With high bunker prices, the fuel savings are large enough to make it worthwhile employing more ships. One could therefore say that the comparison made here is relevant in a market situation with little excess TEU capacity. Our example looks at a single trip, but in the case of container traffic a total service view would make more sense commercially, but this is beyond the scope of this simplified feasibility exercise.[12]

Distance NSR in nm	2500
Distance outside NSR in nm	4700
Speed in NSR in knots	12
Speed outside NSR in knots	14
Fuel consumption at 12 knots tons/day[11]	15.5
Days in the NSR	9
Days outside NSR	14
Total days	23
Fuel consumption in the NSR in tons	135
Fuel outside NSR in tons	344
Total fuel consumption in tons	479
Fuel consumption reduction in tons	345
Increased insurance costs in $	10,600

Table 7.6 General cargo ship Yokohama–Hamburg via the NSR

Distance NSR in nm	2700
Distance outside NSR in nm	5534
Speed in NSR in knots	14
Speed outside NSR in knots	23
Fuel consumption at 14 knots tons/day	37.6
Days in the NSR	8
Days outside NSR	10
Total days	18
Fuel consumption in the NSR in tons	302
Fuel outside NSR in tons	1669
Total fuel consumption in tons	1971
Fuel consumption reduction in tons	1304
Increased insurance costs in $	20,709
Saved Suez canal toll	135,145

Table 7.7 Container ship Shanghai–Hamburg via the NSR.

7.7.4 Northwest Passage comparisons

General cargo ship Yokohama–Hamburg via the Northwest Passage

As described in Chapter 1, there are basically seven different routes to choose through the NWP and the total distance will vary with the actual choices. Since the purpose of this exercise is just to get a feeling for the overall proportions, we will not consider any exact route, but just assume that the distance from the Bering Strait to the southwest coast of Greenland around the Davis Strait is about 3000 nm. The total round trip would then be as indicated in Table 7.8.

If we assume, as in the NSR comparison above, that the ship must reduce speed on the second leg and that the average speed will be 12 knots, then the results are summarized in Table 7.9.

Again using a fuel price of $465 per ton, the savings on fuel amounts to $137,600, and the number of days is reduced from 33 to 25. The total savings would be $178,100 or about $14 per dwt. The NWP thus seems slightly less attractive than the NSR alternative, *ceteris paribus*.

Container ship Shanghai–Hamburg via the Northwest Passage

The total distance when using the NWP on the Shanghai–Hamburg route, is given in Table 7.10. Again we assume slow-steaming on the second leg down to 14 knots and the results are summarized in Table 7.11.

At a fuel price of $465 per ton the fuel savings amount to $595,500 and total savings are about $710,000, or about $167 per TEU, only slightly less than for the NSR alternative. The saving of time is only one day compared to the Suez alternative.

7.7.5 Transpolar Passage comparisons

General cargo ship Yokohama–Hamburg Transpolar

The transpolar alternative seems a little far-fetched today. We will have to see substantial changes in global affairs before the Central Arctic Ocean becomes navigable without icebreaker support. We will therefore assume more ice in this alternative than the others, reflected in even slower speeds on the Central Arctic Ocean leg. The cost savings will then be converted to willingness to pay for icebreaker support. Since the Central Arctic Ocean belongs to the High Seas, no single country will ever set up a support infrastructure for the TPP, but one could assume that some private business could offer icebreaker support on commercial terms, so the calculations will indicate the order of magnitude of potential revenues for such a service. We assume distances for this alternative as in Table 7.12.and the results are summarized in Table 7.13.

With a fuel price of $465 per ton, the fuel savings amount to about $196,500 and the total savings are $237,000, which corresponds to about $19 per dwt. If a company were to charge, say, $15 per dwt for icebreaker support, it will correspond to almost $20,000 per day. This is hardly enough to attract any private investor into setting up an icebreaker support company.

Yokohama to the Bering Strait	2700
Bering Strait to the Davis Strait	3000
Davis Strait to Hamburg	2300
Total	8000

Table 7.8 Distances in nm Yokohama–Hamburg via the NWP.

Distance NWP in nm	3000
Distance outside NSR in nm	5000
Speed in NWP in knots	12
Speed outside NWP in knots	14
Fuel consumption at 12 knots tons/day	15.5
Days in the NWP	10
Days outside NWP	15
Total days	25
Fuel consumption in the NWP in tons	161
Fuel outside NWP in tons	366
Total fuel consumption in tons	528
Fuel consumption reduction in tons	296
Increased insurance costs in $	10,600
Saved Suez canal toll	51,168

Table 7.9 General cargo ship Yokohama–Hamburg via the NWP.

Shanghai to the Bering Strait	3200
Bering Strait to the Davis Strait	3000
Davis Strait to Hamburg	2300
Total	8500

Table 7.10 Distances in nm Shanghai–Hamburg via the NWP.

Distance NWP in nm	3000
Distance outside NWP in nm	5500
Speed in NWP in knots	14
Speed outside NWP in knots	23
Fuel consumption at 14 knots tons/day	37.6
Days in the NWP	9
Days outside NWP	10
Total days	19
Fuel consumption in the NWP in tons	335
Fuel outside NWP in tons	1659
Total fuel consumption in tons	1994
Fuel consumption reduction in tons	1281
Increased insurance costs in $	20,709
Saved Suez canal toll	135,145

Table 7.11 Container ship Shanghai–Hamburg via the NWP.

Yokohama to Bering Strait	2700
Bering Strait–Svalbard TPP	2300
Svalbard–Hamburg	1600
Total	6600

Table 7.12 Distances in nm Yokohama–Hamburg via the TPP.

Container ship Shanghai–Hamburg Transpolar

We assume distances for this alternative as in Table 7.14.

The results are summarized in Table 7.15. With the same fuel price of $465 per ton, then the fuel savings amount to $761,800 and total savings are $876,200, or $206 per TEU. An icebreaker fee of, say, $175 per TEU would imply a daily income of around $93,000/day for the support provider, almost five times that of the general cargo ship case. One could then imagine convoys of ships being supported by icebreakers that will make a private icebreaker support service something to be investigated out there in the far future.

7.7.6 Some comments on some of the assumptions

This numerical exercise is based on a set of assumptions. Some of these assumptions could

Distance TPP in nm	2300
Distance outside NSR in nm	4300
Speed in TPP in knots	10
Speed outside TPP in knots	14
Fuel consumption at 10 knots tons/day	9
Days in the TPP	10
Days outside TPP	13
Total days	22
Fuel consumption in the TPP in tons	86
Fuel outside TPP in tons	315
Total fuel consumption in tons	401
Fuel consumption reduction in tons	423
Increased insurance costs in $	10,600
Saved Suez canal toll	51,158

Table 7.13 General cargo ship Yokohama–Hamburg TPP.

Shanghai to the Bering Strait	3200
Bering Strait to Svalbard	2300
Svalbard to Hamburg	1600
Total distance	7100

Table 7.14 Distances in nm Shanghai–Hamburg via the TPP.

be debatable. This is clearly the case with the stipulated distances on the three passages. We have assumed that in the distant future ships might be able to navigate these passages without icebreaker support. That does not mean that the passages will be free of ice. A ship observing large ice floes will try to avoid them and navigate around them. The actual sailing distance could then easily increase so much that the advantages are eroded. This is particularly the case for transpolar routes. This should be kept in mind when interpreting the results.

The average speed through the Passages are rather arbitrarily set, and one could argue that if the general cargo ship is going at 12 knots, how come the container vessel could do 14? The reasoning is simple – there will be stretches without any ice at all, and then the container vessel can speed up more than the general cargo vessel, so the average speed will be higher. The actual numbers chosen are still without any empirical foundation.

Distance TPP in nm	2300
Distance outside TPP in nm	4800
Speed in TPP in knots	12
Speed outside TPP in knots	23
Fuel consumption at 12 knots tons/day	23.6
Days in the TPP	8
Days outside TPP	9
Total days	17
Fuel consumption in the TPP in tons	189
Fuel outside TPP in tons	1448
Total fuel consumption in tons	1637
Fuel consumption reduction in tons	1638
Increased insurance costs in $	20,709
Saved Suez canal toll	135,145

Table 7.15: Container ship Shanghai-Hamburg TPP.

	NEP	NWP	TPP
General cargo ship Yokohama–Hamburg	42%	36%	51%
Container ship Shanghai–Hamburg	40%	39%	50%

Table 7.16 Fuel consumption reductions when using the Arctic passages.

	NEP	NWP	TPP
Days saved	11	9	12
Fuel cost savings for fuel price of $465 per ton, $	160,300	137,600	196,500
Total savings, $	732,200	178,100	237,000
Total savings per dwt, $	16	14	19

Table 7.17 General cargo vessel Yokohama–Hamburg (nil icebreaker fees).

	NEP	NWP	TPP
Days saved	3	1	3
Fuel cost savings for fuel price of $465 per ton, $	616,700	595,500	761,800
Total savings, $	732,200	710,000	876,200
Total savings per container, $	172	167	206

Table 7.18 Container vessel Shanghai–Hamburg (nil icebreaker fees).

We have mechanically used the Admiralty formula when calculating the fuel consumption effect of slow-steaming. This is at best an approximation that probably will not reflect the actual savings for the specific ships used as reference. With a lack of empirical data for this, this approximation is our best choice.

We have absolutely no empirical basis for our assumption that insurance costs will triple for ships sailing the Arctic passages. But even a tripling in costs is a cost disadvantage of almost negligible order of magnitude.

7.7.7 Advantages not quantified

The two dominant cost savings factors in the examples above are the fuel savings and the saved Suez Canal toll. On the Yokohama–Hamburg route, the time savings are substantial. A reduction in sailing time from 34 to 22 days will free up capacity that has a value for the ship owner as the ship will be more quickly available for new contracts. This value would have been explicit if we had chosen a total yearly service approach. The value is difficult to stipulate, however, as it will totally depend on the actual market situation at the time of the sailing.

Time savings could also have a value for the cargo owners. Commodities in transport tie up capital, which is an implicit cost for the cargo owners. For high value cargo, this cost element could be significant. Since we have not considered any particular cargo in our examples, we also ignore this advantage.

A reduction in fuel consumption implies a reduction in emissions of greenhouse gases and other substances. Using the Arctic routes rather than the Suez route contributes to more sustainable transport networks, *ceteris paribus*. Emissions will be proportional to the actual fuel consumption. Table 7.16 indicates the percentage reductions in fuel consumption and could be indicative of the reductions in CO_2. In a future where environmental concerns are more prominent, this is also an advantage that could be used commercially.

7.7.8 The Passages in comparison

We have used some simple calculations to compare the three different Arctic passages NEP, NWP and TPP, just to get a feel for the order of magnitude of cost savings. Some main results are summarized in Tables 7.17 and 7.18.

Not surprisingly, the TPP comes out as the best alternative, simply because crossing the Pole is the shortest route if ice is not too much of an obstacle. This is also by far the least likely route to become accessible without icebreaker support in the near future.

The NSR seems marginally better than the NWP, but then there is the question of ice-breaker fees. The average fee rose to $23 in 2003, and this level will kill the economics for both cases. The fee must be way below $15 per ton if commercial activities want to seek an Arctic solution, given current oil prices. The latest official data for fees that we have been able to obtain dates back to 2003 (see Section 5.5.1). The current situation is non-transparent, but the Beluga experience in 2009 clearly indicates that fees are negotiable.

The actual numerical results are of course very dependent on the oil price chosen. To see the sensitivity of the results for both lower and higher oil prices, Table 7.19 shows the total cost savings for a fuel price of $175, which reflects the 2002–2003 price level and $750, which reflects the mid-2008 record high. Most experts seem to believe we are facing much higher oil prices in the future. This will make use of Arctic passages more attractive.

7.7.9 Endnote: The commercial development of the NSR

In the summer season of 2011 there were no less than 34 passages of the NSR, up from six in 2010 and two in 2009, as documented in Chapter 5. Although it is difficult to obtain detailed information about the economics of these transits, some conclusions can be drawn from the transits of the last couple of years.

The icebreaker fees do not follow the official tariffs. Both the Beluga experience in 2009 and the Nordic Barents trip in 2010 indicate that the fees can be negotiated to a level where the total sum approximately corresponds to the Suez Canal toll, which means some $5–6 per ton carried. Thist means that, compared to using the Suez route, the main savings when using the NSR are reduced bunker costs. For ships in the 40,000–60,000 dwt range, this means $300,000–400,000 per trip (using average bunker prices of 2009–2010). In addition comes the potential advantage of the days saved by using a shorter route. This has typically been 18–22 days for most of the transits.

Although the majority of the NSR trips in 2011 involved Russian interests either as cargo owners or operators, the route now attracts quite an international mix of operators. We also see non-cargo vessels using the route, the seismic vessel *Polarcus Alima* and the passenger ship *Akademik Shokalskiy* being examples.

	NEP	NWP	TPP
Yokohama–Hamburg (fuel price $175), in $/dwt	8	7	9
% change in advantage	–51	–48	–52
Yokohama–Hamburg (fuel price $750), in $/dwt	25	21	28
% change in advantage	45	48	48
Shanghai–Hamburg (fuel price $175), in $/TEU	81	80	94
% change in advantage	–52	–52	–54
Shanghai–Hamburg (fuel price $750), in $/TEU	257	253	316
% change in advantage	52	51	53

Table 7.19 Cost savings sensitivity with respect to fuel prices of using Arctic passages.

In the 2011 season, we saw for the first time a ship using the NSR without any icebreaker support when the ship *Polar* transported iron ore along the NSR. This could be an indication that the assumption made in this section about a future where ice breaker assistance will not be needed in Arctic waters may not be so far away after all.

If the Russian authorities continue to price icebreaker assistance so that the fees do not exceed the corresponding Suez Canal toll, we will no doubt experience increased use of the NSR. With increased activity, there is also an increased risk of accidents, so operators need to focus on all security and safety aspects. The last thing international shipping needs is a major oil spill in Arctic waters.

Notes

1 This is one of the two ships that made a NSR transit in September 2009
2 The calculator is available on http://www.suezcanal.gov.eg/calc.aspx
3 SDR – Special Drawing Rights in IMF – the International Monetary Fund. Current exchange rate (April 2010) was used, extracted from the IMF site: http://www.imf.org/external/np/fin/data/rms_five.aspx
4 Drewry (2008), p. 130
5 Jacobson (2010), p. 5
6 SCNT = 1.1* (GRT+NRT)/2
7 P&I – Protection and Indemnity is the self insurance against third party liabilities
8 TEU – Twenty foot equivalent unit – the no. of loaded 8x8x20-feet containers the ship can carry
9 According to http://navigatemag.ru/bunker/, 28.4.2010
10 On 9.10.2008 the IMO decided on new restrictions regarding sulphur contents in marine bunker. At the time of the decision, the maximum allowed level of sulphur was 4.5% by weight. By 2012 it was reduced to 3.5% and by 2020 it should be 0.5%. This will make bunker oil more expensive
11 We have used the Admiralty formula as an approximation. This says that bunker consumption (Cons) varies with the cube of the speed – Cons = k*speed^3
12 See Liu & Kronbak (2009) and Verny & Grigentin (2009) for a system approach to this

Bibliography

AAPG Explorer, 2008.

ACIA (2004), *Arctic Climate Impact Assessment*, Nøkkelfunn: Konsekvenser av klimaendringer i Arktis.

ACIA (2004), *Arctic Climate Impact Assessment,* Impacts of a Warming Arctic, Cambridge University Press 4.

ACIA (2005), *Arctic Climate Impact Assessment*, Cambridge University Press, Cambridge.

Althingi (2011), *A Parliamentary Resolution on Iceland's Arctic Policy*. Approved by the Althingi at the 139th Legislative Session, 28 March 2011.

Althoff, W.F. (2007), *Drift Station. Arctic Outposts of Superpower Science*, Potomac Books, Washington D.C.

AMAP (2002), *Arctic Monitoring and Assessment Programme*, Factsheet: Transport of Contaminants Exposure of Arctic Humans and their Fate, ACIA 2002, Cambridge University Press, Cambridge.

AMAP (2007), *Arctic Monitoring and Assessment Programme, Arctic Oil and Gas*, Oslo 2007.

AMSA (2008), *Arctic Marine Shipping Assessment*, Report Draft, 14 November 2008.

AMSA (2009), *Arctic Marine Shipping Assessment,* Report, PAME, Arctic Council, Terragraphica, Anchorage, April 2009.

ARCDEV (1998), *Arctic Demonstration Voyage,* Final Public Report of the ARCDEV Project. By the European Commission Under the Transport RTD Programme of the 4th Framework Programme, 1998.

ARCOP (2002), *Arctic Operational Platform*, Working Paper D3.6.1, User Requirements for VTMIS. By Jens Froese, Karsten Bruns-Schüler Maritime Logistics – ISSUS, R.N. Chernyaev, Central Marine Research and Design Institute, 2004.

ARCOP (2003a), *Arctic Operational Platform*, Working Paper D3.7.1, Report on Current Crew Training Practices. By Anniek Platzer, Wagenborg Shipping, Finland 2003.

ARCOP (2003b), *Arctic Operational Platform*, Workshop 1. Legal and Administrative Issues. By Anders Backlund, Liisa Laiho, Britta Jourio, Ministry of Trade and Industry, Finland 2003.

ARCOP (2004a), *Arctic Operational Platform*, Working Paper D2.2.1, Report on Current Regulations and Practices and Impact of IMO Guidelines. By L.G. Tsoy, Central Marine Research and Design Institute.

ARCOP (2004b), *Arctic Operational Platform*, Working Paper D 2.4.1, Marine Insurance Coverage for Oil and LNG Tankers. By E. Gold, Fridtjof Nansen Institute.

ARCOP (2005a), *Arctic Operational Platform*, Working Paper D1.1a, NSR Ice Information State of the Art: Russian Ice Information System for supporting the NSR navigation. By V.Smirnov, Arctic and Antarctic Research Institute.

ARCOP (2005b), *Arctic Operational Platform*, Working Paper D4.2.3, Environmental protection: Biology and potential effects of oil spills on the Arctic sea ice. By Johanna Ikävalko, Finnish Institute of Marine Research, Finland.

ARCOP (2005c), *Arctic Operational Platform*, Working Paper D3.7.2. Report on Existing Courses and Facilities. By Anniek Platzer, Wagenborg Shipping, Sergey Rodionov, Central Marine Research and Design Institute, Leif Baarman, Meriturva, Finland 2005.

ARCOP (2006a), *Arctic Operational Platform*, Workshop 6, Industry Interests in the NSR. By Liisa Laiho, Piia Rahiikanen, Brita Jourio, Sebastian Sala, Ministry of Trade and Industry, Finland.

ARCOP (2006b), *Arctic Operational Platform*, Workshop 7. Legal and Administrative Issues on Arctic Transportation. By Liisa Laiho, Piia Nordstrom, Brita Jourio, Sebastian Sala, Ministry of Trade and Industry, Finland.

ARCOP (2006c), *Arctic Operational Platform*, Working Paper D2.4.2, Marine Insurance Coverage for the Sea Carriage of Oil and Other Energy Materials on the Northern Sea Route: Moving from Theory to Reality. By E. Gold and L.Wright, Fridtjof Nansen Institute.

ARCOP (2006d), *Arctic Operational Platform*, ARCOP Final Report. By Kimmo Juurmaa, Aker Finnyards Inc., Finland.

Arctic Bridge (1994), An Overview of Trade Opportunities between Russia and Canada via the Arctic Ports of Churchill and Murmansk, Final Report, August 1994, Caribou Ventures Limited, Winnipeg, Manitoba.

Arctic Bulletin (1996), Declaration on the establishment of the Arctic Council, No. 4, September 1996.

Arctic Marine Transport Workshop (2004), 28–30 September 2004, Scott Polar Research Institute, Cambridge University, Cambridge.

Arctic Pollution Issues. A State of the Arctic Environment, Arctic Monitoring and Assessment Program, Arctic Council Oslo, 1998.

Arikainen, A. and Chubakov, K.N. (1987), The A and the Z for Ice Navigation in *Transport*.

Armstrong, T (1981), The Northern Sea Route, 1980, *Polar Record*, 20:128.

ASPPR (1978), *Arctic Shipping Pollution Prevention Regulations*.

AWPPA (1970), *Arctic Waters Pollution Prevention Act*.

Baker, Betzy (1978) Law, science, and the continental shelf. The Russian Federation and the promise of Arctic cooperation, *American University Law Review*, 25:2.

Bakharin, B. (2006), Russia's nuclear icebreaker fleet, *Science and Global Security*, 14, 25–31.

Bambulyak, A. and Frantzen, B. (2005), Oil transport in the Barents region. In: The Third International Conference of the Oil and Gas Industry and Sustainable Development of the Barents Region. Murmansk, Russia, 9–11 November 2005. Svanhovd Environmental Center, Norway.

Bambulyak, A and Frantzen, B. (2009), Oil transport from the Russian part of the Barents region. Status per January 2009. Akvaplan-niva AS, The Norwegian Barents Secretariat.

Barabolya, P.D., Bakhov, A.S., Ivannashcehnko, L.A., Kolesnik, D.N.,. Logunov, V.D., Molodtsov, S.V. and Nasinovskiy, Ye. N. (1966), *Manual of International Maritime Law*, Military Publishing House of the Ministry of Defense of the USSR, Moscow, 1966. Translation by Naval Intelligence Command, Department of Transportation, United States Coast Guard.

Barry, R.G. and Ives, J.D. (1974), *Arctic and Alpine Environments*, Methuen & Co, London, pp. 1–13.

Baskin, A., Egorov, L., et al. (1998), The NSR Simulation Study Work Package 1: Routes and Associated Operational Infrastructure. INSROP Working Paper 108-1998, FNI, Lysaker.

Beauregard-Tellier, F. (2008), *The Arctic: Hydrocarbon Resources*, Library of Parliament, 24 October 2008.

BBC News (2009), Dispute hits Europe gas supplies, 1 July 2009.

Beluga Group (2009), Beluga Shipping Masters First Commercial Transit of the Northeast-Passage, Internal Paper written by the Beluga Group, Bremen.

Bergesen, H., Moe, A., and Østreng, W. (1987), *Soviet Oil and Security Interests in the Barents Sea*, Frances Pinter, London.

Boswell, R. (2009), Fishing ban report places pressure on Canada, Canada. com, 14 February 2009.

Boyle, R. and Lyon, W. (1998), Arctic ASW: Have we lost?, *Naval Institute Proceedings*, 124: 6/1, 144.

BP Statistical Review of World Energy (2005), Natural Gas and Putting Energy in the Spotlight, *BP Statistical Review of World Energy 2005*: Oil.

BP Statistical Review of World Energy, June 2008.

BP Statistical Review of World Energy, June 2009.

BP Statistical Review of World Energy, June 2011.

Brigham, L (1995), Arctic Rendevouz. *U.S. Naval Institute Proceedings*, January 1995.

Brigham, L. (1999), The natural environment, ice navigation and ship technology. In W. Østreng (Ed.) (1999), *The Natural and Societal Challenges of the Northern Sea Route. A Reference Work*. Kluwer Academic Publishers, Dordrecht.

Brigham, L. (2006), The Arctic Ocean. In *The Oxford Encyclopedia of Maritime History*, Oxford University Press, Oxford.

Brigham, L. (2007), Thinking about the Arctic's future: Scenarios for 2040, *The Futurist*, September–October.

Brubaker, D. (2000), Regulation of navigation and vessel-source pollution in the Northern Sea Route: Article 234 and state practice. In Davor Vidas (Ed.) *Protecting the Polar Marine Environment. Law and Policy for Pollution Prevention,* Cambridge University Press, Cambridge, pp. 221–243.

Brubaker, D. (2002), Environmental Protection of Arctic Waters. Specific Focus on the Russian Northern Sea Route. Sth. 2002.

Brubaker, D. (2005), *The Russian Arctic Straits*, Martinus Nijhoff Publishers, Leiden, Boston.

Brubaker, D. and Østreng, W. (1999), The Northern Sea Route regime: Exquisite superpower subterfuge? *Ocean Development and International Law*, 30.

Budzik, P. (2009), Arctic Oil and Natural Gas Potential, US Energy Information Administration, Office of Integrated Analysis and Forecsating, Oil and Gas Division.

Bulatov, V. (1997), Historical and Current Uses of the Northern Sea Route, Part IV: The Administration of the Northern Sea Route. INSROP Working Paper 84-1997, FNI, Lysaker.

Butler, W.E. (1978), *International Straits of the World: Northeast Arctic Passage*, Sijthoff and Noordhof, Alphen aan Rijen.

Calvert, J. (1960), *Surface at the Pole. The Extraordinary Voyages of the* USS Skate, New York.

Canada Today (1974) Pingoes and the Northwest Passage, 5:2.

Canwest Interactive, Canwest Media Works Publications, canada.com, August 2008.

CARA (2008), Circum Arctic Resource Appraisal. Estimates of Undiscovered Oil and Gas North of the Arctic Circle, USGA Fact Sheet 2008-3049, July 2008, U.S. Department of the Interior, U.S. Geological Survey.

CASA (2007), *Canadian Arctic Shipping Assessment*, The Mainport Group Ltd. for Transport Canada, June.

Castel, J.-G. (1976), *International Law. Chiefly Interpreted and Applied in Canada*, Butterworths Canadian Legal Casebook Series, Butterworths, Toronto.

CBC News (2009), Mackenzie pipeline gets green light from panel, 30 December 2009.

Chance, N. (2008), Alaska Natives and the Land Claims Settlement Act of 1971, no date.

Chircop, Aldo (2011), The emergence of China as a Polar-capable state, *Canadian Naval Review*, 7:1.

Christopher, J. (2008), The Arctic: Transportation, Infrastructure and Communication, Library of Parliament, (Canada), PRB 08-08E, 24 October 2008.

Coté, F. and Dufresne, R. (2008), The Arctic: Canada's Legal Claims. Parliamentary Information and Research Service. 2008. PRB 08-05E5.

Council of the European Union (2010), Council Conclusions on Arctic Issues. 2985th Foreign Affairs Council Meeting, Brussels, 8 December 2009.

Dagbladet (2010), Frykter arktisk splittelse, 31 March 2010.

Dagsavisen (2009), Kina kommer i nord, 12 May 2009.

Dahle, S. (2010), Flytende på isen I politisk farvann", *Chronicle*, Fiskeribladet Fiskaren, 23 April 2010.

Dallmann, W.K. Indigenous peoples of the Russian North, Tromsø, no date

Dewar, B. (2009), Nunavut and the Nunavut Land Claims agreement – An unresolved relationship, *Policy Options*, July/August 2009.

Diatchkova, G. (2001), Indigenous peoples of Russia and political history, *The Canadian Journal of Native Studies* XXI, 2, 217–233.

Digital Chosunilbo (English Edition) (2008), Daily News in English about Korea: Http://english.chosun.com. 10.09.08.

Doyle, A. (2008), Arctic ice bigger in 2007, but thawing long-term, 30 July 2008, blogs.reuters.com/environment

Drewry Consultants (2007), Ship Operating Costs. Annual Report and Forecasts 2007/2008, London.

Drivenes, E.A. and Harald D.J. (Eds) (2004), *Norsk Polarhistorie*, Norsk Gyldendal Forlag, Oslo.

Dunbar, M. (1966), The Arctic setting. In R. Macdonald (Ed.) *The Arctic Frontier*, University of Toronto Press, Toronto, 1966, pp. 3–25.

Dyson, J.L. (1963), *The World of Ice*, London.

Eger, Karl Magnus and Kristiansen, Øystein Russ (2011), Sjøsikkerhet i nordlige isdekte farvann. Utarbeidet av forskningsinstituttet Ocean Futures på oppdrag fra det Norsk Sjømannsforbund Oslo, 15 March 2011.

Eide, A. Indigenous self-government in the Arctic, and their right to land and natural resources. No date.

Eide, A. (2006), Rights of indigenous peoples – Achievements in international law during the last quarter of a century. In *Netherlands Yearbook of International Law Vol. XXXVII*, pp. 155–212.

Elferink, A. (1994), *The Law of Maritime Boundary Delimitation: A Case Study of the Russian Federation*, Publications on Ocean Development, University of Utrecht, Utrecht.

Energy Information Administration (2008), *Analysis of Crude Oil Production in the Arctic National Wildlife Refuge*, Report#: SR-OIAF72008-03, May 2008. Energy Information Administration.

European Union (2008), Communication from the Commission to the European Parliament and the Council: The European Union and the Arctic Region. EU, Brussels, 20 November 2008.

European Union (2009), EU Council Conclusions on Arctic Issues, 8 December 2009, EU, Brussels.

European Union (2010), Draft Report on a Sustainable EU Policy for the High North 2009 (2214(INI)), European Parliament, Committee on Foreign Affairs, Brussels.

Evensen, J. (1978), Speech in the Storting, 9 March 1978. In UD-informasjon, no. 11/1978, Oslo, The Foreign Ministry.

Falkingham, J. (2004), Sea ice in the Canadian Arctic. Talk at Arctic Marine Transport Workshop, Scott Polar Research Institute, 28–30 September 2004.

Fenge, T. (2008), Inuit and the Nunavut Land Claims Agreement: Supporting Canada's Arctic sovereignty, *Political Options*, December 2007–January 2008.

Fenge, T. (2008), Implementing comprehensive land claims agreements, *Policy Options*, July–August 2008.

Fleischer, C.A. (2007), The new international law of the sea and Svalbard.

Fløistad. B. (2007), The controversy over the applicable regime outside Svalbard's territorial sea, UiO 2007.

Food and Agriculture Organization of the United Nations (2006), *The State of the World's Fisheries and Aquaculture 2006*, FAO: Rome.

Food and Agriculture Organization of the United Nations (2007), *The State of the World's Forests 2007*, FAO: Rome.

Franckx, E. (1988), Non-Soviet shipping in the Northeast Passage and the legal status of Prov Vil'kitskogo, *Polar Record*, 24:151.

Franckx, E. (1993), *Maritime Claims in the Arctic: Canadian and Russian Perspectives,* Martinus Nijhoff Publishers, Dordrecht, Boston, London.

Frolov, I.E. and Krutskih, B.A. (Eds) (2008), *Hydrometeorological Supplying of Navigation in XX and Beginning of XXI Centuries*, Arctic and Antarctic Research Institute, St. Petersburg.

Frolov, I.E., Borodachev, V.Ye and Alexandrov, V.Yu. (2007), Section 1.4: The Soviet-Russian system for supporting navigation, In Johannessen, Alexandrov, Frolov, Sandven, Pettersson, Bobylev, Kloster, Smirnov, Mironov and Babich (2007)

Fuhs, P. (1992), Marco Polo in the 21st Century. In Simonsen Henning (Ed), *Proceedings from the Northern Sea Route Expert Meeting*, 13–14 October 1992.

Fujita, Y., Narita, H. and Kitagawa, H. (1986), Design study of a 200,000 dwt icebreaking tanker. In: *Proceedings of the Fourth Offshore Mechanics and Engineering Symposium*, Vol. 4, ASME.

Glasby, G.P. and Voytekhovsky, Yu. L. (2009), Arctic Russia: Minerals and Mineral Resources, *Geochemical News*, July 2009.

Glomsrød, S. and Aslaksen, J. (Eds) (2009), *The Economy of the North*, Oslo-Kongsvinger; Statistics, Norway.

Gold, E. (1999), Economy and Commercial Viability. In W. Østreng (Ed.) (1999), *The Natural and Societal Challenges of the Northern Sea Route. A Reference Work*. Kluwer Academic Publishers, Dordrecht.

Gold, E. (2000), Transiting the Northern Sea Route: Shipping and marine insurance interests. In C.L. Ragner (Ed.), *The 21st Century: Turning Point for the Northern Sea Route?* Proceedings of the Northern Sea Route User Conference, Kluwer, Dordrecht.

Golovnev, A.V. and Osherenko, G. (1999), *Siberian Survival. The Nenets and their Story*, Cornell University Press, Ithaca.

Gordienko, P.A. (1960), The Arctic Ocean, *Ocean*, 2.

Gorshkovsky, A.G. Rules to be followed on the Northern Sea Route.

Gould, L.M. (1958), *The Polar Regions and their Relations to Human Affairs*, American Geographical Society, New York.

Granberg, A. (1992), The Northern Sea Route and the policy of the new Russia, *International Challenges*, 12:1.

Granovsky, A.E. (1989), Soviet approaches to security in the Arctic. In W.A. Hurlburt (Ed.), *The Arctic. Choices for Peace and Security*, Gordon Soules Book Publishers, Seattle.

Griffiths, Franklyn and George, I. (2007), Canadian Arctic Sovereignty: Time to take Yes for an Answer on the Northwest Passage. Paper prepared for Aboriginal and Circumpolar Affairs, Foreign Affairs Canada, Mimeographed paper, 15 October 2007.

Griffiths, Franklyn (2012), Stewardship as Concept and Practice in an Arctic Context, Draft article, 20 February 2012.

Grishchenko, V (1999), The natural environment, ice navigation and ship technology. In W. Østreng (Ed.) *The Natural and Societal Challenges of the Northern Sea Route. A Reference Work*. Kluwer Academic Publishers, Dordrecht.

Heininen, Lassi (2011) *Arctic Strategies and Policies. Inventory and Comparative Study,* Northern Research Forum and University of Lappland Press, Akureyri.

Hernesniemi, H. (2006), Industrial Clusters in Northwest Russia, Lappeenranta University of Technology, Northern Dimension Research Center, Publication 36, 2006.

Hoel, A. (1928), Suverænitetsspøsmaalene i polartraktene, *Nordmands-forbundet*, 4:5, 21.

Homeland Security (2009), US National Presidential Directive and Homeland Security Presidential Directive, The White House, President George W. Bush, January 2009.

Hønneland, G.B. (1997), Navigating the Straits of the Northern Sea Route. INSROP Working Paper 82-1997, FNI, Lysaker.

Hurley, M.C. Aboriginal self-government. Parlimentary Information and Research Service, Canada, 96-2E.

Icelandic Ministry for Foreign Affairs (2006), North Meets North. Navigation and the Future of the Arctic. Report of a Working Group of the Ministry for Foreign Affairs, Reykjavik, Iceland.

Icelandic Ministry of Foreign Affairs (2010), An Executive Summary of the Report of Øssur Skarphedinsson to Althingi, Minister of Foreign Affairs, 14 May 2010, Reykyavik, Iceland.

ICIA, *Impacts of a Warming Arctic: Arctic Climate Impact Assessment*, Cambridge University Press, Cambridge.

IEA (2005), *Resources to Reserves*, IEA/OECD.

IEA (2008), *World Energy Outlook 2008 Executive Summary.*

IEA (2009), *World Energy Outlook 2009 Executive Summary.*

IMF (2010), *World Economic Outlook Update*, 26 January 2010.

IMO (2000), Study of greenhouse gas emissions from ships, Issue no. 2, 31 March 2000.

IMO Guidelines (2002), Guidelines for ships operating in Arctic ice-covered waters. IMO Doc. SC/Circ.1056 MEPC/Circ.399, 23 December 2002.

IMO COMSAR (2006), IMO Sub-Committee on Radio Communications and Search and Rescue, 10th Session, Report to the Maritime Safety Committee, COMSAR 10/16. London 2006.

Information Canada (1972) A voyage without end, Information Canada, Ottawa.

INSROP (1994), International Sea Route Programme, Working Paper 2. *Routing Communication and IT-Customizing*. By N. Kjerstad, August 1994.

INSROP (1996a), International Sea Route Programme, Working Paper 28, *Historical and Current Uses of the Northern Sea Route: Part 1*. By T. Armstrong, January 1996.

INSROP (1996b), International Sea Route Programme, Working Paper 58, *Regional Cooperation in Northeast Asia: New Opportunities for the Russian Arctic?* By H. Simonsen, 1996.

INSROP (1996c), International Sea Route Programme, Working Paper 59, *The Economics of the NSR. A feasibility study of the Northern Sea Route as an alternative to the international shipping market*. By T.R. Ramsland and S. Hedels, September 1996.

INSROP (1996d), International Sea Route Programme, Working Paper 64, *Requirements to NSR Shore Reception Facilities*. By G. Semanov, V. Molchanov. S. Lothukov, A. Stepanov and L. Gagieva, October 1996.

INSROP (1996e), International Sea Route Programme, Working Paper 93. *The Northern Sea Route and Native Peoples. Lessons from the 20th Century for the 21st Century*. By G. Osherenko, 1997.

INSROP (1996f), International Sea Route Programme, Working Paper 95. *Using the INSROP Phase 1 Data in Transport Evaluation Process*. By A. Backlund, E. Gold and J. Kivela, January 1998.

INSROP (1997a), International Sea Route Programme, Working Paper 76, *Coastal Pollution Emergency Plan. Part I*. By G. Semanov, V. Volkov, V. Somkin and D. Iljushenko-Krylov, February 1997.

INSROP (1997b), International Sea Route Programme, Working Paper 81, *Navigating the Straits of the Northern Sea Route*. By G. May Hønneland, 1997.

INSROP (1997c), International Sea Route Programme, Working Paper 84, *Historical and Current Uses of the Northern Sea Route, Part IV: The Administration of the Northern Sea Route (1917–1991)*. By V. Bulatov, 1997.

INSROP (1997d), International Sea Route Programme, Working Paper 90, *Indigenous Peoples of the Northern Part of the Russian Federation and their Environment. Atlas and Historical/Ethnographical Background Information*, By W.K. Dallmann, 1997.

INSROP (1997e), International Sea Route Programme, Working Paper 93, *The Northern Sea Route and Native Peoples. Lessons from the 20th Century for the 21st*. By Gail Osherenko, Debra L. Scindler, Alexander I. Pika and Dimitri Bogoyavlensky, 1997.

INSROP (1998a), International Sea Route Programme, Working Paper 99, *The NSR Environmental Atlas*. By O.W. Brude, K.A. Moe, V. Bakken, R. Hansson, L.H. Larsen, J. Thomassen and Ø. Wiig, May 1998.

INSROP (1998b), International Sea Route Programme, Working Paper 111, *Assessment of Social and Cultural Impacts on Indigenous Peoples of Expanded Use of the Northern Sea Route*. By Z. Sokolova, and A. Yakolev, 1998.

INSROP (1998c), International Sea Route Programme, Working Paper 113, *Historical and Current Uses of the Northern Sea Route, Part II: The Period 1745–1855*. By E. Okhuizen, 1998.

IPCC (2007), Intergovernmental Panel on Climate Change, Working Group 1, *The Physical Science Basis of Climate Change*, AR4 Report, 2007.

Jacobson, Linda (2010), China prepares for an ice-free Arctic, *SIPRI Insights on Peace and Security*, No. 2010/2, March 2010.

Jakobsson, Thor (2004), The Northern Sea Route – Interest in Iceland, talk at Arctic Marine Transport Workshop, Scott Polar Research Institute, 28–30 September 2004.

Jensen, Ø. (2007), The IMO Guidelines for Ships Operating in Arctic Ice-covered Waters. From Voluntary to Mandatory Tool for Navigation Safety and Environmental Protection? Fridtjof Nansen Institute, Report 2/2007, Oslo 2007.

Jensen, Ø. (2008), Arctic shipping guidelines: towards a legal regime for navigational safety and environmental protection? *Polar Record*, 44(229), 107–114.

Johannessen, O.M. (2008), Arctic sea ice mirrors increasing CO2 on decadal time scale, *Atmospheric and Oceanic Science Letters*, 1:1.

Johannessen, O.M, Alexandrov, V.Y., Frolov, I.Y., Sandven, S. Pettersson, L. Bobylev, L.P., Kloster, K., Smirnov, V.G., Mironov, Y.U. and N.G. Babich (2007), *Remote Sensing of Sea Ice in the Northern Sea Route. Studies and Applications*, Springer Verlag/PraxisPublishing, Berlin, Heidelberg, Chichester.

Johnston, D.M. (Ed.) (1981), *Arctic Ocean Issues in the 1980s*, Proceedings, Workshop, 10–12 June 1981, Machinac Island, Michigan, Law of the Sea Institute, University of Hawaii and Dalhousie Ocean Studies Programme, Dalhousie University, Honolulu, 1982.

Joint Research Center of the OECD and the International Transport Forum, *Transport Outlook 2009, Globalisation, Crisis and Transport,* Discussion paper No 2008-12, May 2009, preliminary version.

Joling, D. (2009), Council moves to bar commercial fishing in the Arctic, report filed from Anchorage late January or early February 2009, Associated Press.

Jones, R.S (1981), Alaska Natives Claims Settlement Act of 1971. History and Analyses Together with Subsequent Amendments.

Jørgensen-Dahl, A. and Østreng, W. (Eds) (1991), *The Antarctic Treaty System in World Politics*, Macmillan, London.

Jørgensen, T.S. (1991), Sea ice – Nature's major challenge for human enterprise in the Arctic. In *Festscrift to Willy Østreng: Challenges of a Changing World*, Fridtjof Nansen Institute, Lysaker.

Karczynski, V.M. (2007), US–Russian Bering Sea marine border dispute: Conflict over strategic assets fisheries and energy resources. *Russia Analytical Digest*, No 20, May 2007.

Kassel, Bernard M. (1959), Soviet logistics in the Arctic. In *US Naval Institute Proceedings*, 85: 2, February 1959.

Kilman,W. and Whiteman, A. (2006), The Russian forestry sector in the global forest product market: Trends, outlook and opportunities for development, FAO, Rome. Presentation to the VII International Forestry Forum, 10–13 October 2006, St Petersburg.

Kitagawa, H. et al. (1995), Northern sea route: Future and perspectives. In *The Proceedings of INSROP Symposium*,1–6 October 1995, Ship and Ocean Foundation, Tokyo.

Kitagawa, H. (2001), *The Northern Sea Route: The Shortest Sea Route Linking East Asia and Europe*. Tokyo: Ship & Ocean Foundation. Retrieved from Ship & Ocean Foundation.

Kitagawa, H. (Ed.) (2006), *New Era in the Far East Russia and Asia*, Ocean Policy and Research Foundation, Tokyo, February 2006.

Kolodkin, A.L. and Kolosov, M.E. (1990), The legal regime of the Soviet Arctic: Major issues, *Marine Policy*, Vol. 14.

Krauss, C. (2009), New way to tap gas may expand global supplies, *The New York Times*, October 2009

Lakthine,V.L. (1928), Retten til de nordlige polarområder. En undersøkelse av de nordlige polarområders politiske, økonomiske og rettslige stilling i forbindelse med utviklingen av luftfarten og de transarktiske flyvninger Foreign Commissariat, Moscow. Confidential translation into Norwegian. In the private archive of the author.

Lasserre, F. and Pelletier, S. (2011), Polar super seaways? Maritime transport in the Arctic: An analysis of shipowners' intentions. *Journal of Transport Geography*, 19.

Lensky, L. Damage Statistics on Ships Sailing the Northern Sea Route, undated mimeographed paper.

Liu, M. and Kronbak, J. (2009), The Potential Economic Viability of Using the Northern Sea Route as an Alternative Route between Asia and Europe. Paper presented at the IAME conference, Copenhagen 2009.

Loukacheva, N. (2007), *The Arctic Promise. Legal and Political Autonomy for Greenland and Nunavut*, University Press of Toronto, Toronto, Buffalo, London, 2007.

Løvas, S.M. and Brude, O.W. (1996), INSROP GIS – User Guide and System Documentation. INSROP Working Paper 47-1996, FNI, Lysaker.

Lundgren, K (2010), Cairn shares rise on Greenland plans, Bloomberg.com, 23 March 2010.

Lyon, Waldo (1963), The submarine and the Arctic Ocean, *The Polar Record*, 11(75).

Lysaker, D.I. (2007), "Framstredet", e-mail from Statens Kart verk, geodesidivisjonen, Hønefoss, 9 June 2009 to Willy Østreng.

Mariport Group for Transport Canada (2007), Canadian Arctic Shipping Assessment, Main Report, June 2007.

Marr, J. (2001), Impact of Climate Change in the Arctic on Ship Operations and Support Systems: A Mariners Perspective, Canadian Maritime Law Association, 50th AGM, June 15–16 2001, Montreal.

Maugeri, L (2009), Squeezing more oil from the ground, *Scientific American*, October.

Mauritzen, Cecilie and Kolstad, E.W. (2011), The Arctic Ocean: An ocean in transition. In John Grue and Roy H. Gabrielsen (Eds), *Marine Transport in the High North*, Novus Forlag, Det norske Videnskapsakademi og Norges Tekniske Vitenskapsakademi, Oslo.

McCarthy, S. (2009), In Mackenzie Valley, frustration and a sense of foreboding, *The Globe and Mail*, 30 August 2009.

Mejlænder-Larsen, M. and Espeland, Ø. (2009), ARCON-Arctic Container, Det Norske Veritas, Arctic Container Project, September 2009.

Ministry for Foreign Affairs (2006), North Meets North. Navigation and the Future of the Arctic. Report of a Working Group of the Ministry for Foreign Affairs, Reykjavik, 2006.

Ministry for Foreign Affairs (2010), An Executive Summary of the Report of Øssur Skarphedinsson, Minister for Foreign Affairs to Althingi, 14 May 2010.

Molenaar, E.J. and Corell, R. (2009), Background paper. *Arctic Shipping*.

Molloy, A.E. (1969), The Arctic Ocean and the marine science in the North Polar Region, *Oceans*, 1:2.

Mouawad, J. (2010), China's growth shifts the geopolitics of oil, *The New York Times*, 19 March 2010.

Mutanen, A., Viitanen, J., Toppinen, A. Hanninen, R. and Holopainen, P. (2009), Forest Resorces, Production and Exports of Roundwood and Sawnwood from Russia. Working Papers of the Finnish Forest Research Institute 9.

Natsuhiko, O. (2006), Transport infrastructure in the Russian Far East. In: H. Kitagawa (Ed.) *New Era in the Far East Russia and Asia*, Ocean Policy and Research Foundation, Tokyo, February 2006.

New York Times, 20 February 2004.

NOU (2003), *Norsk Offentlig Utredning,* Mot Nord! Utfordringer og muligheter i Nordområdene, NOU 2003: 32, Statens forvaltningstjeneste, Oslo.

Norsk Polarinstitutt (2009), Best Practices in Ecosystem-based Ocean Management in the

Arctic. Rapportserie no.129. Tromsø, April.

Nystø, S.R. (2010), Uro og lederskap, Chronicle, Fiskeribladet Fiskaren, 5 March 2010

Ocean Futures (2006), Fakta om Nordområdene. By Vik, J. Kvarving, S. Sawhill, O. Thommessen, and W. Østreng, for NHO "Oppdrag Nord", Oslo, January 2006.

Ocean Futures (2007), Focus-North 4 - 2007. Ocean Futures, Oslo.

Ocean Futures (2008), The Northern Sea Route's Role in the System of International Transport Corridors, by Peresypkin, V. and Yakovlev, A., produced by the Reasearch Company Ocean Futures in cooperation with the Norwegian Atlantic Committee, Focus North 2-2008, Oslo 2008.

Ocean Futures (2007), The Environment and Maritime Transport, produced by Ocean Futures for The Norwegian Atlantic Committee, Focus North 4-2007, Oslo 2007.

Ocean Futures (2007), Oil and Gas, produced by Ocean Futures for The Norwegian Atlantic Committee, Focus North 5-2007, Oslo 2007.

Ocean Futures (2009), Vurdering av andre uåpnete områder i nord. By B. Fløistad, A. Jørgensen-Dahl, and W. Østreng, for Oljeindustriens Landsforening (OLF), Oslo, February 2009.

OECD (2009), Joint Research Center of the OECD and the International Transport Forum, Transport Outlook 2009, Globalisation, Crisis and Transport, Discussion paper No 2008-12, May 2009, preliminary version.

Office of Naval Research (2001), Naval Operations in an Ice-free Arctic, *Final Report of a Symposium*, 17–18 April 2001, Office of Naval Research, Naval Ice Center, Oceanographer of the Navy and the Arctic Research Commission, Whitney, Bradley & Brown, Washington 2001.

Office of the Press Secretary (2009), US National Security Presidential Directive/NSPD – 66, The Arctic Region. The White House, Washington DC, 9 January 2009.

Oljedirektoratet (2007), Petroleumsressursene på norsk kontinentalsokkel 2007.

Olenikoff, S. (1972), *Territorial Waters of the Arctic: The Soviet Position*, Rand Corporation, Santa Monica.

O'Rourke, R. (2010), Changes in the Arctic. Issues for the Congress. Congressional Research Service, 10 March 2010.

Osherenko, G. and Oran, Y. (1989), *The Age of the Arctic. Hot Conflicts and Cold Realities*, Studies in Polar Research, Cambridge University Press, Cambridge.

Osherenko, G., Schindler, D., Pika, A. and Bogayavlensky, D. (1997) The Northern Sea Route and Native Peoples. Lessons from the 20th Cenetury to the 21st. INSROP Working Paper 93-1997, FNI, Lysaker.

Østerud, Ø (1996), Statsvitenskap. Innføring i politisk analyse, Universitetsforlaget, Oslo.

Østreng, W. (1977), The strategic balance and the Arctic Ocean, *Cooperation and Conflict*, 1.

Østreng, W. (1979), Polhavet i internasjonal politikk, Mimeographed, Studie AA:H012, FNI, Lysaker 1979.

Østreng, W. (1982), Sovjet i Nordlige Farvann. Atomstrategien, Nordflåten og norsk sikkerhet, Gyldendal Norsk Forlag, Oslo, 1982.

Østreng, W. (1982), The strategic balance in the Arctic Ocean – Soviet options. In: William Gutteridge (Ed.), *European Security, Nuclear Weapons and Public Confidence*, Macmillian Press, London, Basingstoke.

Østreng, W. (1991), Den nordlige sjørute: En ny æra i sovjetisk politikk? Økonomiske, folkerettslige og sikkerhetspolitiske aspekter", Den norske Atlanterhavskomite, Oslo, Det sikkerhetspolitiske bibliotek,nr. 9/1991

Østreng, W. (1992), Political-Military Relations among the Ice-States: The Conceptual Basis of State Behaviour. In Franklyn Griffiths (Ed.) *Arctic Alternatives: Civility or Militarism in the Circumpolar North*, Science for Peace/Samuel Stevens, Toronto.

Østreng, W. (Ed.) (1999), *National Security and International Environmental Cooperation in the Arctic – the Case of the Northern Sea Route*, Kluwer, Dordrecht/Boston/London.

Østreng, W. (Ed.) (1999), *The Natural and Societal Challenges of the Northern Sea Route. A Reference Work*. Kluwer Academic Publishers, Dordrecht.

Østreng, W. (2005), Transport og miljøutfordringer I nordområdene, *Horisont. Næringspolitisk tidsskrift*, 3:6.

Østreng, W. (2009), Trenger Arktis en egen traktat? (Is the Arctic in need of a separate treaty?), *Chronicle*, Fiskeribladet Fiskaren, 22 May 2009.

Østreng, W. (2009), Security and climate change in the global and Pan-Arctic context: A historic dynamic perspective. In: Lassie Heininen and Heather Nicol (Eds), *Climate Change and Human Security – From a Northern Point of View*, Ottawa University Press.

Østreng, Willy (2010), *Science without Boundaries. Interdisciplinarity in Research, Society and Politics*, American University Press, New York.

Østreng, Willy. The Conception of a Race for Arctic Resources, mimeographed to be published by University of Alberta and edited by Professor Mark Nuttall (forthcoming).

Østreng, W. The politics and law of sea ice (working title), *International Development and International Law* (forthcoming).

Østreng, W. and Simonsen, H. (1992), The Barents Region and the Northern Sea Route,

International Challenges, 12:4.

Otsuka, N. (2006), Transport infrastructure in the Russian Far East. In: H. Kitagawa (Ed.) *New Era in Far East Russia and Asia*, Ocean Policy Research Foundation, Tokyo.

Overenskomst mellom Kongeriket Norges regjering på den ene siden og Kongeriket Danmarks regjering sammen med Grønlands landsstyre på den annen side om avgrensningen av kontinentalsokkelen og fiskerisonene i området mellom Grønland og Svalbard av 20 February 2006.

Oxman, B.H. (1985), From Cooperation to Conflict; The Soviet Union and the United States at the Third U.N. Conference on the Law of the Sea, Mimeographed paper, University of Washington, Seattle, 1985.

Painter, J. (1995), *Politics, Geography and "Political Geography": A Critical Perspective*, Arnold, London.

Parrott, B. (1988), Soviet national security under Gorbachev, *Problems of Communism*, 37:10.

Peresypkin, V. and Yakovlev, A. (2008), The Northern Sea Routes in the System of Transport Corridors. Ocean Future/Norwegian Atlantic Committee, Oslo.

Petroleum Intelligence Weekly (2009), 5 January 2009.

Pharand, Donat (1973) *The Law of the Sea of the Arctic with Special Reference to Canada*, University of Ottawa Press, Ottawa.

Pharand, D. (2007), The Arctic Waters and the Northwest Passage: A final revisit, *Ocean Development and International Law*, 38, 3–69.

Pollack, H. and Anderson, P. (1973), United States policy for the Arctic, *Arctic Bulletin*, 2

Polunin, N. (1951), The real Arctic: Suggestions for its limitation, subdivision and characterization, *Journal of Ecology*, 39, 308–315.

Ragner, C.L. (Ed.) (2000a), The 21st century: Turning point for the Northern Sea Route? In: *Proceedings of the Northern Sea Route User Conference*, Kluwer, Dordrecht/ Boston/London.

Ragner, C.L. (2000b), Northern Sea Route Cargo Flows and Infrastructure – Present State and Future Potential, The Fridtjof Nansen Institute Report 13/2000, Lysaker, Norway.

Ramsland, T.R. (1999), Economic Evaluation of the NSR Commercial Shipping, INSROP Working Paper 140-1999, FNI, Lysaker.

Resources to Reserves, IEA/OECD, 2005.

Riska, Kaj, Challenges and possibilities in Arctic Marine operations. In: John Grue and Roy H. Gabrieløsen (Eds), *Marine Transport in the High North*, Novus Forlag, Det Norske Videnskabakademi and Norges Tekniske Viotenskapsakademi, Oslo.

Rogers, T.G. and Rowley, G. (1978), *The Circumpolar North*, Methuen & Co., London.

Rompkey, W. (2009a), Rising to the Arctic Challenge: Report on the Canadian Coast Guard. Report of the Standing Senate Committee on Fisheries and Oceans, Ottawa, April 2009.

Rompkey, W. (2009b), Canada's Arctic Waters: Role of the Canadian Coast Guard. Report of the Standing Senate Committee on Fisheries and Oceans, Ottawa, December 2009.

Russian Federation (1990), Regulations for Navigation on the Seaways of the Northern Sea Route, 14 September 1990.

Russian Federation (1991), Regulations for Navigation on the Seaways of the Northern Sea Route (1991), issued by the Head Department of Navigation and Oceanography, USSR Ministry of Defence, 1991, published in *International Challenge*, 12:1, Special Issue on the Northern Sea Route.

Russian Federation (1996a), Guide to Navigating through the Northern Sea Route. English version published by the Head Department of Navigation and Oceanography of the Ministry of Defence of the Russian Federation, St. Petersburg.

Russian Federation (1996b), Regulations for Icebreaker and Pilot Guiding of Vessels through the Northern Sea Route. English version published in *Guide to Navigation* (1996), 84–89.

Russian Federation (1996c), Requirements for the Design, Equipment and Supplies of Vessels Navigating the Northern Sea Route. English version published in *Guide to Navigation* (1996), 317–323.

Russian Federation (1998), Federal Law of Internal Sea Waters, Territorial Sea and Contiguous Zone of July 1998, no. 155-F3, and The Regulations for Navigation on the Sea Ways of the Northern Sea Route.

Russian Federation (2009), Fundamentals of the Russian Federation's Policy in the Arctic for the Period up to 2020 and Beyond, Security Council of Russian Federation, Moscow, 27 March 2009

Russian Federation (2010), Decree of the President of the Russian Federation of 17 January about the Concept of the Federal Programme of World Oceans, no. 11, Sub-programme 9: Conquering and Utilising the Arctic. The translation from Russian has been done by Anne Berteig.

Said, E.W. (1993), *Culture and Imperialism*, Vintage Books, London.

Saksina, T. (2008), Arctic change: New management challenges, *WWF Arctic Bulletin*, 3.

Samuelson, U. (1970), The Arctic and its Role in World Affairs. Unpublished manuscript, Royal Swedish Navy, London.

Sater, J.E. (1969), The Arctic Basin, Washington D.C..

Sawhill, S. (2008), Defining the North in Norwegian Politics, produced by Ocean Futures for the Norwegian Atlantic Committee, Focus North 5, 2008.

Solheim, A., Hauge, O. and Leknes, E. (2004), Northern Maritime Corridor. Interreg III B at sea. In: Louis von Flotow (Ed.) *Innovative City and Business Regions, Structural Changes in Europe 3*, Lou Hagbart Publisher, Bollschweil.

Scrivener, D. (1989), Gorbachev's Murmansk Speech: The Soviet Initiative and Western Responses, The Norwegian Atlantic Committee, Oslo, 1989.

Simone, T. (2008), The Arctic: Northern Aboriginal Peoples, Library of Parliament – Parliamentary Information and Research Service. October 2008 (PRB 08-10E).

Simonson, H. (1996), Regional Cooperation in Northeast Asia: New Opportunities for the Russian Arctic? INSROP Working Paper 58-1996, FNI, Lysaker.

Skagestad, O.G. (1975), *Norsk Polarpolitikk. Hovedtrekk og utviklingslinjer 1905–1974*, Dreyer, Oslo.

Smith, R.W. (1966), Sovereignty in the North: The Canadian aspect of an international problem. In: R.St. J. Macdonald (Ed.), *The Arctic Frontier*, University of Toronto Press, Toronto, pp.194–255.

Smith, R.W. (1984), National claims and the geography of the Arctic. In: *Proceedings of the Law of the Sea Institute*.

SOLAS (1974), International Convention for the Safety of Life at Sea, London, 1 November 1974, 1334 *U.N.T.S.* 2, and Protocols of 17 February, 1978, 1276 *U.N.T.S.* and 11 November 1988, U.S.Treaty Doc. 102-2.

STCW (2010), DE 54/13 30 June 2010, Section B - V/g "Guidance regarding training of masters and officers for ships operating in polar waters". Development of a mandatory code for ships operating in polar waters. Outcome of the 2010 Manila conference.

Stephenson, Scott R., Smith, L.C. and Agnew, J.A. (2011), Divergent long-term trajectories of human access to the Arctic, *Nature Climate Change*, Vol. 1, June 2011/ www. Nature.com/natureclimatechange

Stokke, K. (1999), *Politisk Geografi*, Tano Aschehoug, Oslo.

Stortinget (2009), St. Meld. Nr. 15. Interesser, ansvar og muligheter. Hovedlinjer i norsk utenrikspolitikk! Det Kgl Norske utenriksdepartement, Oslo, 2009

Støre, Jonas Gahr (2008), Å gjøre en forskjell. Refleksjoner fra en utenriksminister, Cappelen Damm, Oslo, 2008.

Swartztrauber, S.A. (1965), Alaska and Siberia: A Strategic Analysis, Naval Review, US Naval Institute, 1965.

Sæther, R. (2000), What do we need? The shipping industry's view on the Northern Sea Route's potential and problems. In: Claes Lykke Ragner (Ed.), *The 21st Century –*

Turning Point for the Northern Sea Route?, Kluwer Academic Publishers, Dordrecht, Boston, London.

Tamvakis, M., Alexander G. Granberg and Gold, G. (1999), Economy and commercial viability. In Willy Østreng (Ed.), *The Natural and Societal Challenges of the Northern Sea Route. A Reference Work*, Kluwer Academic Publishers, Dordrecht, Boston, London.

The Arctic Council (2011), The Agreement on Cooperation on Aeronautical and Maritime Search and Rescue in the Arctic. Ministerial meeting in Nuuk, 12 May 2011.

The Fundamentals of State Policy of the Russian Federation in the Arctic in the Period 2020 and Beyond (2009). Adopted by the President of the Russian Federation, D. Medvedev, 18 September 2008. Promulgated 20 March 2009, translated from *Rossiyskaya Gazeta*.

The Northern Maritime Corridor, Ocean Futures Maps, Oslo, 2006.

Theutenberg, B.J. (1984), *The Evolution of the Law of the Sea. A Study of Resources and Strategy with Special Regard to the Polar Areas*, Tycooly International Publishing Limited, Dublin.

Timtchenko, L. (1997), The Russian Arctic sectoral concept: Past and present, *Arctic*, 50:1, 29–35.

Tsoy, L.G. (2004), Northern Ports of Russia, Hydrometeorological and Ice Conditions of Navigation in White Sea, South East Region of Barents Sea, South Western Region of Kara Sea, Ob and Yenisei Gulfs. Prepared by CNIIMF for Det Norske Veritas. St. Petersburg.

Ugryumov, A. and Korovin, V. (2005), På isflak mot nordpolen, Forlaget Nord, Tromsø.

UNCTAD (2009), Review of Maritime Transport 2009.

US Department of Energy (2007), Energy Information Administration, US Crude Oil, Natural Gas, and Natural Gas Liquids Reserves, 2007 Annual Report.

US Department of Energy (2008), US Energy Information Administration, Country Analysis Briefs, Russia Natural Gas, May 2008.

US Department of the Interior (2005), *US Geological Survey Minerals Yearbook*.

US Department of the Interior (2007), USGS, *2007 Minerals Yearbook, Canada* (Advance Release), August 2009.

US Department of the Interior (2009), US Geological Survey, Assessment of Undiscovered Petroleum Resources of the Barents Sea Shelf, Fact Sheet 2009-3037, June 2009.

US Department of the Interior (2009), USGS, *2006 Minerals Yearbook: Alaska*, July 2009.

US Department of the Interior (2009), USGS, *2006 Minerals Yearbook, Commonwealth of Independent States, Russia,* July 2009.

US Department of the Interior (2006), Minerals Management Service, MMS Fact Sheet RED-2006-01b, February 2006, Assessment of Undiscovered Technically Recoverable Oil and Gas Resources of the Nation's Outer Continental Shelf.

US Energy Information Administration (2009), Country Analysis Briefs 2009, Canada Oil-Natural Gas.

US Energy Information Administration (2009), Country Analysis Briefs, Russia Oil, May 2008.

US Energy Information Administration (2008), Country Analysis Briefs, Russia – Oil Exports, May 2008.

US Energy Information Administration (2008), Russia – Natural Gas, May 2008.

US Geological Services (2008), USGS Arctic Oil and Gas Report: Estimates of Undiscovered Oil and Gas North of the Arctic Circle. US Geological Services, Washington DC.

US Navy (2009), *US Navy Arctic Roadmap*. United States Navy, Chief of Naval Operations, sponsored by Task Force Climate Change/Oceanography of the Navy, Washington DC.

Valkky, E., Nousiainen, H., and Karjalainen, T. (2008), Facts and Figures of the Barents Forest Sector. Working Paper of the Finnish Forest Research Institute 78.

Van der Zwaag, D.L. (Ed.) (2008), *Governance of Arctic Marine Shipping*, Marine & Environmental Law Institute, Dalhousie University, Halifax, October 2008.

Vartanov, E., Roginko, A. and Kolossov, V. (1999), Russian Security Policy 1945–1996. In W. Østreng (lEd.) *National Security and International Environmental Cooperation in the Arctic – The Case of the Northern Sea Route*, Kluwer Academic Publishers, Dordrecht, Boston, London.

Verny, J. and Grigentin, C. (2009), Container shipping of the Northern Sea Route, *International Journal of Production Economics*, November 2009, 107–117.

Vogt, Carl Emil (2011) *Fridtjof Nansen. Mannen og verden*, Cappelen Damm, Oslo.

Vylegjanin, A.N. (2007), *Future Problems of International Law in the High North: Review of the Russian Legal Literature*.

Wadhams, P. (2004), Arctic Sea Ice Thickness Changes. Presentation at Arctic Marine Transport Workshop, Scott Polar Research Institute, 28–30 September 2004

Waldichuk (1990), Histopathology in winter flounder (Pleuronectes americanus) living sdjacent to a pulp and paper mill, *Archives of Environmental Contamination and Toxicology*, 26:1.

Watson, G.G. (1991), Technical aspects of ice navigation and port construction in Soviet

Arctic. In: Lawson Brigham (Ed.), *The Soviet Maritime Arctic*, Belhaven Press, Scott Polar Research Institute, Cambridge.

Weller, G.E. (2000), Climate change and its impact on the Arctic environment. In: Henry P. Huntington (Ed.), *Impacts of Changes in Sea Ice and Other Environmental Parameters in the Arctic*, Report of the Marine Mammal Commission Workshop, Girdwood, Alaska, 15–17 February 2000.

Wergeland, T. (1992), The Northern Sea Route – Rosy prospects for commercial shipping? *International Challenges*, 12:1.

Winjolst, N. and Wergeland, T. (2008), Shipping Innovation, Rotterdam 2008, presentation by Lloyd's Register Fairplay Research, SAI Executive Meeting, Gothenburg,

The World Bank Group 2009.

Yakovlev, A., Kossov, O. and Ushakov, A. (1994), Political Aspects of International Shipping along the Northern sea Route, INSROP Discussion Paper, IV.2.2., St. Petersburg, December 1994.

Yamaguchi, H. (1995), Experimental voyage through the Northern Sea Route. In: H. Kitagawa (Ed.), *Northern Sea Route; Future and Perspectives*, SOF, Tokyo.

Yeong-Seok, Ha (2006), Transportation sSystem of natural tesources in Korea. In: Hiromitzu Kitagawa (Ed.), *New Era in Far East Russia and Asia*, OPRF, Tokyo.

Other sources

Data retrieved from websites

Aleut International: www.aleut-international.org/files/Download/2008_1st_2nd_Quarter%20Newsletter.pdf -

Archangelsk Sea Commercial Port: http://www.ascp.ru/en_htm/9.htm

http://www.ascp.ru/en_htm/3.htm

Arctic Centre, University of Lapland: http://arcticcentre.ulapland.fi/barentsinfo/

Arctic Council Oil and Gas Assessment: http://arctic-council.org/article/2008/1/release_of_the_arctic_council_oil_and_gas_assessment

Arctic Economics:
http://benmuse.typepad.com/arctic_economics/2008/10/russias-shallow-arctic-seas-and-straits.html#more

Arctic Focus:
http://arcticfocus.com/2009/07/13/coast-guard-needs-more-icebreakers-for-arctic/

Athropolis Map of the Arctic:
http://www.interklasa.pl/portal/dokumenty/ue005/map-circled.gif

Baffin Land:
www.baffinland.com

Barents Observer:
-http://www.barentsobserver.com/norilsk-nickel-boosts-arctic-independence.4647401-116320.html

-http://www.barentsobserver.com/murmansk-port-and-sovcomflot-on-privatisation-list.4656621-116320.html

-http://www.barentsobserver.com/murmansk-port-and-sovcomflot-on-privatisation-list.4656621-116320.html

-http://www.barentsobserver.com/one-step-closer-to-deep-water-port-in-arkhangelsk.4609737.html

-http://www.barentsobserver.com/10-million-tons-shipped-from-varandey-oil-terminal.4725304-116321.html

http://www.barentsobserver.com/first-oil-shipment-planned-for-northern-sea-route.4752806-116320.html

http://barentsobserver.com/more-traffic-along-the-northern-sea-route.4947500-16149.html

http://www.barentsobserver.com/34-vessels-in-transit-on-northern-sea-route.4991248.html

Bellona:
http://www.bellona.org/articles/articles_2009/atomflot_prays_for_sun

Cairn Energy:
http://www.cairn-energy.plc.uk

Canadian Arctic Resources Committee:
http://www.carc.org/pubs/v18no3/1.htm

Canadian Broadcasting Corporation:
http://www.casr.ca/id-arctic-empires-2.htm

Canadian Coast Guard:
www.ccg-gcc.gc.ca

Canadian Defence Policy, Foreign Policy and Canadian-US Relations:
http://www.casr.ca/id-arctic-empires-2.htm
http://www.casr.ca/id-iqaluitport2.htm

The Canadian Environmental Assessment Agency (CEAA):
http://a100.gov.bc.ca/appsdata/epic/documents/p239/d21939/1151444602531_f314149f768a4109abd3ab91484841ea.pdf

Canadian Ice Service:
http://ice-glaces.ec.gc.ca/IceGraph/IceGraph-GraphdesGlaces.jsf

Canadian Ministers reveal resources management strategy" 28 July 2009:

www.mineweb.com/mineweb/view

Canwest News Service:
http://www2.canada.com/topics/news/features/arcticambitions/story.html?id=de0a569c-1ec6-478c-9fd0-a45622e2fe9a

China Babel International Investment Group, "Progress stalled on Russian gas deal with China", 12.04.2009:
http://en.worldmr.net/shownews

Det Norske Veritas:
http://www.dnv.com/industry/maritime/publicationsanddownloads/publications/special_ships/2009/01_2009/arctictrendsincreasingdemandsonarcticseafarers.asp

Environment Canada:
http://www.weatheroffice.gc.ca/marine/index_e.html

Federal State Statistics Service, Russia:
www.gks.ru/bgd/regl/

The Financial Express: Blomberg, Mark Dean, *The Financial Express*, "The West's share of global GDP set to fall", 3 June 2009
www.financialexpress.com

Free Republic:
http://www.freerepublic.com/focus/f-news/2234272/posts

Gazprom, Release, "On meeting dedicated to Shtokman field development", 28 January 2010
www.gazprom.com/press/news/2010/january

Gazprom, Release, "Gazprom launches Portovaya compressor station construction, 15 January 2010

www.gazprom.com/press/news/

Gazprom, "Shtokman Development AG Board of Directors approve way forward on Shtokman" 8 February 2010
www.gazprom.com/press/news/february/2010

Gulf Oil and Gas:
www.gulfoilandgas.com

International Group of P&I Clubs:
http://www.igpandi.org/

IMO (2000), Study of greenhouse gas emissions from ships", Issue no. 2, 31 March 2000:
IMO-study-2000-ghg-emissions.pdf

Kosich, Dorothy, "Canadian Ministers reveal resources management strategy" 28 July 2009:
www.mineweb.com/mineweb/view

Library of the Canadian Parliament:
http://www.parl.gc.ca/information/library/PRBpubs/prb0561-e.htm

Lloyd's List:
http://www.lloydslist.com/ll/news/scf-to-test-northern-sea-route/20017751993.htm;.5fa4e8cc80be35e2653c9f87d8b8be45bf6ba69a

Lloyd's Register Fairplay database, extracted via:
www.sea-web.com

The National Academic Press:
http://www.nap.edu/openbook.php?record_id=12024&page=119

National Oceanic and Atmospheric Administration:
http://www.lme.noaa.gov/LMEWeb/Publications/tm208.pdf

The New York Times, "Europe's Recovery Comes to Near Halt", 13 February 2010:
www.nytimes.com/2010/02/13

Norilsk Nickel: Building an Independent Transport Chain:
http://img0.custompublish.com/getfile.php/1051433.900.pcpceveesd/BUILDING_INDEPENDENT_TRANSPORT_CHAIN.pdf?return=www.barents.no

OzCoasts Australian Coastal Information:
http://www.ozcoasts.org.au/indicators/benthic_inverts.jsp

Port of Churchill:
http://www.portofchurchill.ca/

Port of Murmansk:
www.portmurmansk.ru

Port of Vitino:
http://www.vitino.ru/

Ria Ru:
http://ria.ru/arctic_news/20111128/500547773.html

River Lake Group:
https://www.riverlake.ch/news.php?newsid=87

Russian Ministry of Transport:
http://www.mintrans.ru

Russian stats on sailing:
http://rus-shipping.ru/ru/stats/?id=53

Science Daily, "World oil reserves at tipping point", University of Oxford, 26 March 2010:
www.sciencedaily.com/releases/2010/03/100324225511.htm

Sovcomflot:
www.scf-group.com/

Statoil, Facts about Snøhvit, 23 November 2009:
www.statoil.com/en/OurOperations/

St.Meld. 22 (White Paper 22), 2008–2009, extracted from: http://www.regjeringen.no/nb/dep/jd/dok/regpubl/stmeld/2008-2009/stmeld-nr-22-2008-2009-/10/1/4.html?id=555011

Trade and Environment Database (TED) - Case Studies (2009), *Russia Oil Spill*
http://74.125.77.132/search?q=cache:yw31lu7C87IJ:http://www.american.edu/ted/KOMI.HTM+oil%20spill+russia+financial

Transport Canada:
http://www.tc.gc.ca/marinesafety/debs/arctic/acts-regulations/images/zones2.JPG

The World Icebreaker and Icebreaking Supply Vessel Fleet:
http://portal.fma.fi/portal/page/portal/fma_fi/merenkulun_palvelut/talvimerenkulku/icebreakerfleets.pdf

Wikipedia:
http://en.wikipedia.org/wiki/MS_Explorer
http://en.wikipedia.org/wiki/Canadian_Coast_Guard
http://en.wikipedia.org/wiki/Nuclear_powered_icebreaker
http://en.wikipedia.org/wiki/Pevek#Port
http://en.wikipedia.org/wiki/Ecosystem
http://en.wikipedia.org/wiki/Arctic_ocean
http://en.wikipedia.org/wiki/Large_marine_ecosystem
http://en.wikipedia.org/wiki/Oil_reserves

World Port Source:
http://www.worldportsource.com/ports/RUS_Port_of_Arkhangelsk_1532.php
http://www.worldportsource.com/ports/portCall/RUS_Port_of_Dikson_2549.php
http://www.worldportsource.com/ports/portCall/RUS_Port_of_Dudinka_2550.php
http://www.worldportsource.com/ports/portCall/RUS_Port_of_Igarka_2551.php
http://www.worldportsource.com/ports/RUS_Port_of_Tiksi_2574.php
http://www.worldportsource.com/ports/RUS_Port_of_Pevek_2547.php
http://www.worldportsource.com/shipping/ports/RUS_Port_of_Providenija_2548.php

Interviews/conversations

Canadian Coast Guard, 14 January 2010

Canadian Polar Commission, 13 January 2010

Donat Pharand, 11 January 2010

Edgar Gold, 22 December 2009

Foreign Affairs and International Trade Canada, 13 January 2010

Representative from the Beluga Group, 2009

Transport Canada, 9 and 15 January 2010

Index

D

P

U

MIX
Papier aus verantwortungsvollen Quellen
Paper from responsible sources
FSC® C105338

Printed by Books on Demand, Germany